Library of America, a nonprofit organization,
champions our nation's cultural heritage
by publishing America's greatest writing in
authoritative new editions and providing resources
for readers to explore this rich, living legacy.

RACHEL CARSON

Rachel Carson

Silent Spring
&
Other Writings
on the Environment

Sandra Steingraber, *editor*

THE LIBRARY OF AMERICA

Rachel Carson:
Silent Spring & Other Writings on the Environment
is published and kept in print with a gift from

THE GOULD FAMILY FOUNDATION

to the Guardians of American Letters Fund,
established by the Library of America
to ensure that every volume in the series
will be permanently available.

Contents

List of Illustrations
(following page 372)

Introduction

BY SANDRA STEINGRABER

I was three years old when biologist Rachel Carson published *Silent Spring* in 1962. To me, *Silent Spring* was the green book that all the grown-ups carried around. My father, a high school teacher, used *Silent Spring* as a course text for his business class. When he arrived home each afternoon, he opened his briefcase and laid the book on the coffee table.

Green. With squiggly lines.

Because of *Silent Spring*, Dad put in a compost pile. Even more exciting: ladybugs and praying mantises began arriving at our house by mail order, in pin-holed boxes, to serve as pest-hunting predators in the backyard garden. Mulch was for weed control. Soon enough, my sister and I were selling surplus tomatoes at the end of our driveway. *Organic* tomatoes.

Narrowly speaking, *Silent Spring* is a book about the toxicological properties of nineteen pesticides. Among them: aldrin, dieldrin, endrin, DDT, lindane, chlordane, and heptachlor. As alien as this roll call of chemicals may sound now, their names were well-known to readers in 1962. "DDT is now so universally used," Carson noted in her second chapter, "that in most minds the product takes on the harmless aspect of the familiar."

Ten years after *Silent Spring*'s publication, DDT was outlawed in the United States. One by one, a half dozen other pesticides featured in its chapters met the same fate—or were heavily restricted for most uses. If it had accomplished only this—the banishing of seven common, persistent, highly toxic, organochlorine chemicals—*Silent Spring* would be responsible for more substantive reform than any other environmental book before or since. But it is rightly credited with much more. The first book to gather together all existing evidence on the risks of pesticides for public health and the environment, *Silent Spring* was composed in such an artful and compelling way that it aroused public awareness far beyond its immediate subject matter.

In the ways that it provoked an awakening of public consciousness about the interconnections of human beings and their ecological world—and a public outcry against ecological abuses of all kinds—*Silent Spring* became a founding force of modern American environmentalism. In an era of better living through chemistry, it brought the environment into the body politic. *Silent Spring* is the book that led to the establishment of the U.S. Environmental Protection Agency in 1970, tasked with regulating pesticides and protecting food safety. It inspired major national environmental legislation, including the Clean Air Act (1963), the Wilderness Act (1964), the National Environmental Policy Act (1970), the Clean Water Act (1972), and the Endangered Species Act (1973).

Above and beyond the individual chemicals it indicted, *Silent Spring* revealed that the attempt to conquer and master the natural world is a fool's game. Human actions, insisted Carson, have environmental consequences. These, in turn, affect us. Wars waged against the web of life—as, say, when plotting the eradication of insect pests via the broadcast spraying of chemical poisons—will, sooner or later, transform the cells of our own bodies into battlefields. "[O]ur heedless and destructive acts enter into the vast cycles of the earth and in time return to bring hazard to ourselves," as Carson testified before a Senate subcommittee in 1963.

Silent Spring prompted not only congressional hearings but an invitation to its author to meet with the president's science advisers. Transcending genre and disciplinary boundaries, it won medals of honor from both scientific and literary societies. In 1992, *Silent Spring* was named the single most influential book of the past fifty years by a panel of twenty-two distinguished Americans that included a former president, a supreme court justice, and a senator. Like Harriet Beecher Stowe's *Uncle Tom's Cabin*, to which it is often compared, *Silent Spring* is a book that altered the course of history.[1] But to me, a biologist and author who has followed the path that Carson blazed, the power of *Silent Spring* is contained within this single fact: it's the book that made my father—a World War II veteran, a lifelong Republican, and a man of unchangeable habit—lay down his spray gun and go shopping for ladybugs.

Organized into seventeen interlocking chapters, *Silent Spring* makes four big points. First, Carson contends, we are all being contaminated, without our permission, by inherently toxic chemicals in the form of pesticides. These chemicals, Carson pointed out, largely came into being as weapons of war during the 1940s and had subsequently been turned to purposes for which they were never intended. Most famously, DDT halted a typhus epidemic among our troops in Italy by killing the lice that transmitted the pathogen. Immediately after the war ended, pesticide research was declassified and, with an assist from Madison Avenue advertising agencies, these "economic poisons" were then sold to the general citizenry—from housewives to farmers—not because of some unmet demand for insect and weed control but because abundant wartime production capacity was in need of domestic markets. In this way, DDT and its chemical brethren were repositioned for peacetime duty without any advance testing of their long-term safety or efficacy. (An abundance of former military planes that could be cheaply converted into spray planes—and an abundance of former military pilots eager to fly them—helped seal the deal.)[2]

Second, pesticides don't even work terribly well to solve pest problems. Whether broadcast over the entirety of Long Island to eradicate gypsy moths or added to kitchen floor wax to kill houseflies on contact, pesticides, in the long run, tend to make the original problem worse. Just as the indiscriminate use of antibiotics breeds antibiotic-resistant bacteria, routine use of pesticides induces evolved resistance among populations of weeds and pest insects, which then require ever higher doses to knock them back. Pesticides also kill the natural enemies of the insects they target—spiders, birds, frogs, wasps—and the absence of such creatures likewise fuels future pest outbreaks. "Thus the chemical war is never won," Carson explains, "and all life is caught in its violent crossfire."

Third, at the very least, the public has the right to know about the risks they are being asked to assume in the form of nonconsensual exposures to inherently toxic substances—exposures that begin in the womb and continue throughout life. Of particular concern: pesticide residues that linger on food crops; contamination of surface and ground water; and the suffering and death of songbirds, eagles, fish, and other wildlife that are the unintended casualties of mass spraying programs. Out of the basic right to know about all of these issues flows a moral obligation to speak up and take action.

Fourth, the consequent risks to our health and the health of other species are needless risks because so many elegant, non-toxic methods of pest control are available if we only have eyes to see—and research dollars to study and develop them. These various alternatives involve biological, rather than chemical, approaches and look to the natural world, rather than petrochemical engineering, for inspiration. Sex attractants that lure pests into traps. Hormones that prevent their metamorphosis from larvae to adults. Insect sterilization. Habitats for their natural enemies.

The evidence that Carson gathered to support her arguments came from no one journal or field of study. Rather, she cites hundreds of scientific studies across many disciplines, ranging from pharmacology to wildlife ecology. She treats each study like a piece of a puzzle—and, in her private correspondence, often refers to them this way. No single study demonstrates absolute proof, but, when assembled together in the right configuration, they create a bigger picture that, Carson argues, we ignore at our peril. Her task, she felt, was something like a trial lawyer's. Sifting through masses of often arcane evidence—crop statistics, forestry reports, dissertations on ornithology—she would pull out telling details, translating data into narrative and making a case.

Silent Spring was a publishing sensation when it appeared in September 1962. It rocketed to the top of best-seller lists and stayed there for months. It was named a main selection of the Book-of-the-Month Club. It was the subject of an hour-long primetime *CBS Reports* documentary, watched by millions. And it would go on to sell millions of copies. Among the secrets to its long and ongoing success are these: the science was brilliant, the writing beautiful, the messenger right, the timing perfect, and the public advocacy bold.

The veracity of its science is the central genius of *Silent Spring*. Carson had worked as a federal Fish and Wildlife Service official in the 1930s and 1940s, giving her access to field reports, internal documents, and technical journals that, altogether, clearly indicated many unintended and troubling consequences of chemical spraying programs that attempted to eradicate specific insect pests. Even by midcentury, she had understood that the scientific case against chlorinated pesticides like DDT was damning. Study after study was beginning to show that pesticides were triggering population explosions in insect pests and mass die-offs among the birds, fish, and mammals that fed on them. Pesticides lingered as a residue in the food crops that we ate and, as fat-soluble substances, concentrated in our bodies over time. By 1950, pesticide residues were detected in baby food. By 1951, pesticides were known to pass from mother to child through breast milk.

In 1952, now a best-selling author for her book *The Sea Around Us*, Carson was able to retire from her post and turned her hand to other book projects, including *The Edge of the Sea* (1955). But she continued to keep an eye on the pesticide debate raging within government agencies. When she returned to the subject in earnest in 1958, she knew exactly whom to call to ferret out important, closely held information. She knew some of the frontline researchers personally.

In the end, Carson's list of principal sources for *Silent Spring* filled more than fifty pages in the published book. But her research was more than just thorough or accumulative. Carson had an uncanny ability, when she discovered gaps in the data, to see across to the far shore and make predictions that have turned out to be correct. Which is not to say that she was prescient. She just knew how to extrapolate. She read across the scientific disciplines and could trace the connections between findings published in widely divergent reports. She sought out the authors of these studies—visiting labs, or calling, or writing—both to seek out more data and to fact-check her conclusions. Researchers she visited remember her as unerringly well-informed and curious, and she clearly loved to find herself on the leading edge of discovery. Here, for example, is part of a 1959 letter to a colleague in the sciences, sent when she was in the thick of her research:

> Now that I have tabulated all the pesticidal chemicals that disturb cytochrome oxidase, a great light is breaking in my mind. I have never seen this connection fully pursued. . . . [U]ncoupling phosphorylation from respiration may lead to the same end. Here again, the list of chemicals disturbing phosphorylation is impressive. And Weinbach's work on pentachlorophenol (to which you led me) seems to tie up with a paper presented at the American Chemical Society's recent meetings. . . . To me, it all seems to fit together.

Ahead of her time, Carson saw the first glimmers of the mechanisms by which pesticides disrupt enzyme systems, putting cells on a path to tumor formation.

Consider how, in *Silent Spring*, Carson lays out evidence linking pesticide exposure to cancer risk. First, she demonstrates that the synthetic chemical era, ushered in by World War II, has created countless new cancer-causing chemicals—carcinogens—for which we have no naturally occurring defenses. Next, she documents the rise in pesticidal chemical usage during the postwar period, the ubiquitous presence of these chemicals in the general environment, and the myriad routes of exposure by which we all have intimate, regular contact with them. She traces the concurrent rise of cancer's prevalence within the general population and presents troubling cases of cancer within pesticide-exposed subpopulations. (Behold the spray-gun-toting housewives with leukemia. Behold the farmers with bone marrow degeneration.) She supports these case studies with evidence from laboratory and veterinary studies, showing that both lab animals and livestock exposed to pesticides develop cancer significantly more often than those not so exposed. (Behold the sheep with nasal tumors.) And, finally, she takes the reader inside the unseen molecular machinery within the cell itself to explore how pesticides may sabotage certain metabolic processes, resulting in genetic injury and runaway cell growth.

All of these lines of inquiry and the conclusions Carson drew from them have stood the test of time. Many more puzzle pieces have been added to those Carson put together herself, but her ability to reach correct inferences based on the data points available to her is nothing less than brilliant. At the time of *Silent Spring*'s publication, of course, molecular biology was in its infancy and our understanding of DNA's role in heredity, growth, and development less than a decade old. (Watson and Crick's classic double-helix paper—based on X-ray images from Rosalind Franklin—was published in 1953.) Knowledge of DNA's role in the story of human cancer was younger still.

To put a finer point on Carson's scientific achievement in *Silent Spring*: she was able to document troubling links between pesticide exposures and cancer risk even though she was working without state-based cancer registries (which came into being c. 1970–90) and without a working model of endocrine disruption (c. 1996) or epigenetics (c. 2008), both of which elucidate how toxic chemicals can alter DNA activity even in the absence of mutations or other genetic injury. When she was drafting her chapters, it was not yet accepted toxicological wisdom that the timing of exposure matters, with exposures during pregnancy, infancy, and early childhood mattering the most. Carson also lacked geographic information systems (GIS) and computer mapping programs that can generate visually compelling pictures of potential cancer clusters and other temporo spatial patterns for statistical analysis. And yet, through careful, painstaking sleuthing, Carson believed that she was seeing the early signs of a cancer epidemic in slow motion. She was especially concerned with early-life exposures and the apparent rise in cancers among children. And she was right.

Silent Spring opens not with data, nor even with structural formulas of organochlorine compounds (those arrive in Chapter 2), but with a lyrical fable and a masterful narrative voice. If she wanted to keep her readers turning the pages of what she privately referred to as her "poison book," Carson knew she needed to weave a spell, bring the science to life, and keep readers focused on the beauty of the living world as much as the chemical assaults hurled at it. To show how harmful chemicals enter groundwater, food chains, bloodstreams, cells, and chromosomes, she relied on imagery and metaphor. Rivers of death. Elixirs of poison. The mute testimony of the dead ground squirrels. Along the way, her writer's eye captures panoramic views of entire ecosystems, descends into subterranean aquifers, rises with crop dusters into the skies, and then travels inside the human body to make visible the damage on subcellular levels.

As a young woman, before the biological sciences attracted her interest, Carson hoped she might become a poet or fiction writer, and her book demonstrates a powerful literary imagination as well as scientific insight. Woven within her descriptive narrative are allusions to classical tragedy, mythology, and romatic poetry.[3] To conclude *Silent Spring*, she distills our dire and complicated situation with regard to national pest control

practices into a resonant, memorable image, borrowed from
Robert Frost's familiar poem "The Road Not Taken":

> The road we have long been traveling is deceptively easy, a
> smooth superhighway on which we progress with great speed,
> but at its end lies disaster. The other fork of the road—the one
> "less traveled by"—offers our last, our only chance to reach a
> destination that assures the preservation of our earth.

Carson's largest literary accomplishment in *Silent Spring* lies
not only in how she tells her story, however, but in the ways she
deliberately retells one. Chemicals had come home from the war
as heroes. Welcomed into our kitchens, schools, hospitals, and
workplaces, they would, the citizenry was assured, make everyone's lives better. DDT in particular had saved the lives of Allied
troops, both in Europe and the Pacific theater, and its inventor
had won the Nobel Prize for Medicine. Along with other modern technological solutions, it was promoted as a miracle worker.
DDT was said to be, at the same moment, a lethal assassin to
undesirable insect pests (the new enemy of the home front) and
a completely benign helpmate, safe enough to mothproof the
baby's blanket.

Taking issue with these positive associations, Carson revises
and upends powerful cultural narratives about science, nature,
and our relationship to both. Confronting values and assump-
tions fundamental to technological modernity, she emphasized
that command-and-control operations, when directed at the
natural world, are neither enlightened nor advanced but brutish,
ignorant, arrogant, and regressive:

> The "control of nature" is a phrase conceived in arrogance,
> born of the Neanderthal age of biology and philosophy,
> when it was supposed that nature exists for the convenience
> of man. . . . It is our alarming misfortune that so primitive
> a science has armed itself with the most modern and terrible
> weapons, and that in turning them against the insects it has also
> turned them against the earth.

By 1955, Rachel Carson was already a celebrated, best-selling nature writer, winner of the John Burroughs Medal and the National
Book Award. Her reputation and scientific credibility provided

her a reserve of goodwill that made her a perfect messenger for the narrative contained in *Silent Spring*—a darker, more oppositional narrative than those she had previously attempted.

Her earlier trilogy of books about the sea—*Under the Sea-Wind* (1941), *The Sea Around Us* (1951), and *The Edge of the Sea* (1955)—had offered her readers accounts of a world in which the human race scarcely appears. The first two books function like submarine cameras, revealing the recent findings of oceanography and marine biology in cinematic detail. The third, her "shore book," explores the dynamic boundary between oceanic and terrestrial life. Altogether, with their focus on how complicated webs of direct and indirect interactions among many species create organized communities, Carson's earlier books are credited by contemporary ecologists as having anticipated complexity theory as a central tool for understanding ecosystem functioning.[4] Certainly, Carson's ability to see patterns in emerging data is on display. By 1951, she had already perceived early signs of climate change—although she did not know that rising levels of heat-trapping greenhouse gases were driving the trend. From *The Sea Around Us*:

> The evidence that the top of the world is growing warmer is to be found on every hand. The recession of the northern glaciers is going on at such a rate that many smaller ones have already disappeared. If the present rate of melting continues others will soon follow them.[5]

While Carson's sea books occasionally allude to environmental threats, they call for no particular action. Carson's theory of social change here, if she had one, seems to rest on helping her readers visualize the oceanic world below the waves, or teeming shore life—full of communities of interacting creatures that possessed agency and distinct personalities—and so inculcate a sense of wonder and humility. Wonder and humility, she believed, "do not exist side by side with a lust for destruction."[6]

Behind the scenes, however, Carson's pre–*Silent Spring* concerns about the despoilment of natural systems were fully articulated, if not yet publicly expressed. By 1945, she was sufficiently troubled by the Department of Agriculture's plan to allow commercial distribution of DDT—in the face of nascent evidence showing harm to birds and nontarget insects—that she pitched the *Reader's Digest* for a story assignment. She was turned down. In 1952, with McCarthyism ascendant, Carson became sufficiently disgusted by the summary replacement of the long-standing director of the U.S. Fish and Wildlife Service with a pro-business appointee that she dashed off a bold letter to the editor of *The Washington Post*: "It is one of the ironies of our time, that while concentrating on the defense of our country against enemies from without, we should be so heedless of those who would destroy it from within." In 1954, her rage at the "blind and wilful destructiveness" behind a plan to erect a dam within Dinosaur National Monument reached such a boiling point that she urged her beloved, Dorothy Freeman, not only to join her in wiring the president but also to organize her entire Garden Club to action. In these and other passages, we see the impulse to confrontation and redress of grievance that later found full-throated expression in *Silent Spring*.

In January 1958, a gardener and bird-watcher named Olga Owens Huckins sent Carson a letter full of painful details about a mosquito-control campaign that had resulted in a mass death of songbirds in her Massachusetts backyard. Dead birds lay piled around her DDT-contaminated birdbath, frozen in postures of convulsion. This letter, which had been published in the *Boston Herald*, moved Carson from quiet concern to action. "Knowing what I do," she wrote to Dorothy in June 1958, "there would be no future peace for me if I kept silent."[7]

Carson prominently credits Huckins's letter as the starting point for her book at the beginning of *Silent Spring*'s acknowledgments, but she was more circumspect, perhaps strategically, about the political organizing in which Huckins and others were deeply involved. Huckins was indeed a bird lover, but she was also what we would now call a grassroots environmental activist, her letter part of a citizen campaign to halt the aerial spraying of pesticides via lawsuits and protests. Banding themselves together into the ad hoc Committee Against Mass Poisoning, she and others wrote hard-hitting letters to the editors of many New England and Long Island newspapers. These defined pesticides as poisons and urged a precautionary approach. (Said one: "Stop the spraying of poisons everywhere until all the evidence, biological and scientific, immediate and long run, of the effects upon wildlife and human beings is known.") Further, the committee took a human rights approach to environmental harm, pointing to the absence of informed consent, a call that was soon taken up by Carson. Huckins condemned aerial spraying of pesticides on the grounds that it was "inhumane, undemocratic, and probably unconstitutional." Carson echoed this assertion in *Silent Spring*:

> If the Bill of Rights contains no guarantee that a citizen shall be secure against lethal poisons distributed either by private individuals or by public officials, it is surely only because our forefathers, despite their considerable wisdom and foresight, could conceive of no such problem.

The Committee Against Mass Poisoning was crucial in other ways, as well. The citizen lawsuit filed by the group, with ornithologist Robert Cushman Murphy as the lead plaintiff, wended its way to the Supreme Court. (It lost on a technicality.) Along the way, the suit became a magnet for media attention, which certainly didn't hurt when Carson proposed writing about pesticides for *The New Yorker*. At the same time, Marjorie Spock, another plaintiff in the Long Island lawsuit and the younger sister of celebrity pediatrician Benjamin Spock, responded prodigiously to Carson's early requests for information, sending her stacks of reports and studies, trial transcripts, and names of experts crucial for her research.

In short, an environmental campaign in the 1950s opened a space in the culture for a considered analysis of synthetic chemicals and their possible dangers and helped to open a space within the publishing industry for environmental writing. Carson was as transformed by the power of political advocacy that preceded *Silent Spring* as much as the nascent environmental movement was magnified by its publication. So, while Carson is rightly lauded as a solitary figure whose writing inspired modern environmentalism, it is also the case that environmental activism inspired the writing—and that the accompanying legal action provided the author key research materials. As Carson herself might say, the relationship between the book and the movement was interconnected and reciprocal.

By the time the final draft of *Silent Spring* was submitted for publication, Carson was two frustrating years behind schedule.

But, as she later confessed to Dorothy, the long delay was actually fortuitous. The timing for the book's release, she believed, was now considerably better. And, about this, she was certainly proved right.

President Kennedy, who took office in January 1961, would emerge as a champion of *Silent Spring*. His mention of Carson's work in response to a reporter's question about the possible harms of pesticides, in August 1962, prompted an investigation by the Life Sciences Panel of the President's Science Advisory Committee. Its final report was a vindication of *Silent Spring*'s central argument and endorsed Carson's call for public health protections. These recommendations, in turn, prompted full-blown congressional hearings in June 1963, at which Carson testified.

At the same time, the international arms race was a source of mounting social anxiety in the early 1960s, as the continued testing of atomic weapons brought increased attention to the public health threats of nuclear fallout. In 1961, a group of Boston doctors founded Physicians for Social Responsibility to address these and related concerns. Meanwhile, in St. Louis, researchers with the Committee for Nuclear Information (CNI) were documenting the presence of radioactive strontium in the baby teeth of children. These levels had spiked sharply with the advent of the atomic age, and they rose and fell in correlation with the timing of ongoing bomb tests. (See Carson's correspondence with pathologist and CNI science advisor Walter C. Bauer.) The notion that radioactive contamination from nuclear detonations could drift across entire continents, insinuate itself into the food chain (from grass to cows to milk), and find its way into the bones and teeth of children was deeply unnerving. The spectre of wholesale global destruction through nuclear warfare was even more so.

The sheer folly and terror of the arms race was a significant factor in pushing Carson beyond the reverential mode of her earlier books toward the critical engagements of *Silent Spring* and her later public statements. Here is how she explained her change of heart to Dorothy in February 1958:

> I suppose my thinking began to be affected soon after atomic science was firmly established. Some of the thoughts that came

were so unattractive to me that I rejected them completely, for the old ideas die hard. . . . It was comforting to suppose that the stream of life would flow on through time in whatever course that God had appointed for it—without interference by one of the drops of the stream—man. And to suppose that, however the physical environment might mold Life, that Life could never assume the power to change drastically—or even destroy—the physical world. These beliefs have almost been part of me for as long as I have thought about such things. To have them even vaguely threatened was so shocking that, as I have said, I shut my mind—refused to acknowledge what I couldn't help seeing. But that does no good, and I have now opened my eyes and my mind. I may not like what I see, but it does no good to ignore it. . . . So it seems time someone wrote of Life in the light of the truth as it now appears to us.

Later, as a rhetorical strategy, Carson made the parallels between radioactive fallout and drifting pesticides explicit, urging her readers to direct the same sense of outrage and urgency to the recklessness of the pesticide crisis as they did to the atomic one.[8] "We are rightly appalled by the genetic effects of radiation," she wrote in *Silent Spring*. "[H]ow then, can we be indifferent to the same effect in chemicals that we disseminate widely in our environment?"

In the summer of 1962, prior to the book's fall launch, *The New Yorker* serialized *Silent Spring*, running generously long excerpts in three consecutive issues. For this, the magazine was threatened with lawsuits by the makers of DDT and other pesticides. But did not blink. Her book publisher, Houghton Mifflin, was likewise threatened and also stood firm. When *CBS Reports* broadcast an hour-long program about *Silent Spring*, two major corporate sponsors threatened to withdraw their support. And did. But the show went on. Bold and courageous advocacy—and the unflinching willingness of publishers, editors, and media outlets to stand against a deep-pocketed, well-lawyered industry—became another factor in the unprecedented success of *Silent Spring*.

Led by American Cyanamid, Monsanto, and Velsicol, and joined by the Department of Agriculture, the countercampaign to discredit *Silent Spring* and its author was immediate, well financed, relentless, and deeply personal. Carson was accused of sentimentality and hysteria. She was called a spinster and a Communist. She was defined as existing outside the scientific community altogether.[9] On television, Robert White Stevens, employee of American Cyanamid, led the charge: "Miss Carson maintains that the balance of nature is a major force in the survival of man, whereas the modern chemist, the modern biologist and scientist believes that man is steadily controlling nature." In print, Monsanto's *Silent Spring* parody, "The Desolate Year," imagined a world in which ticks, maggots, lice, beetles, worms, and rats, "freed from pesticidal opposition," take over a disarmed America that has been stripped of its chemical weapons. Behind the scenes, anticipating that readers of *Silent Spring* would turn to their doctors with questions on the possible health impacts of pesticides, the chemical industry convinced the American Medical Association to disseminate propaganda from the National Agricultural Chemicals Association.

In response, Carson stood firm on the data and remained confident. The many prizes, medals, and awards garnered by *Silent Spring* each occasioned acceptance speeches, and she used these and other public appearances as platforms to push back against her enemies in industry by reiterating her book's main arguments and providing new, corroborating evidence that had emerged since *Silent Spring*'s release.

Even as she kept the spotlight firmly trained on the evidence for harm, Carson was unafraid to identify the crude, ad hominem tactics of those attacking her, to whom she referred as "the masters of invective and insinuation." And she was unafraid of biting sarcasm. "I am a bird lover—a cat lover—a fish lover—a priestess of nature—a devotee of a mystical cult having to do with laws of the universe which my critics consider themselves immune to." Most importantly, she made clear that she would not allow herself to be attacked for things she never said. From her address to the Women's National Press Club:

> I *do not* advocate complete abandonment of chemical control . . . I criticize the modern chemical method not because it *controls* harmful insects, but because it controls them *badly* and *inefficiently* and creates many dangerous side effects in doing so. I criticize the present methods because they are based

on a rather low level of scientific thinking. We are capable of much greater sophistication in our solution of the problem.

Finally, Carson went on the offensive, pointing out the hidden conflicts of interest and growing liaisons between industry trade groups and scientific societies that, she argued, were corrupting both university research—especially at land grant institutions—and scientific communication. The American Entomological Society, she noted, now included a pesticide manufacturer and chemical companies among their "sustaining associates."

Pointedly, in a speech before the Garden Club of America, she asked, who speaks for science and why?

> When the scientific organization speaks, whose voice do we hear—that of science? or of the sustaining industry? It might be a less serious situation if this voice were always clearly identified, but the public assumes it is hearing the voice of science.

She dug deep. That reassuring report on the impact of pesticides from the National Academy of Sciences? Its authors were hardly disinterested individuals nor were they even members of the Academy. Instead, as Carson revealed, they consisted of representatives of nineteen different chemical companies and four trade organizations, including the National Agricultural Chemical Association.

In her 1963 speech "The Pollution of Our Environment," delivered to a convocation of physicians and medical professionals, Carson dismantled the claim that no viable alternatives to chemical poisons exist. More broadly, she asked why, given the existence of safer alternatives and in the face of overwhelming evidence of human harm, do we continue to pollute? Why do we behave "not like people guided by scientific knowledge, but more like the proverbial bad housekeeper who sweeps dirt under the rug in the hope of getting it out of sight?" To that rhetorical question, she offered three explanations for our collective reluctance to give up on entrenched, poisonous practices. First, she said, we wait too long to evaluate the risks. Once a new technology is deployed and a vast economic and political commitment has been made, dislodging it is impossible. Second, we fail to acknowledge that nature invariably has its own unpredictable way with harmful pollutants. Within living ecosystems, there are no compartments. Third, we act as though the evidence for harm in other animals does not apply to us even though we share biological ancestry and are thus clearly susceptible to damage from the same forces.

Throughout all of these speeches, the theme of human rights is front and center. More specifically, Carson urges a shift from one set of rights to another. The right to "wage war on other organisms"—thereby pushing "whole species over the brink of extinction"—needs to be rescinded, along with the right to despoil the earth for "generations unborn." Their replacement would be a "basic human right" to live without being poisoned and a right to know about our own chemical exposures. As scientific societies become fronts for the chemical industry, this right to know is fundamentally undermined.

Shortly after *Silent Spring*'s thirtieth anniversary in 1992, breast cancer patients took to the streets carrying signs proclaiming, "Rachel Carson Was Right!" and holding aloft copies of her book. The research findings of the day demonstrated a rising incidence of breast cancer and suggested emerging links between breast cancer and pesticide exposure. A seminal paper published in 1994 then showed that women born in the United States between 1947 and 1958 suffered breast cancer at almost three times the rate their great-grandmothers had at the same age.[10] Reading *Silent Spring*—many for the first time—women born in this era began to recognize that pesticides now banned had been used freely and ubiquitously during their infancies and childhoods. Some had firsthand memories of chasing DDT fogging trucks through their neighborhoods as young girls. In their own lives, years later, they saw the effects of the disaster-in-the-making that Carson had predicted.

These women insisted that more breast cancer research funding be directed down environmental pathways with an eye toward identifying preventable causes. To a small extent, it was. The Silent Spring Institute in Newton, Massachusetts, for example, opened its doors in 1994 after breast cancer activists on Cape Cod called for "a laboratory of their own." The Institute continues to conduct original research on the effects of modern chemicals on breast development and breast cancer and curates two major databases, including the Mammary Carcinogens Review Database, used by researchers around the globe.[11] The results of these research efforts have largely confirmed Carson's concerns. To cite just one: a study of 15,000 California mothers and daughters—conducted over fifty years—finds that women exposed in the womb to high levels of DDT have a fourfold risk of developing breast cancer later in life.[12]

As scientists directed more research down the lines of inquiry that Carson elucidated, many more pieces were added to the environmental health jigsaw puzzle. Animal studies showed that ten common pesticides are associated with increases in mammary gland tumors.[13] The Long Island Breast Cancer Study Project, which began in the early 1990s with a focus on pesticides, uncovered a causal link between breast cancer and exposure to air pollution, a connection that is now well documented in the field of epidemiology.[14] At the same time, the field of endocrinology has substantiated Carson's observation that chlorinated pesticides appear to be hormonally active at very low levels of exposure and, among other issues, exert demasculinizing effects during male development.

The phenomenon of endocrine disruption eventually reached the public's attention in 1996 with the publication of *Our Stolen Future*, a book hailed as a sequel to *Silent Spring*.[15] Scientific attention to the issue fanned out from there. We now know that some pesticides can, as Carson hypothesized, mimic estrogen outright. They can also, variously, block male hormones, magnify the sensitivity of cells to female hormones, alter breast development, disrupt menstrual cycling, or otherwise interfere with hormone metabolism. We know that endocrine disruptors include not just pesticides but also certain synthetic ingredients in personal care products, food packaging materials, and home furnishings. We know that there is no such thing as a safe dose: even vanishingly small exposures to hormonally active agents—at levels insufficient to induce acute poisoning—can exert potent, irreversible effects when those exposures occur during key moments of development, including pregnancy and puberty. We know that multiple hormonal pathways can be disrupted—not just those involving reproduction—and that the potential impacts are manifold. Strong evidence suggests that endocrine disruptors play a role in infertility, birth defects, asthma, obesity, diabetes, and learning disabilities.[16]

Of particular note: forty years after *Silent Spring*, epidemiologists discovered that DDT's own endocrine-disrupting abilities include the power to shorten human gestation. Exposure to trace levels of DDT during pregnancy raises the risk for preterm birth, a leading cause of infant death and disability.[17]

In a June 1962 commencement address at Scripps College, Carson looked beyond the just-finished chapters of *Silent Spring* to broader, even existential questions. How is it that environmental harms, committed in the name of dominance and control, continue to accumulate in spite of the abundant evidence of the threats they pose to public health, even humanity itself? It's a recurring, self-destructive pattern of human behavior that is enabled, Carson asserted, by a proclivity for "taking refuge in ignorance" and "a curious unwillingness" to be guided by scientific expertise. The result: "there is all too little awareness that man is *part* of nature, and that the price of conquest may well be the destruction of man himself." This is Carson speaking in her most prophetic voice. And it seems to hint strongly that, of the two diverging roads to the future that head out from *Silent Spring*'s final chapter, she was placing her bets on the superhighway to ruin rather than the less traveled path of ecological enlightenment.

It is difficult, as evidence of environmental damage continues to mount, to say that she was wrong. Pollution in various forms now kills nine million people annually and is responsible for 16 percent of all deaths worldwide—claiming more lives each year than malaria, AIDS, tuberculosis, war, and violence combined.[18] The combustion of fossil fuels is responsible for a large number of these casualties; exposure to chemical pollution, including pesticides, is responsible for many others.

The price of conquest is very high indeed. In fact, it's been monetized. The *Lancet* estimates the worldwide costs attributable to pollution-related disease at $4.6 trillion per year, or 6.2 percent of global economic output, and rising. An earlier economic analysis pegged the public health costs of pesticides alone, and just in the United States, at $11 billion annually.[19]

In other words, the alarming trends Carson identified in *Silent Spring* have, by and large, continued trending in more or less the same alarming direction. Carson documented a fivefold increase in the U.S. production of chemical pesticides from 1947 to 1960. Since that time, production has increased by a factor of four, and pesticide use has doubled, with more than 20,000 different pesticidal products now on the world market. While the environmental advocacy organizations that sprang up in the wake of *Silent Spring*'s publication enjoyed some early, hard-won victories, they did not succeed in moving pest control off the chemical treadmill and into the realm of ecologically inspired solutions.

Nor did they prevent the return of pesticides to the battlefield as chemical weapons. Originally developed during World War II by the U.S. military to destroy enemy crops, the herbicide formulation known as Agent Orange was, two decades later, deployed in Vietnam to defoliate forests suspected of harboring enemy combatants. The U.S. Department of Veterans Affairs now attributes fourteen different illnesses to wartime exposure to Agent Orange. Among them: leukemia, lymphoma, diabetes, heart diseases, and birth defects to the children of Vietnam War veterans.[20]

On the home front, domestic bans on individual organochlorine pesticides mostly provoked a switch to organophosphate and carbamate pesticides, which are less persistent—they linger in soil for months rather than generations—but more toxic. Beginning in the 1980s, neurotoxic pesticides called neonicotinoids came on the market and quickly became popular. Neonicotinoids, like the organochlorine chemicals of yore, can persist for years in water, dust, and soil—and are detectable in commonly consumed foods. Famous for their ability to derange the behavior of pollinators, neonicotinoids have become a leading suspect in colony collapse disorder among honeybees.

Meanwhile, as the parade of ever-changing pesticides marches by, the fundamental problem that Carson identified—indiscriminateness—remains. It is still the case that 99.99 percent of pesticides applied, whatever their names, miss their target pests. Birds, fish, wildlife, and beneficial insects and plants are still caught in the crossfire. Pesticide resistance also remains an intractable problem, with more than 1,000 species of pests now impervious to the chemicals intended to poison them and with pest resistance to biotech crops also surging.[21]

Carson noted that pesticide residues were, by 1960, thoroughly distributed across the globe. This is still the case. The U.S. Geological Survey reports the presence of pesticides in every stream it monitors. Pesticides continue to be detected in all environmental media tested, from groundwater drawn up from deep beneath the earth's surface to ambient air, household dust, rain, fog, snowflakes, soil, and the upper atmosphere. In the United States, according to the U.S. Department of Agriculture's Pesticide Data Program, pesticides are found in more than 70 percent of fruits and vegetables, more than 60 percent of wheat samples, and 99 percent of milk samples. The body fluids of nearly all U.S. adults and children contain pesticides.[22] U.S. farmworkers and their families continue to suffer particularly high exposures, both at work and at home, and in ways that contribute to acute pesticide poisonings, chronic pesticide-related illnesses, and neuropsychological impairments.[23] While *Silent Spring* roused and continues to rouse many to advocacy and action, it is also still the case that corporate influence over public agencies, research institutions, scientific societies, and governments distorts policy-making and drives research and development down chemical pathways.

The Stockholm Convention on Persistent Organic Pollutants, an international treaty signed in 2001 and enacted in 2004, eliminated the worldwide production and use of twelve highly toxic, long-lasting chemicals. All are characterized by their ability to evaporate and travel long distances, sift back down to earth, ascend the food chain, concentrate in body fat, and impair biological processes. These substances are considered inherently unmanageable under any regulatory framework. Eight of the twelve are pesticides singled out by Rachel Carson. Indeed, during the treaty's negotiations, Klaus Töpfer, then executive director of the United Nations Environmental Program, referred to the Convention as "the unfinished business of *Silent Spring*."

While most chemicals on the Convention's list are now banned around the globe, DDT received a special exemption for use in public health emergencies involving malaria—in keeping with Carson's recommendation. Body burdens of the pesticides banned are, happily, now in decline. Because of these bans, lives have been saved and suffering prevented. This, too, is the legacy of *Silent Spring*.

At the same time, developments in the fields of agroecology and integrated pest management have demonstrated that pesticides are not necessary to maintain high yields. We don't have to poison ourselves to eat. The 2009 United Nations report *Agriculture at a Crossroads* found that small-scale farming reliant on ecological rather than chemical pest control is a more sustainable method to feed the world. By protecting crucial ecological resources, such farming may be the only way to prevent the eventual collapse of agricultural systems altogether. Similarly, a major study that examined yields on organic and conventional farms around the world found that ecological farming methods that do not rely on pesticides could produce sufficient food to sustain the current world's population without the need to plow up more acreage.[24] Further, pesticide-free soils, which teem with living organisms, can hold more moisture, are more resilient, and can outperform chemically treated fields during conditions of drought and flooding. In light of these and other findings, a number of United Nations agencies have called for a reduced dependency on pesticides and a shift toward an ecosystem-based approach to pest management as a cornerstone of global food policy and as a strategy to address climate change, with its attendant water crises. In short, the best available evidence shows that replacing chemical spraying with biological solutions to keep pests at bay leads to greater food security in the long run. The ongoing call for biological solutions is also an abiding legacy of *Silent Spring*.[25]

Carson was diagnosed with breast cancer in 1960, at the midpoint of her work on *Silent Spring*. Falsely reassured by her first physician about the seriousness of her situation, she insisted on full disclosure from her second surgeon, but her cancer rapidly metastasized. Surgery and radiation treatments left her immobilized and exhausted, and slowed her writing. Angina, an ulcer, and recurring eye infections further vexed her. As the cancer spread to the vertebrae of her neck, her writing hand went numb. She leaned heavily on her personal assistant, Jeanne Davis, for help with research, typing, and correspondence. At the same time, out of a desire to yield her enemies in industry no further ground from which to launch personal attacks and to maintain the appearance of scientific objectivity, Carson strictly forbade public conversation about her illness.[26] Even to Dorothy, her intimate letters reveal few details about what clearly must have qualified, at some point, as grievous physical suffering. Toward the end, she expressed relief that she had seen *Silent Spring* through to completion, bemusement that she was unable to seize opportunities that her success had, at last, afforded her, and hope for the remission that would not come.

Carson began *Silent Spring* out of a sense of outrage and moral responsibility; seeing and recognizing a horrible wrong, she simply felt she had to do *something*. That she completed it at all, through the nausea, fatigue, and disability of her final illness, and as the single mother of a recently adopted grandnephew on top of that, is astonishing—surely no one would have faulted her for simply letting it go, or giving up. The book's very existence stands as testimony to her strength of character and the depth of her commitments. In a speech delivered to the National Audubon Society a few months before her death, at age fifty-six, she challenged us to continue the environmental revolution to which she had contributed even more than she knew:

> And so the effort must and shall go on. Though the task will never be ended we must engage in it with a patience that refuses to be turned aside, with determination to overcome obstacles, and with pride that it is our privilege to contribute so greatly.

---

1. See, for example, Al Gore, "Rachel Carson and *Silent Spring*," in Peter Matthiessen, ed., *Courage for the Earth: Writers, Scientists, and Activists Celebrate the Life and Writing of Rachel Carson* (Houghton Mifflin, 2007), 63–78.

2. For more on the coevolution of chemical warfare and chemical pest control practices, see Edmund Russell, *War and Nature: Fighting Humans and Insects with Chemicals from World War I to* Silent Spring (Cambridge University Press, 2001).

3. For more on the literary devices of *Silent Spring*, see John Elder, "Withered Sedge and Yellow Wood: Poetry in *Silent Spring*," in Peter Matthiessen, ed., *Courage for the Earth* (Houghton Mifflin, 2007), 79–95.

4. John H. Vandermeer and Ivette Perfecto, *Ecological Complexity and Agro-ecology* (Routledge, 2018), 8–10.

5. Rachel Carson, *The Sea Around Us* (Oxford University Press, 1951), 143.

6. Rachel Carson, "Speech Accepting the John Burroughs Medal, April 1952," in Linda Lear, ed., *Lost Woods: The Discovered Writing of Rachel Carson* (Beacon Press, 1998), 94.

7. Martha Freeman, ed., *Always, Rachel: The Letters of Rachel Carson and Dorothy Freeman, 1952–1964* (Beacon Press, 1995), 259.

8. For more on the ways in which Carson linked the dangers of pesticide spraying with the threats of radioactive fallout, see William Souder, *On a Farther Shore: The Life and Legacy of Rachel Carson* (Crown Publishers, 2012).

9. For an in-depth analysis of the public relations campaigns to discredit *Silent Spring*, see Michael Smith, "'Silence, Miss Carson!': Science, Gender, and the Reception of *Silent Spring*," in Lisa H. Sideris and Kathleen Dean Moore, eds., *Rachel Carson: Legacy and Challenge* (State University of New York Press, 2008), 168–87.

10. Devra Lee Davis et al., "Decreasing Cardiovascular Disease and Increasing Cancer among Whites in the United States from 1973 through 1987: Good News and Bad News," *Journal of the American Medical Association* 271 (1994): 431–37.

11. I worked as a biologist on two other research initiatives that were the indirect result of *Silent Spring*–inspired advocacy efforts: Cornell University's Program on Breast Cancer and Environmental Risk Factors and the California Breast Cancer Research Program.

12. Barbara A. Cohn et al., "DDT Exposure *in Utero* and Breast Cancer," *Journal of Clinical Endocrinology & Metabolism* 100 (2015): 2865–72.

13. Ruthann A. Rudel et al., "Chemicals Causing Mammary Gland Tumors in Animals Signal New Directions for Epidemiology, Chemicals Testing, and Risk Assessment for Breast Cancer Prevention," *Cancer* 109, S12 (2007): 2635–66.

14. Marilie D. Gammon et al., "Environmental Toxins and Breast Cancer on Long Island: I. Polycyclic Hydrocarbon DNA Adducts," *Cancer Epidemiology, Biomarkers and Prevention* 11 (2002): 677–85; Julia Green Brody and Ruthann A. Rudell, "Environmental Pollutants and Breast Cancer: The Evidence from Animal and Human Studies," *Breast Diseases: A Year Book Quarterly* 19 (2008): 17–19.

15. Theo Colborn, Dianne Dumanoski, and John Preston Myers, *Our Stolen Future: Are We Threatening Our Fertility, Intelligence, and Survival? A Scientific Detective Story*, (Dutton, 1996); Carol F. Kwaitkowski et al., "Twenty-five

Years of Endocrine Disruption Science: Remembering Theo Colburn," *Environmental Health Perspectives* 124 (2012): A151–A54; Robert K. Musil, *Rachel Carson and Her Sisters: Remarkable Women Who Have Shaped America's Environment* (Rutgers University Press, 2015), 225–51.

16. R. Thomas Zoeller et al., "Endocrine-Disrupting Chemicals and Public Health Protection: A Statement of Principles from the Endocrine Society," *Endocrinology* 153 (2012): 4097–110.

17. Matthew P. Longnecker et al., "Association between Maternal Serum Concentration of the DDT Metabolite DDE and Preterm and Small-for-Gestational-Age Babies at Birth," *The Lancet* 358 (2001): 110–14.

18. Philip J. Landrigan et al., "The *Lancet* Commission on Pollution and Health," thelancet.com, accessed November 7, 2017.

19. David Pimentel, "Environmental and Economic Costs of the Application of Pesticides Primarily in the United States," *Environment, Development, and Sustainability* 7 (2005): 229–52.

20. Arthur H. Westing, ed., *Herbicides in War: The Long-Term Ecological and Human Consequences* (Stockholm International Peace Research Institute, 1984). See also Viet Thanh Nguyen and Richard Hughes, "The Forgotten Victims of Agent Orange," *The New York Times*, September 16, 2017.

21. Bruce E. Tabashnik, "Surge in Insect Resistance to Transgenic Crops and Prospects for Sustainability," *Nature Biotechnology* 35 (2017): 926–35.

22. Julia G. Brody et al., eds., "Identifying Gaps in Breast Cancer Research: Addressing Disparities and Roles of Physical and Social Environment," California Breast Cancer Research Program, Special Research Initiative (2007), www.cbcrp.org/files/other-publications/GAPS_full.pdf, accessed December 12, 2017.

23. Thomas A. Acury et al., "Lifetime and Current Pesticide Exposure among Latino Farmworkers in Comparison to Other Latino Immigrants," *American Journal of Industrial Medicine* 57 (2014): 776–87; Maria T. Muñoz-Quezada et al., "Chronic Exposure to Organophosphate (OP) Pesticides and Neuropsychological Functioning in Farm Workers: A Review," *International Journal of Occupational and Environmental Health* 22 (2016): 68–79.

24. Catherine Badgley et al., "Organic Agriculture and the Global Food Supply," *Renewable Agriculture and Food Systems* 22 (2007): 86–108.

25. This call dates back to at least 1919 when entomologist Harry Scott Smith coined the term "biological control" in a paper published in the *Journal of Economic Entomology*. At the time, the idea of introducing or otherwise encouraging the natural enemies of particular agricultural scourges—like the cottony-cushion scale that had devastated California's citrus crops—was poised to become the road "*most* traveled by." Research programs on biological pest

control were thriving at some universities and experimental stations, most notably at the University of California, Riverside. Commercial production was likewise ascendant. But the marketing of chemical pesticides after World War II—and their ease of application—made biological methods, with their prerequisite need for ecological knowledge, seem obsolete. Worse, it sabotaged them outright. By 1947, entomologist Paul DeBach—whose research Carson references in *Silent Spring*—warned that DDT residues were harming the living organisms needed for biological control methods that were already established and working effectively. At the same time, banks, betting big on chemical pesticides, began refusing to finance biological control purchases. Commercial insectories, which had provided farmers with predatory and parasitic insects, were defunded. And biological control, as an avenue to the safe and successful management of weed and insect pests, fell into disuse. For more on this history, see Paul DeBach and David Rosen, *Biological Control by Natural Enemies*, 2nd ed. (Cambridge University Press, 1991); C.B. Huffaker and P.S. Messenger, *Theory and Practice of Biological Control* (Academic Press, 1976); Richard C. Sawyer, *To Make a Spotless Orange: Biological Control in California* (Iowa State University Press, 1996). Contemporary calls for biological solutions arise from the field of agroecology and the food sovereignty movement. Beyond the recruitment of natural enemies to serve as substitutes for chemical poisons, practitioners of agroecology employ a myriad of techniques—crop rotation, intercropping with multiple species, nutrient recycling, permaculture, pasturing, promotion of genetic diversity—that seek to mimic the interspecies complexity of natural systems and obviate the need for chemical inputs. See Peter M. Rosset and Miguel A. Altieri, *Agroecology: Science and Politics* (Fernwood Books, 2017).

26. For more on Carson's lived experience as a cancer patient, see Linda Lear, *Rachel Carson: Witness for Nature* (Houghton Mifflin, 1997), and Sandra Steingraber, *Living Downstream: An Ecologist's Personal Investigation of Cancer and the Environment* (Perseus Books Group, 2010), 17–34.

SILENT SPRING

To Albert Schweitzer
who said

"Man has lost the capacity to foresee
and to forestall. He will end by
destroying the earth."

The sedge is wither'd from the lake,
  And no birds sing.

KEATS

I am pessimistic about the human race because it is too ingenious for its own good. Our approach to nature is to beat it into submission. We would stand a better chance of survival if we accommodated ourselves to this planet and viewed it appreciatively instead of skeptically and dictatorially.

E. B. WHITE

Acknowledgments

In a letter written in January 1958, Olga Owens Huckins told me of her own bitter experience of a small world made lifeless, and so brought my attention sharply back to a problem with which I had long been concerned. I then realized I must write this book.

During the years since then I have received help and encouragement from so many people that it is not possible to name them all here. Those who have freely shared with me the fruits of many years' experience and study represent a wide variety of government agencies in this and other countries, many universities and research institutions, and many professions. To all of them I express my deepest thanks for time and thought so generously given.

In addition my special gratitude goes to those who took time to read portions of the manuscript and to offer comment and criticism based on their own expert knowledge. Although the final responsibility for the accuracy and validity of the text is mine, I could not have completed the book without the generous help of these specialists: L. G. Bartholomew, M.D., of the Mayo Clinic, John J. Biesele of the University of Texas, A. W. A. Brown of the University of Western Ontario, Morton S. Biskind, M.D., of Westport, Connecticut, C. J. Briejèr of the Plant Protection Service in Holland, Clarence Cottam of the Rob and Bessie Welder Wildlife Foundation, George Crile, Jr., M.D., of the Cleveland Clinic, Frank Egler of Norfolk, Connecticut, Malcolm M. Hargraves, M.D., of the Mayo Clinic, W. C. Hueper, M.D., of the National Cancer Institute, C. J. Kerswill of the Fisheries Research Board of Canada, Olaus Murie of the Wilderness Society, A. D. Pickett of the Canada Department of Agriculture, Thomas G. Scott of the Illinois Natural History Survey, Clarence Tarzwell of the Taft Sanitary Engineering Center, and George J. Wallace of Michigan State University.

Every writer of a book based on many diverse facts owes much to the skill and helpfulness of librarians. I owe such a debt to many, but especially to Ida K. Johnston of the

Department of the Interior Library and to Thelma Robinson of the Library of the National Institutes of Health.

As my editor, Paul Brooks has given steadfast encouragement over the years and has cheerfully accommodated his plans to postponements and delays. For this, and for his skilled editorial judgment, I am everlastingly grateful.

I have had capable and devoted assistance in the enormous task of library research from Dorothy Algire, Jeanne Davis, and Bette Haney Duff. And I could not possibly have completed the task, under circumstances sometimes difficult, except for the faithful help of my housekeeper, Ida Sprow.

Finally, I must acknowledge our vast indebtedness to a host of people, many of them unknown to me personally, who have nevertheless made the writing of this book seem worthwhile. These are the people who first spoke out against the reckless and irresponsible poisoning of the world that man shares with all other creatures, and who are even now fighting the thousands of small battles that in the end will bring victory for sanity and common sense in our accommodation to the world that surrounds us.

RACHEL CARSON

Author's Note

I HAVE NOT WISHED to burden the text with
footnotes but I realize that many of my readers
will wish to pursue some of the subjects discussed.
I have therefore included a list of my principal
sources of information, arranged by chapter and
page, in an appendix which will be found at the
back of the book.

R.C.

ONE

A Fable for Tomorrow

THERE WAS ONCE a town in the heart of America where all life seemed to live in harmony with its surroundings. The town lay in the midst of a checkerboard of prosperous farms, with fields of grain and hillsides of orchards where, in spring, white clouds of bloom drifted above the green fields. In autumn, oak and maple and birch set up a blaze of color that flamed and flickered across a backdrop of pines. Then foxes barked in the hills and deer silently crossed the fields, half hidden in the mists of the fall mornings.

Along the roads, laurel, viburnum and alder, great ferns and wildflowers delighted the traveler's eye through much of the year. Even in winter the roadsides were places of beauty, where countless birds came to feed on the berries and on the seed heads of the dried weeds rising above the snow. The countryside was, in fact, famous for the abundance and variety of its bird life, and when the flood of migrants was pouring through in spring and fall people traveled from great distances to observe them. Others came to fish the streams, which flowed clear and cold out of the hills and contained shady pools where trout lay. So it had been from the days many years ago when the first settlers raised their houses, sank their wells, and built their barns.

Then a strange blight crept over the area and everything began to change. Some evil spell had settled on the community: mysterious maladies swept the flocks of chickens; the cattle and sheep sickened and died. Everywhere was a shadow of death. The farmers spoke of much illness among their families. In the town the doctors had become more and more puzzled by new kinds of sickness appearing among their patients. There had been several sudden and unexplained deaths, not only among adults but even among children, who would be stricken suddenly while at play and die within a few hours.

There was a strange stillness. The birds, for example—where had they gone? Many people spoke of them, puzzled and disturbed. The feeding stations in the backyards were deserted. The few birds seen anywhere were moribund; they trembled violently and could not fly. It was a spring without voices. On the mornings that had once throbbed with the dawn chorus of robins, catbirds, doves, jays, wrens, and scores of other bird voices there was now no sound; only silence lay over the fields and woods and marsh.

On the farms the hens brooded, but no chicks hatched. The farmers complained that they were unable to raise any pigs— the litters were small and the young survived only a few days. The apple trees were coming into bloom but no bees droned among the blossoms, so there was no pollination and there would be no fruit.

The roadsides, once so attractive, were now lined with browned and withered vegetation as though swept by fire. These, too, were silent, deserted by all living things. Even the streams were now lifeless. Anglers no longer visited them, for all the fish had died.

In the gutters under the eaves and between the shingles of the roofs, a white granular powder still showed a few patches; some weeks before it had fallen like snow upon the roofs and the lawns, the fields and streams.

No witchcraft, no enemy action had silenced the rebirth of new life in this stricken world. The people had done it themselves.

*

This town does not actually exist, but it might easily have a thousand counterparts in America or elsewhere in the world. I know of no community that has experienced all the misfortunes I describe. Yet every one of these disasters has actually happened somewhere, and many real communities have already suffered a substantial number of them. A grim specter has crept upon us almost unnoticed, and this imagined tragedy may easily become a stark reality we all shall know.

What has already silenced the voices of spring in countless towns in America? This book is an attempt to explain.

TWO

The Obligation to Endure

THE HISTORY OF LIFE on earth has been a history of inter-action between living things and their surroundings. To a large extent, the physical form and the habits of the earth's vegetation and its animal life have been molded by the environment. Considering the whole span of earthly time, the opposite effect, in which life actually modifies its surroundings, has been relatively slight. Only within the moment of time represented by the present century has one species—man—acquired significant power to alter the nature of his world.

During the past quarter century this power has not only increased to one of disturbing magnitude but it has changed in character. The most alarming of all man's assaults upon the environment is the contamination of air, earth, rivers, and sea with dangerous and even lethal materials. This pollution is for the most part irrecoverable; the chain of evil it initiates not only in the world that must support life but in living tissues is for the most part irreversible. In this now universal contamination of the environment, chemicals are the sinister and little-recognized partners of radiation in changing the very nature of the world—the very nature of its life. Strontium 90, released through nuclear explosions into the air, comes to earth in rain or drifts down as fallout, lodges in soil, enters into the grass or corn or wheat grown there, and in time takes up its abode in the bones of a human being, there to remain until his death. Similarly, chemicals sprayed on croplands or forests or gardens lie long in soil, entering into living organisms, passing from one to another in a chain of poisoning and death. Or they pass mysteriously by underground streams until they emerge and, through the alchemy of air and sunlight, combine into new forms that kill vegetation, sicken cattle, and work unknown harm on those who drink from once pure wells. As Albert Schweitzer has said, "Man can hardly even recognize the devils of his own creation."

It took hundreds of millions of years to produce the life that now inhabits the earth—eons of time in which that developing and evolving and diversifying life reached a state of adjustment and balance with its surroundings. The environment, rigorously shaping and directing the life it supported, contained elements that were hostile as well as supporting. Certain rocks gave out dangerous radiation; even within the light of the sun, from which all life draws its energy, there were short-wave radiations with power to injure. Given time—time not in years but in millennia—life adjusts, and a balance has been reached. For time is the essential ingredient; but in the modern world there is no time.

The rapidity of change and the speed with which new situations are created follow the impetuous and heedless pace of man rather than the deliberate pace of nature. Radiation is no longer merely the background radiation of rocks, the bombardment of cosmic rays, the ultraviolet of the sun that have existed before there was any life on earth; radiation is now the unnatural creation of man's tampering with the atom. The chemicals to which life is asked to make its adjustment are no longer merely the calcium and silica and copper and all the rest of the minerals washed out of the rocks and carried in rivers to the sea; they are the synthetic creations of man's inventive mind, brewed in his laboratories, and having no counterparts in nature.

To adjust to these chemicals would require time on the scale that is nature's; it would require not merely the years of a man's life but the life of generations. And even this, were it by some miracle possible, would be futile, for the new chemicals come from our laboratories in an endless stream; almost five hundred annually find their way into actual use in the United States alone. The figure is staggering and its implications are not easily grasped—500 new chemicals to which the bodies of men and animals are required somehow to adapt each year, chemicals totally outside the limits of biologic experience.

Among them are many that are used in man's war against nature. Since the mid-1940's over 200 basic chemicals have been created for use in killing insects, weeds, rodents, and other organisms described in the modern vernacular as "pests"; and they are sold under several thousand different brand names.

These sprays, dusts, and aerosols are now applied almost universally to farms, gardens, forests, and homes—nonselective chemicals that have the power to kill every insect, the "good" and the "bad," to still the song of birds and the leaping of fish in the streams, to coat the leaves with a deadly film, and to linger on in soil—all this though the intended target may be only a few weeds or insects. Can anyone believe it is possible to lay down such a barrage of poisons on the surface of the earth without making it unfit for all life? They should not be called "insecticides," but "biocides."

The whole process of spraying seems caught up in an endless spiral. Since DDT was released for civilian use, a process of escalation has been going on in which ever more toxic materials must be found. This has happened because insects, in a triumphant vindication of Darwin's principle of the survival of the fittest, have evolved super races immune to the particular insecticide used, hence a deadlier one has always to be developed—and then a deadlier one than that. It has happened also because, for reasons to be described later, destructive insects often undergo a "flareback," or resurgence, after spraying, in numbers greater than before. Thus the chemical war is never won, and all life is caught in its violent crossfire.

Along with the possibility of the extinction of mankind by nuclear war, the central problem of our age has therefore become the contamination of man's total environment with such substances of incredible potential for harm—substances that accumulate in the tissues of plants and animals and even penetrate the germ cells to shatter or alter the very material of heredity upon which the shape of the future depends.

Some would-be architects of our future look toward a time when it will be possible to alter the human germ plasm by design. But we may easily be doing so now by inadvertence, for many chemicals, like radiation, bring about gene mutations. It is ironic to think that man might determine his own future by something so seemingly trivial as the choice of an insect spray.

All this has been risked—for what? Future historians may well be amazed by our distorted sense of proportion. How could intelligent beings seek to control a few unwanted species by a method that contaminated the entire environment and brought the threat of disease and death even to their own kind? Yet this is precisely what we have done. We have done it, moreover, for reasons that collapse the moment we examine them. We are told that the enormous and expanding use of pesticides is necessary to maintain farm production. Yet is our real problem not one of *overproduction*? Our farms, despite measures to remove acreages from production and to pay farmers *not* to produce, have yielded such a staggering excess of crops that the American taxpayer in 1962 is paying out more than one billion dollars a year as the total carrying cost of the surplus-food storage program. And is the situation helped when one branch of the Agriculture Department tries to reduce production while another states, as it did in 1958, "It is believed generally that reduction of crop acreages under provisions of the Soil Bank will stimulate interest in use of chemicals to obtain maximum production on the land retained in crops."

All this is not to say there is no insect problem and no need of control. I am saying, rather, that control must be geared to realities, not to mythical situations, and that the methods employed must be such that they do not destroy us along with the insects.

The problem whose attempted solution has brought such a train of disaster in its wake is an accompaniment of our modern way of life. Long before the age of man, insects inhabited the earth—a group of extraordinarily varied and adaptable beings. Over the course of time since man's advent, a small percentage of the more than half a million species of insects have come into conflict with human welfare in two principal ways: as competitors for the food supply and as carriers of human disease.

Disease-carrying insects become important where human beings are crowded together, especially under conditions where sanitation is poor, as in time of natural disaster or war or in situations of extreme poverty and deprivation. Then control of some sort becomes necessary. It is a sobering fact, however, as we shall presently see, that the method of massive chemical control has had only limited success, and also threatens to worsen the very conditions it is intended to curb.

Under primitive agricultural conditions the farmer had few insect problems. These arose with the intensification of agriculture—the devotion of immense acreages to a single crop. Such a system set the stage for explosive increases in specific insect populations. Single-crop farming does not take advantage of the principles by which nature works; it is agriculture as an engineer might conceive it to be. Nature has introduced great variety into the landscape, but man has displayed a passion for simplifying it. Thus he undoes the built-in checks and balances by which nature holds the species within bounds. One important natural check is a limit on the amount of suitable habitat for each species. Obviously then, an insect that lives on wheat can build up its population to much higher levels on a farm devoted to wheat than on one in which wheat is intermingled with other crops to which the insect is not adapted.

The same thing happens in other situations. A generation or more ago, the towns of large areas of the United States lined their streets with the noble elm tree. Now the beauty they hopefully created is threatened with complete destruction as disease sweeps through the elms, carried by a beetle that would have only limited chance to build up large populations and to spread from tree to tree if the elms were only occasional trees in a richly diversified planting.

Another factor in the modern insect problem is one that must be viewed against a background of geologic and human history: the spreading of thousands of different kinds of organisms from their native homes to invade new territories. This worldwide migration has been studied and graphically described by the British ecologist Charles Elton in his recent book *The Ecology of Invasions.* During the Cretaceous Period, some hundred million years ago, flooding seas cut many land bridges between continents and living things found themselves confined in what Elton calls "colossal separate nature reserves." There, isolated from others of their kind, they developed many new species. When some of the land masses were joined again, about 15 million years ago, these species began to move out into new territories—a movement that is not only still in progress but is now receiving considerable assistance from man.

The importation of plants is the primary agent in the modern spread of species, for animals have almost invariably gone along with the plants, quarantine being a comparatively recent and not completely effective innovation. The United States Office of Plant Introduction alone has introduced almost 200,000 species and varieties of plants from all over the world. Nearly half of the 180 or so major insect enemies of plants in the United States are accidental imports from abroad, and most of them have come as hitchhikers on plants.

In new territory, out of reach of the restraining hand of the natural enemies that kept down its numbers in its native land, an invading plant or animal is able to become enormously abundant. Thus it is no accident that our most troublesome insects are introduced species.

These invasions, both the naturally occurring and those dependent on human assistance, are likely to continue indefinitely. Quarantine and massive chemical campaigns are only extremely expensive ways of buying time. We are faced, according to Dr. Elton, "with a life-and-death need not just to find new technological means of suppressing this plant or that animal"; instead we need the basic knowledge of animal populations and their relations to their surroundings that will "promote an even balance and damp down the explosive power of outbreaks and new invasions."

Much of the necessary knowledge is now available but we do not use it. We train ecologists in our universities and even employ them in our governmental agencies but we seldom take their advice. We allow the chemical death rain to fall as though there were no alternative, whereas in fact there are many, and our ingenuity could soon discover many more if given opportunity.

Have we fallen into a mesmerized state that makes us accept as inevitable that which is inferior or detrimental, as though having lost the will or the vision to demand that which is good? Such thinking, in the words of the ecologist Paul Shepard, "idealizes life with only its head out of water, inches above the limits of toleration of the corruption of its own environment . . . Why should we tolerate a diet of weak poisons, a home in insipid surroundings, a circle of acquaintances who are not quite our enemies, the noise of motors with just enough relief to prevent insanity? Who would want to live in a world which is just not quite fatal?"

Yet such a world is pressed upon us. The crusade to create a chemically sterile, insect-free world seems to have engendered a fanatic zeal on the part of many specialists and most of the so-called control agencies. On every hand there is evidence that those engaged in spraying operations exercise a ruthless power. "The regulatory entomologists . . . function as prosecutor, judge and jury, tax assessor and collector and sheriff to enforce their own orders," said Connecticut entomologist Neely Turner. The most flagrant abuses go unchecked in both state and federal agencies.

It is not my contention that chemical insecticides must never be used. I do contend that we have put poisonous and biologically potent chemicals indiscriminately into the hands of persons largely or wholly ignorant of their potentials for harm. We have subjected enormous numbers of people to contact with these poisons, without their consent and often without their knowledge. If the Bill of Rights contains no guarantee that a citizen shall be secure against lethal poisons distributed either by private individuals or by public officials, it is surely only because our forefathers, despite their considerable wisdom and foresight, could conceive of no such problem.

I contend, furthermore, that we have allowed these chemicals to be used with little or no advance investigation of their effect on soil, water, wildlife, and man himself. Future generations are unlikely to condone our lack of prudent concern for the integrity of the natural world that supports all life.

There is still very limited awareness of the nature of the threat. This is an era of specialists, each of whom sees his own problem and is unaware of or intolerant of the larger frame into which it fits. It is also an era dominated by industry, in which the right to make a dollar at whatever cost is seldom challenged. When the public protests, confronted with some obvious evidence of damaging results of pesticide applications, it is fed little tranquilizing pills of half truth. We urgently need an end to these false assurances, to the sugar coating of unpalatable facts. It is the public that is being asked to assume the risks that the insect controllers calculate. The public must decide whether it wishes to continue on the present road, and it can do so only when in full possession of the facts. In the words of Jean Rostand, "The obligation to endure gives us the right to know."

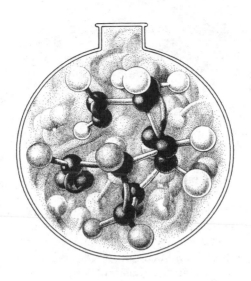

THREE

Elixirs of Death

FOR THE FIRST TIME in the history of the world, every human being is now subjected to contact with dangerous chemicals, from the moment of conception until death. In the less than two decades of their use, the synthetic pesticides have been so thoroughly distributed throughout the animate and inanimate world that they occur virtually everywhere. They have been recovered from most of the major river systems and even from streams of groundwater flowing unseen through the earth. Residues of these chemicals linger in soil to which they may have been applied a dozen years before. They have entered and lodged in the bodies of fish, birds, reptiles, and domestic and wild animals so universally that scientists carrying on animal experiments find it almost impossible to locate subjects free from such contamination. They have been found in fish in remote mountain lakes, in earthworms burrowing in soil, in the eggs of birds—and in man himself. For these chemicals are now stored in the bodies of the vast majority of human beings, regardless of age. They occur in the mother's milk, and probably in the tissues of the unborn child.

All this has come about because of the sudden rise and prodigious growth of an industry for the production of man-made
or synthetic chemicals with insecticidal properties. This industry is a child of the Second World War. In the course of
developing agents of chemical warfare, some of the chemicals
created in the laboratory were found to be lethal to insects.
The discovery did not come by chance: insects were widely
used to test chemicals as agents of death for man.

The result has been a seemingly endless stream of synthetic
insecticides. In being man-made—by ingenious laboratory ma-
nipulation of the molecules, substituting atoms, altering their
arrangement—they differ sharply from the simpler inorganic
insecticides of prewar days. These were derived from naturally
occurring minerals and plant products—compounds of arsenic,
copper, lead, manganese, zinc, and other minerals, pyrethrum
from the dried flowers of chrysanthemums, nicotine sulphate
from some of the relatives of tobacco, and rotenone from leguminous plants of the East Indies.

What sets the new synthetic insecticides apart is their enormous biological potency. They have immense power not
merely to poison but to enter into the most vital processes of
the body and change them in sinister and often deadly ways.
Thus, as we shall see, they destroy the very enzymes whose
function is to protect the body from harm, they block the oxidation processes from which the body receives its energy, they
prevent the normal functioning of various organs, and they
may initiate in certain cells the slow and irreversible change
that leads to malignancy.

Yet new and more deadly chemicals are added to the list
each year and new uses are devised so that contact with
these materials has become practically worldwide. The production of synthetic pesticides in the United States soared
from 124,259,000 pounds in 1947 to 637,666,000 pounds in
1960—more than a fivefold increase. The wholesale value of
these products was well over a quarter of a billion dollars. But
in the plans and hopes of the industry this enormous production is only a beginning.

A Who's Who of pesticides is therefore of concern to us all.
If we are going to live so intimately with these chemicals—
eating and drinking them, taking them into the very marrow of

our bones—we had better know something about their nature and their power.

Although the Second World War marked a turning away from inorganic chemicals as pesticides into the wonder world of the carbon molecule, a few of the old materials persist. Chief among these is arsenic, which is still the basic ingredient in a variety of weed and insect killers. Arsenic is a highly toxic mineral occurring widely in association with the ores of various metals, and in very small amounts in volcanoes, in the sea, and in spring water. Its relations to man are varied and historic. Since many of its compounds are tasteless, it has been a favorite agent of homicide from long before the time of the Borgias to the present. Arsenic was the first recognized elementary carcinogen (or cancer-causing substance), identified in chimney soot and linked to cancer nearly two centuries ago by an English physician. Epidemics of chronic arsenical poisoning involving whole populations over long periods are on record. Arsenic-contaminated environments have also caused sickness and death among horses, cows, goats, pigs, deer, fishes, and bees; despite this record arsenical sprays and dusts are widely used. In the arsenic-sprayed cotton country of southern United States beekeeping as an industry has nearly died out. Farmers using arsenic dusts over long periods have been afflicted with chronic arsenic poisoning; livestock have been poisoned by crop sprays or weed killers containing arsenic. Drifting arsenic dusts from blueberry lands have spread over neighboring farms, contaminating streams, fatally poisoning bees and cows, and causing human illness. "It is scarcely possible . . . to handle arsenicals with more utter disregard of the general health than that which has been practiced in our country in recent years," said Dr. W. C. Hueper, of the National Cancer Institute, an authority on environmental cancer. "Anyone who has watched the dusters and sprayers of arsenical insecticides at work must have been impressed by the almost supreme carelessness with which the poisonous substances are dispensed."

Modern insecticides are still more deadly. The vast majority fall into one of two large groups of chemicals. One, represented by DDT, is known as the "chlorinated hydrocarbons." The other group consists of the organic phosphorus insecticides, and is represented by the reasonably familiar malathion and parathion. All have one thing in common. As mentioned above, they are built on a basis of carbon atoms, which are also the indispensable building blocks of the living world, and thus classed as "organic." To understand them, we must see of what they are made, and how, although linked with the basic chemistry of all life, they lend themselves to the modifications which make them agents of death.

The basic element, carbon, is one whose atoms have an almost infinite capacity for uniting with each other in chains and rings and various other configurations, and for becoming linked with atoms of other substances. Indeed, the incredible diversity of living creatures from bacteria to the great blue whale is largely due to this capacity of carbon. The complex protein molecule has the carbon atom as its basis, as have molecules of fat, carbohydrates, enzymes, and vitamins. So, too, have enormous numbers of nonliving things, for carbon is not necessarily a symbol of life.

Some organic compounds are simply combinations of carbon and hydrogen. The simplest of these is methane, or marsh gas, formed in nature by the bacterial decomposition of organic matter under water. Mixed with air in proper proportions, methane becomes the dreaded "fire damp" of coal mines. Its structure is beautifully simple, consisting of one carbon atom to which four hydrogen atoms have become attached:

$$\begin{array}{ccc} H & & H \\ & \diagdown \; \diagup & \\ & C & \\ & \diagup \; \diagdown & \\ H & & H \end{array}$$

Chemists have discovered that it is possible to detach one or all of the hydrogen atoms and substitute other elements. For example, by substituting one atom of chlorine for one of hydrogen we produce methyl chloride:

$$\begin{array}{ccc} H & & Cl \\ & \diagdown \; \diagup & \\ & C & \\ & \diagup \; \diagdown & \\ H & & H \end{array}$$

Take away three hydrogen atoms and substitute chlorine and we have the anesthetic chloroform:

$$\begin{array}{ccc} H & & Cl \\ & \diagdown C \diagup & \\ & \diagup \diagdown & \\ Cl & & Cl \end{array}$$

Substitute chlorine atoms for all of the hydrogen atoms and the result is carbon tetrachloride, the familiar cleaning fluid:

$$\begin{array}{ccc} Cl & & Cl \\ & \diagdown C \diagup & \\ & \diagup \diagdown & \\ Cl & & Cl \end{array}$$

In the simplest possible terms, these changes rung upon the basic molecule of methane illustrate what a chlorinated hydrocarbon is. But this illustration gives little hint of the true complexity of the chemical world of the hydrocarbons, or of the manipulations by which the organic chemist creates his infinitely varied materials. For instead of the simple methane molecule with its single carbon atom, he may work with hydrocarbon molecules consisting of many carbon atoms, arranged in rings or chains, with side chains or branches, holding to themselves with chemical bonds not merely simple atoms of hydrogen or chlorine but also a wide variety of chemical groups. By seemingly slight changes the whole character of the substance is changed; for example, not only what is attached but the place of attachment to the carbon atom is highly important. Such ingenious manipulations have produced a battery of poisons of truly extraordinary power.

DDT (short for dichloro-diphenyl-trichloro-ethane) was first synthesized by a German chemist in 1874, but its properties as an insecticide were not discovered until 1939. Almost immediately DDT was hailed as a means of stamping out insect-borne disease and winning the farmers' war against crop destroyers overnight. The discoverer, Paul Müller of Switzerland, won the Nobel Prize.

DDT is now so universally used that in most minds the product takes on the harmless aspect of the familiar. Perhaps the myth of the harmlessness of DDT rests on the fact that one of its first uses was the wartime dusting of many thousands of soldiers, refugees, and prisoners, to combat lice. It is widely believed that since so many people came into extremely intimate contact with DDT and suffered no immediate ill effects the chemical must certainly be innocent of harm. This understandable misconception arises from the fact that—unlike other chlorinated hydrocarbons—DDT *in powder form* is not readily absorbed through the skin. Dissolved in oil, as it usually is, DDT is definitely toxic. If swallowed, it is absorbed slowly through the digestive tract; it may also be absorbed through the lungs. Once it has entered the body it is stored largely in organs rich in fatty substances (because DDT itself is fat-soluble) such as the adrenals, testes, or thyroid. Relatively large amounts are deposited in the liver, kidneys, and the fat of the large, protective mesenterics that enfold the intestines.

This storage of DDT begins with the smallest conceivable intake of the chemical (which is present as residues on most foodstuffs) and continues until quite high levels are reached. The fatty storage depots act as biological magnifiers, so that an intake of as little as $1/10$ of 1 part per million in the diet results in storage of about 10 to 15 parts per million, an increase of one hundredfold or more. These terms of reference, so commonplace to the chemist or the pharmacologist, are unfamiliar to most of us. One part in a million sounds like a very small amount—and so it is. But such substances are so potent that a minute quantity can bring about vast changes in the body. In animal experiments, 3 parts per million has been found to inhibit an essential enzyme in heart muscle; only 5 parts per million has brought about necrosis or disintegration of liver cells; only 2.5 parts per million of the closely related chemicals dieldrin and chlordane did the same.

This is really not surprising. In the normal chemistry of the human body there is just such a disparity between cause and effect. For example, a quantity of iodine as small as two ten-thousandths of a gram spells the difference between health and disease. Because these small amounts of pesticides are cumulatively stored and only slowly excreted, the threat of chronic poisoning and degenerative changes of the liver and other organs is very real.

Scientists do not agree upon how much DDT can be stored in the human body. Dr. Arnold Lehman, who is the chief pharmacologist of the Food and Drug Administration, says there is neither a floor below which DDT is not absorbed nor a ceiling beyond which absorption and storage ceases. On the other hand, Dr. Wayland Hayes of the United States Public Health Service contends that in every individual a point of equilibrium is reached, and that DDT in excess of this amount is excreted. For practical purposes it is not particularly important which of these men is right. Storage in human beings has been well investigated, and we know that the average person is storing potentially harmful amounts. According to various studies, individuals with no known exposure (except the inevitable dietary one) store an average of 5.3 parts per million to 7.4 parts per million; agricultural workers 17.1 parts per million; and workers in insecticide plants as high as 648 parts per million! So the range of proven storage is quite wide and, what is even more to the point, the minimum figures are above the level at which damage to the liver and other organs or tissues may begin.

One of the most sinister features of DDT and related chemicals is the way they are passed on from one organism to another through all the links of the food chains. For example, fields of alfalfa are dusted with DDT; meal is later prepared from the alfalfa and fed to hens; the hens lay eggs which contain DDT. Or the hay, containing residues of 7 to 8 parts per million, may be fed to cows. The DDT will turn up in the milk in the amount of about 3 parts per million, but in butter made from this milk the concentration may run to 65 parts per million. Through such a process of transfer, what started out as a very small amount of DDT may end as a heavy concentration. Farmers nowadays find it difficult to obtain uncontaminated fodder for their milk cows, though the Food and Drug Administration forbids the presence of insecticide residues in milk shipped in interstate commerce.

The poison may also be passed on from mother to offspring. Insecticide residues have been recovered from human milk in samples tested by Food and Drug Administration scientists. This means that the breast-fed human infant is receiving small but regular additions to the load of toxic chemicals building up in his body. It is by no means his first exposure, however: there is good reason to believe this begins while he is still in the womb. In experimental animals the chlorinated hydrocarbon insecticides freely cross the barrier of the placenta, the traditional protective shield between the embryo and harmful substances in the mother's body. While the quantities so received by human infants would normally be small, they are not unimportant because children are more susceptible to poisoning than adults. This situation also means that today the average individual almost certainly starts life with the first deposit of the growing load of chemicals his body will be required to carry thenceforth.

All these facts—storage at even low levels, subsequent accumulation, and occurrence of liver damage at levels that may easily occur in normal diets, caused Food and Drug Administration scientists to declare as early as 1950 that it is "extremely likely the potential hazard of DDT has been underestimated." There has been no such parallel situation in medical history. No one yet knows what the ultimate consequences may be.

Chlordane, another chlorinated hydrocarbon, has all the unpleasant attributes of DDT plus a few that are peculiarly its own. Its residues are long persistent in soil, on foodstuffs, or on surfaces to which it may be applied, yet it is also quite volatile and poisoning by inhalation is a definite risk to anyone handling or exposed to it. Chlordane makes use of all available portals to enter the body. It penetrates the skin easily, is breathed in as vapor, and of course is absorbed from the digestive tract if residues are swallowed. Like all other chlorinated hydrocarbons, its deposits build up in the body in cumulative fashion. A diet containing such a small amount of chlordane as 2.5 parts per million may eventually lead to storage of 75 parts per million in the fat of experimental animals.

So experienced a pharmacologist as Dr. Lehman has described chlordane as "one of the most toxic of insecticides—anyone handling it could be poisoned." Judging by the carefree liberality with which dusts for lawn treatments by suburbanites are laced with chlordane, this warning has not been taken to heart. The fact that the suburbanite is not instantly stricken has little meaning, for the toxins may sleep long in his body, to become manifest months or years later in an obscure disorder almost impossible to trace to its origins. On the other hand, death may strike quickly. One victim who accidentally spilled a 25 per cent solution on his skin developed symptoms of poisoning within 40 minutes and died before medical help could be obtained. No reliance can be placed on receiving advance warning which might allow treatment to be had in time.

Heptachlor, one of the constituents of chlordane, is marketed as a separate formulation. It has a particularly high capacity for storage in fat. If the diet contains as little as $1/10$ of 1 part per million there will be measurable amounts of heptachlor in the body. It also has the curious ability to undergo change into a chemically distinct substance known as heptachlor epoxide. It does this in soil and in the tissues of both plants and animals. Tests on birds indicate that the epoxide that results from this change is about four times as toxic as the original chemical, which in turn is four times as toxic as chlordane.

As long ago as the mid-1930's a special group of hydrocarbons, the chlorinated naphthalenes, was found to cause hepatitis, and also a rare and almost invariably fatal liver disease in persons subjected to occupational exposure. They have led to illness and death of workers in electrical industries; and more recently, in agriculture, they have been considered a cause of a mysterious and usually fatal disease of cattle. In view of these antecedents, it is not surprising that three of the insecticides that belong to this group are among the most violently poisonous of all the hydrocarbons. These are dieldrin, aldrin, and endrin.

Dieldrin, named for a German chemist, Diels, is about 5 times as toxic as DDT when swallowed but 40 times as toxic when absorbed through the skin in solution. It is notorious for striking quickly and with terrible effect at the nervous system, sending the victims into convulsions. Persons thus poisoned recover so slowly as to indicate chronic effects. As with other chlorinated hydrocarbons, these long-term effects include severe damage to the liver. The long duration of its residues and the effective insecticidal action make dieldrin one of the most used insecticides today, despite the appalling destruction of wildlife that has followed its use. As tested on quail and pheasants, it has proved to be about 40 to 50 times as toxic as DDT.

There are vast gaps in our knowledge of how dieldrin is stored or distributed in the body, or excreted, for the chemists' ingenuity in devising insecticides has long ago outrun biological knowledge of the way these poisons affect the living organism. However, there is every indication of long storage in the human body, where deposits may lie dormant like a slumbering volcano, only to flare up in periods of physiological stress when the body draws upon its fat reserves. Much of what we do know has been learned through hard experience in the antimalarial campaigns carried out by the World Health Organization. As soon as dieldrin was substituted for DDT in malaria-control work (because the malaria mosquitoes had become resistant to DDT), cases of poisoning among the spraymen began to occur. The seizures were severe—from half to all (varying in the different programs) of the men affected went into convulsions and several died. Some had convulsions as long as *four months* after the last exposure.

Aldrin is a somewhat mysterious substance, for although it exists as a separate entity it bears the relation of alter ego to dieldrin. When carrots are taken from a bed treated with aldrin they are found to contain residues of dieldrin. This change occurs in living tissues and also in soil. Such alchemistic transformations have led to many erroneous reports, for if a chemist, knowing aldrin has been applied, tests for it he will be deceived into thinking all residues have been dissipated. The residues are there, but they are dieldrin and this requires a different test.

Like dieldrin, aldrin is extremely toxic. It produces degenerative changes in the liver and kidneys. A quantity the size of an aspirin tablet is enough to kill more than 400 quail. Many cases of human poisonings are on record, most of them in connection with industrial handling.

Aldrin, like most of this group of insecticides, projects a menacing shadow into the future, the shadow of sterility. Pheasants fed quantities too small to kill them nevertheless laid few eggs, and the chicks that hatched soon died. The effect is not confined to birds. Rats exposed to aldrin had fewer pregnancies and their young were sickly and short-lived. Puppies born of treated mothers died within three days. By one means or another, the new generations suffer for the poisoning of their parents. No one knows whether the same effect will be seen in human beings, yet this chemical has been sprayed from airplanes over suburban areas and farmlands.

Endrin is the most toxic of all the chlorinated hydrocarbons. Although chemically rather closely related to dieldrin, a little twist in its molecular structure makes it 5 times as poisonous. It makes the progenitor of all this group of insecticides, DDT, seem by comparison almost harmless. It is 15 times as poisonous as DDT to mammals, 30 times as poisonous to fish, and about 300 times as poisonous to some birds.

In the decade of its use, endrin has killed enormous numbers of fish, has fatally poisoned cattle that have wandered into sprayed orchards, has poisoned wells, and has drawn a sharp warning from at least one state health department that its careless use is endangering human lives.

In one of the most tragic cases of endrin poisoning there was no apparent carelessness; efforts had been made to take precautions apparently considered adequate. A year-old child had been taken by his American parents to live in Venezuela. There were cockroaches in the house to which they moved, and after a few days a spray containing endrin was used. The baby and the small family dog were taken out of the house before the spraying was done about nine o'clock one morning. After the spraying the floors were washed. The baby and dog were returned to the house in midafternoon. An hour or so later the dog vomited, went into convulsions, and died. At 10 P.M. on the evening of the same day the baby also vomited, went into convulsions, and lost consciousness. After that fateful contact with endrin, this normal, healthy child became little more than a vegetable— unable to see or hear, subject to frequent muscular spasms, apparently completely cut off from contact with his surroundings. Several months of treatment in a New York hospital failed to change his condition or bring hope of change. "It is extremely doubtful," reported the attending physicians, "that any useful degree of recovery will occur."

The second major group of insecticides, the alkyl or organic phosphates, are among the most poisonous chemicals in the world. The chief and most obvious hazard attending their use is that of acute poisoning of people applying the sprays or accidentally coming in contact with drifting spray, with vegetation coated by it, or with a discarded container. In Florida, two children found an empty bag and used it to repair a swing. Shortly thereafter both of them died and three of their playmates became ill. The bag had once contained an insecticide called parathion, one of the organic phosphates; tests established death by parathion poisoning. On another occasion two small boys in Wisconsin, cousins, died on the same night. One had been playing in his yard when spray drifted in from an adjoining field where his father was spraying potatoes with parathion; the other had run playfully into the barn after his father and had put his hand on the nozzle of the spray equipment.

The origin of these insecticides has a certain ironic significance. Although some of the chemicals themselves—organic esters of phosphoric acid—had been known for many years, their insecticidal properties remained to be discovered by a German chemist, Gerhard Schrader, in the late 1930's. Almost immediately the German government recognized the value of these same chemicals as new and devastating weapons in man's war against his own kind, and the work on them was declared secret. Some became the deadly nerve gases. Others, of closely allied structure, became insecticides.

The organic phosphorus insecticides act on the living organism in a peculiar way. They have the ability to destroy enzymes— enzymes that perform necessary functions in the body. Their target is the nervous system, whether the victim is an insect or a warm-blooded animal. Under normal conditions, an impulse passes from nerve to nerve with the aid of a "chemical transmitter" called acetylcholine, a substance that performs an essential function and then disappears. Indeed, its existence is so ephemeral that medical researchers are unable, without special procedures, to sample it before the body has destroyed it. This transient nature of the transmitting chemical is necessary to the normal functioning of the body. If the acetylcholine is not destroyed as soon as a nerve impulse has passed, impulses continue to flash across the bridge from nerve to nerve, as the chemical exerts its effects in an ever more intensified manner. The movements of the whole body become uncoordinated: tremors, muscular spasms, convulsions, and death quickly result.

This contingency has been provided for by the body. A

protective enzyme called cholinesterase is at hand to destroy the transmitting chemical once it is no longer needed. By this means a precise balance is struck and the body never builds up a dangerous amount of acetylcholine. But on contact with the organic phosphorus insecticides, the protective enzyme is destroyed, and as the quantity of the enzyme is reduced that of the transmitting chemical builds up. In this effect, the organic phosphorus compounds resemble the alkaloid poison muscarine, found in a poisonous mushroom, the fly amanita.

Repeated exposures may lower the cholinesterase level until an individual reaches the brink of acute poisoning, a brink over which he may be pushed by a very small additional exposure. For this reason it is considered important to make periodic examinations of the blood of spray operators and others regularly exposed.

Parathion is one of the most widely used of the organic phosphates. It is also one of the most powerful and dangerous. Honeybees become "wildly agitated and bellicose" on contact with it, perform frantic cleaning movements, and are near death within half an hour. A chemist, thinking to learn by the most direct possible means the dose acutely toxic to human beings, swallowed a minute amount, equivalent to about .00424 ounce. Paralysis followed so instantaneously that he could not reach the antidotes he had prepared at hand, and so he died. Parathion is now said to be a favorite instrument of suicide in Finland. In recent years the State of California has reported an average of more than 200 cases of accidental parathion poisoning annually. In many parts of the world the fatality rate from parathion is startling: 100 fatal cases in India and 67 in Syria in 1958, and an average of 336 deaths per year in Japan.

Yet some 7,000,000 pounds of parathion are now applied to fields and orchards of the United States—by hand sprayers, motorized blowers and dusters, and by airplane. The amount used on California farms alone could, according to one medical authority, "provide a lethal dose for 5 to 10 times the whole world's population."

One of the few circumstances that save us from extinction by this means is the fact that parathion and other chemicals of this group are decomposed rather rapidly. Their residues on the crops to which they are applied are therefore relatively short-lived compared with the chlorinated hydrocarbons. However, they last long enough to create hazards and produce consequences that range from the merely serious to the fatal. In Riverside, California, eleven out of thirty men picking oranges became violently ill and all but one had to be hospitalized. Their symptoms were typical of parathion poisoning. The grove had been sprayed with parathion some two and a half weeks earlier; the residues that reduced them to retching, half-blind, semiconscious misery were sixteen to nineteen days old. And this is not by any means a record for persistence. Similar mishaps have occurred in groves sprayed a month earlier, and residues have been found in the peel of oranges six months after treatment with standard dosages.

The danger to all workers applying the organic phosphorus insecticides in fields, orchards, and vineyards, is so extreme that some states using these chemicals have established laboratories where physicians may obtain aid in diagnosis and treatment. Even the physicians themselves may be in some danger, unless they wear rubber gloves in handling the victims of poisoning. So may a laundress washing the clothing of such victims, which may have absorbed enough parathion to affect her.

Malathion, another of the organic phosphates, is almost as familiar to the public as DDT, being widely used by gardeners, in household insecticides, in mosquito spraying, and in such blanket attacks on insects as the spraying of nearly a million acres of Florida communities for the Mediterranean fruit fly. It is considered the least toxic of this group of chemicals and many people assume they may use it freely and without fear of harm. Commercial advertising encourages this comfortable attitude.

The alleged "safety" of malathion rests on rather precarious ground, although—as often happens—this was not discovered until the chemical had been in use for several years. Malathion is "safe" only because the mammalian liver, an organ with extraordinary protective powers, renders it relatively harmless. The detoxification is accomplished by one of the enzymes of the liver. If, however, something destroys this enzyme or interferes with its action, the person exposed to malathion receives the full force of the poison.

Unfortunately for all of us, opportunities for this sort of thing to happen are legion. A few years ago a team of Food and Drug Administration scientists discovered that when malathion and certain other organic phosphates are administered simultaneously a massive poisoning results—up to 50 times as severe as would be predicted on the basis of adding together the toxicities of the two. In other words, $\frac{1}{100}$ of the lethal dose of each compound may be fatal when the two are combined.

This discovery led to the testing of other combinations. It is now known that many pairs of organic phosphate insecticides are highly dangerous, the toxicity being stepped up or "potentiated" through the combined action. Potentiation seems to take place when one compound destroys the liver enzyme responsible for detoxifying the other. The two need not be given simultaneously. The hazard exists not only for the man who may spray this week with one insecticide and next week with another; it exists also for the consumer of sprayed products. The common salad bowl may easily present a combination of organic phosphate insecticides. Residues well within the legally permissible limits may interact.

The full scope of the dangerous interaction of chemicals is as yet little known, but disturbing findings now come regularly from scientific laboratories. Among these is the discovery that the toxicity of an organic phosphate can be increased by a second agent that is not necessarily an insecticide. For example, one of the plasticizing agents may act even more strongly than another insecticide to make malathion more dangerous. Again, this is because it inhibits the liver enzyme that normally would "draw the teeth" of the poisonous insecticide.

What of other chemicals in the normal human environment? What, in particular, of drugs? A bare beginning has been made on this subject, but already it is known that some organic phosphates (parathion and malathion) increase the toxicity of some drugs used as muscle relaxants, and that several others (again including malathion) markedly increase the sleeping time of barbiturates.

In Greek mythology the sorceress Medea, enraged at being supplanted by a rival for the affections of her husband Jason, presented the new bride with a robe possessing magic properties. The wearer of the robe immediately suffered a violent death. This death-by-indirection now finds its counterpart in what are known as "systemic insecticides." These are chemicals with extraordinary properties which are used to convert plants or animals into a sort of Medea's robe by making them actually poisonous. This is done with the purpose of killing insects that may come in contact with them, especially by sucking their juices or blood.

The world of systemic insecticides is a weird world, surpassing the imaginings of the brothers Grimm—perhaps most closely akin to the cartoon world of Charles Addams. It is a world where the enchanted forest of the fairy tales has become the poisonous forest in which an insect that chews a leaf or sucks the sap of a plant is doomed. It is a world where a flea bites a dog, and dies because the dog's blood has been made poisonous, where an insect may die from vapors emanating from a plant it has never touched, where a bee may carry poisonous nectar back to its hive and presently produce poisonous honey.

The entomologists' dream of the built-in insecticide was born when workers in the field of applied entomology realized they could take a hint from nature: they found that wheat growing in soil containing sodium selenate was immune to attack by aphids or spider mites. Selenium, a naturally occurring element found sparingly in rocks and soils of many parts of the world, thus became the first systemic insecticide.

What makes an insecticide a systemic is the ability to permeate all the tissues of a plant or animal and make them toxic. This quality is possessed by some chemicals of the chlorinated hydrocarbon group and by others of the organophosphorus group, all synthetically produced, as well as by certain naturally occurring substances. In practice, however, most systemics are drawn from the organophosphorus group because the problem of residues is somewhat less acute.

Systemics act in other devious ways. Applied to seeds, either by soaking or in a coating combined with carbon, they extend their effects into the following plant generation and produce seedlings poisonous to aphids and other sucking insects. Vegetables such as peas, beans, and sugar beets are sometimes thus protected. Cotton seeds coated with a systemic insecticide have been in use for some time in California, where 25 farm laborers planting cotton in the San Joaquin Valley in 1959 were seized with sudden illness, caused by handling the bags of treated seeds.

In England someone wondered what happened when bees made use of nectar from plants treated with systemics. This was investigated in areas treated with a chemical called schradan. Although the plants had been sprayed before the flowers were formed, the nectar later produced contained the poison. The result, as might have been predicted, was that the honey made by the bees also was contaminated with schradan.

Use of animal systemics has concentrated chiefly on control of the cattle grub, a damaging parasite of livestock. Extreme care must be used in order to create an insecticidal effect in the blood and tissues of the host without setting up a fatal poisoning. The balance is delicate and government veterinarians have found that repeated small doses can gradually deplete an animal's supply of the protective enzyme cholinesterase, so that without warning a minute additional dose will cause poisoning.

There are strong indications that fields closer to our daily lives are being opened up. You may now give your dog a pill which, it is claimed, will rid him of fleas by making his blood poisonous to them. The hazards discovered in treating cattle would presumably apply to the dog. As yet no one seems to have proposed a human systemic that would make us lethal to a mosquito. Perhaps this is the next step.

So far in this chapter we have been discussing the deadly chemicals that are being used in our war against the insects. What of our simultaneous war against the weeds?

The desire for a quick and easy method of killing unwanted plants has given rise to a large and growing array of chemicals that are known as herbicides, or, less formally, as weed killers. The story of how these chemicals are used and misused will be told in Chapter 6; the question that here concerns us is whether the weed killers are poisons and whether their use is contributing to the poisoning of the environment.

The legend that the herbicides are toxic only to plants and so pose no threat to animal life has been widely disseminated, but unfortunately it is not true. The plant killers include a large variety of chemicals that act on animal tissue as well as on vegetation. They vary greatly in their action on the organism. Some are general poisons, some are powerful stimulants of metabolism, causing a fatal rise in body temperature, some induce malignant tumors either alone or in partnership with other chemicals, some strike at the genetic material of the race by causing gene mutations. The herbicides, then, like the insecticides, include some very dangerous chemicals, and their careless use in the belief that they are "safe" can have disastrous results.

Despite the competition of a constant stream of new chemicals issuing from the laboratories, arsenic compounds are still liberally used, both as insecticides (as mentioned above) and as weed killers, where they usually take the chemical form of sodium arsenite. The history of their use is not reassuring. As roadside sprays, they have cost many a farmer his cow and killed uncounted numbers of wild creatures. As aquatic weed killers in lakes and reservoirs they have made public waters unsuitable for drinking or even for swimming. As a spray applied to potato fields to destroy the vines they have taken a toll of human and nonhuman life.

In England this latter practice developed about 1951 as a result of a shortage of sulfuric acid, formerly used to burn off the potato vines. The Ministry of Agriculture considered it necessary to give warning of the hazard of going into the arsenic-sprayed fields, but the warning was not understood by the cattle (nor, we must assume, by the wild animals and birds) and reports of cattle poisoned by the arsenic sprays came with monotonous regularity. When death came also to a farmer's wife through arsenic-contaminated water, one of the major English chemical companies (in 1959) stopped production of arsenical sprays and called in supplies already in the hands of dealers, and shortly thereafter the Ministry of Agriculture announced that because of high risks to people and cattle restrictions on the use of arsenites would be imposed. In 1961, the Australian government announced a similar ban. No such restrictions impede the use of these poisons in the United States, however.

Some of the "dinitro" compounds are also used as herbicides. They are rated as among the most dangerous materials of this type in use in the United States. Dinitrophenol is a strong metabolic stimulant. For this reason it was at one time used as a reducing drug, but the margin between the slimming dose and that required to poison or kill was slight—so slight that several patients died and many suffered permanent injury before use of the drug was finally halted.

A related chemical, pentachlorophenol, sometimes known as "penta," is used as a weed killer as well as an insecticide, often being sprayed along railroad tracks and in waste areas. Penta is extremely toxic to a wide variety of organisms from bacteria to man. Like the dinitros, it interferes, often fatally, with the body's source of energy, so that the affected organism almost literally burns itself up. Its fearful power is illustrated in a fatal accident recently reported by the California Department of Health. A tank truck driver was preparing a cotton defoliant by mixing diesel oil with pentachlorophenol. As he was drawing the concentrated chemical out of a drum, the spigot accidentally toppled back. He reached in with his bare hand to regain the spigot. Although he washed immediately, he became acutely ill and died the next day.

While the results of weed killers such as sodium arsenite or the phenols are grossly obvious, some other herbicides are more insidious in their effects. For example, the now famous cranberry-weed-killer aminotriazole, or amitrol, is rated as having relatively low toxicity. But in the long run its tendency to cause malignant tumors of the thyroid may be far more significant for wildlife and perhaps also for man.

Among the herbicides are some that are classified as "mutagens," or agents capable of modifying the genes, the materials of heredity. We are rightly appalled by the genetic effects of radiation; how then, can we be indifferent to the same effect in chemicals that we disseminate widely in our environment?

FOUR

Surface Waters and Underground Seas

O F ALL our natural resources water has become the most precious. By far the greater part of the earth's surface is covered by its enveloping seas, yet in the midst of this plenty we are in want. By a strange paradox, most of the earth's abundant water is not usable for agriculture, industry, or human consumption because of its heavy load of sea salts, and so most of the world's population is either experiencing or is threatened with critical shortages. In an age when man has forgotten his origins and is blind even to his most essential needs for survival, water along with other resources has become the victim of his indifference.

The problem of water pollution by pesticides can be understood only in context, as part of the whole to which it belongs—the pollution of the total environment of mankind. The pollution entering our waterways comes from many sources: radioactive wastes from reactors, laboratories, and hospitals; fallout from nuclear explosions; domestic wastes from cities and towns; chemical wastes from factories. To these is added a new kind of fallout—the chemical sprays applied to croplands and gardens, forests and fields. Many of the chemical agents in this alarming mélange imitate and augment the harmful effects of radiation, and within the groups of chemicals themselves there are sinister and little-understood interactions, transformations, and summations of effect.

Ever since chemists began to manufacture substances that nature never invented, the problems of water purification have become complex and the danger to users of water has increased. As we have seen, the production of these synthetic chemicals in large volume began in the 1940's. It has now reached such proportions that an appalling deluge of chemical pollution is daily poured into the nation's waterways. When inextricably mixed with domestic and other wastes discharged into the same water, these chemicals sometimes defy detection by the methods in ordinary use by purification plants. Most of them are so stable that they cannot be broken down by ordinary processes. Often they cannot even be identified. In rivers, a really incredible variety of pollutants combine to produce deposits that the sanitary engineers can only despairingly refer to as "gunk." Professor Rolf Eliassen of the Massachusetts Institute of Technology testified before a congressional committee to the impossibility of predicting the composite effect of these chemicals, or of identifying the organic matter resulting from the mixture. "We don't begin to know what that is," said Professor Eliassen. "What is the effect on the people? We don't know."

To an ever-increasing degree, chemicals used for the control of insects, rodents, or unwanted vegetation contribute to these organic pollutants. Some are deliberately applied to bodies of water to destroy plants, insect larvae, or undesired fishes. Some come from forest spraying that may blanket two or three million acres of a single state with spray directed against a single insect pest—spray that falls directly into streams or that drips down through the leafy canopy to the forest floor, there to become part of the slow movement of seeping moisture beginning its long journey to the sea. Probably the bulk of such contaminants are the waterborne residues of the millions of pounds of agricultural chemicals that have been applied to farmlands for insect or rodent control and have been leached out of the ground by rains to become part of the universal seaward movement of water.

Here and there we have dramatic evidence of the presence of these chemicals in our streams and even in public water supplies. For example, a sample of drinking water from an orchard area in Pennsylvania, when tested on fish in a laboratory, contained enough insecticide to kill all of the test fish in only four hours. Water from a stream draining sprayed cotton fields remained lethal to fishes even after it had passed through a purifying plant, and in fifteen streams tributary to the Tennessee River in Alabama the runoff from fields treated with toxaphene, a chlorinated hydrocarbon, killed all the fish inhabiting the streams. Two of these streams were sources of municipal water supply. Yet for a week after the application of the insecticide the water remained poisonous, a fact attested by the daily deaths of goldfish suspended in cages downstream.

For the most part this pollution is unseen and invisible, making its presence known when hundreds or thousands of fish die, but more often never detected at all. The chemist who guards water purity has no routine tests for these organic pollutants and no way to remove them. But whether detected or not, the pesticides are there, and as might be expected with any materials applied to land surfaces on so vast a scale, they have now found their way into many and perhaps all of the major river systems of the country.

If anyone doubts that our waters have become almost universally contaminated with insecticides he should study a small report issued by the United States Fish and Wildlife Service in 1960. The Service had carried out studies to discover whether fish, like warm-blooded animals, store insecticides in their tissues. The first samples were taken from forest areas in the West where there had been mass spraying of DDT for the control of the spruce budworm. As might have been expected, all of these fish contained DDT. The really significant findings were made when the investigators turned for comparison to a creek in a remote area about 30 miles from the nearest spraying for budworm control. This creek was upstream from the first and separated from it by a high waterfall. No local spraying was known to have occurred. Yet these fish, too, contained DDT. Had the chemical reached this remote creek by hidden underground streams? Or had it been airborne, drifting down as fallout on the surface of the creek? In still another comparative study, DDT was found in the tissues of fish from a hatchery where the water supply originated in a deep well. Again there was no record of local spraying. The only possible means of contamination seemed to be by means of groundwater.

In the entire water-pollution problem, there is probably nothing more disturbing than the threat of widespread contamination of groundwater. It is not possible to add pesticides to water anywhere without threatening the purity of water everywhere. Seldom if ever does Nature operate in closed and separate compartments, and she has not done so in distributing the earth's water supply. Rain, falling on the land, settles down through pores and cracks in soil and rock, penetrating deeper and deeper until eventually it reaches a zone where all the pores of the rock are filled with water, a dark, subsurface sea, rising under hills, sinking beneath valleys. This groundwater is always on the move, sometimes at a pace so slow that it travels no more than 50 feet a year, sometimes rapidly, by comparison, so that it moves nearly a tenth of a mile in a day. It travels by unseen waterways until here and there it comes to the surface as a spring, or perhaps it is tapped to feed a well. But mostly it contributes to streams and so to rivers. Except for what enters streams directly as rain or surface runoff, all the running water of the earth's surface was at one time groundwater. And so, in a very real and frightening sense, pollution of the groundwater is pollution of water everywhere.

It must have been by such a dark, underground sea that poisonous chemicals traveled from a manufacturing plant in Colorado to a farming district several miles away, there to poison wells, sicken humans and livestock, and damage crops—an extraordinary episode that may easily be only the first of many like it. Its history, in brief, is this. In 1943, the Rocky Mountain Arsenal of the Army Chemical Corps, located near Denver, began to manufacture war materials. Eight years later the facilities of the arsenal were leased to a private oil company for the production of insecticides. Even before the change of operations, however, mysterious reports had begun to come in. Farmers several miles from the plant began to report unexplained sickness among livestock; they complained of extensive crop damage. Foliage turned yellow, plants failed to mature, and many crops were killed outright. There were reports of human illness, thought by some to be related.

The irrigation waters on these farms were derived from shallow wells. When the well waters were examined (in a study in 1959, in which several state and federal agencies participated) they were found to contain an assortment of chemicals. Chlorides, chlorates, salts of phosphonic acid, fluorides, and arsenic had been discharged from the Rocky Mountain Arsenal into holding ponds during the years of its operation. Apparently the groundwater between the arsenal and the farms had become contaminated and it had taken 7 to 8 years for the wastes to travel underground a distance of about 3 miles from the holding ponds to the nearest farm. This seepage had continued to spread and had further contaminated an area of unknown extent. The investigators knew of no way to contain the contamination or halt its advance.

All this was bad enough, but the most mysterious and probably in the long run the most significant feature of the whole episode was the discovery of the weed killer 2,4-D in some of the wells and in the holding ponds of the arsenal. Certainly its presence was enough to account for the damage to crops irrigated with this water. But the mystery lay in the fact that no 2,4-D had been manufactured at the arsenal at any stage of its operations.

After long and careful study, the chemists at the plant concluded that the 2,4-D had been formed spontaneously in the open basins. It had been formed there from other substances discharged from the arsenal; in the presence of air, water, and sunlight, and quite without the intervention of human chemists, the holding ponds had become chemical laboratories for the production of a new chemical—a chemical fatally damaging to much of the plant life it touched.

And so the story of the Colorado farms and their damaged crops assumes a significance that transcends its local importance. What other parallels may there be, not only in Colorado but wherever chemical pollution finds its way into public waters? In lakes and streams everywhere, in the presence of catalyzing air and sunlight, what dangerous substances may be born of parent chemicals labeled "harmless"?

Indeed one of the most alarming aspects of the chemical pollution of water is the fact that here—in river or lake or reservoir, or for that matter in the glass of water served at your dinner table—are mingled chemicals that no responsible chemist would think of combining in his laboratory. The possible interactions between these freely mixed chemicals are deeply disturbing to officials of the United States Public Health Service, who have expressed the fear that the production of harmful substances from comparatively innocuous chemicals may be taking place on quite a wide scale. The reactions may be between two or more chemicals, or between chemicals and the radioactive wastes that are being discharged into our rivers in ever-increasing volume. Under the impact of ionizing radiation some rearrangement of atoms could easily occur, changing the nature of the chemicals in a way that is not only unpredictable but beyond control.

It is, of course, not only the groundwaters that are becoming contaminated, but surface-moving waters as well—streams, rivers, irrigation waters. A disturbing example of the latter seems to be building up on the national wildlife refuges at Tule Lake and Lower Klamath, both in California. These refuges are part of a chain including also the refuge on Upper Klamath Lake just over the border in Oregon. All are linked, perhaps fatefully, by a shared water supply, and all are affected by the fact that they lie like small islands in a great sea of surrounding farmlands—land reclaimed by drainage and stream diversion from an original waterfowl paradise of marshland and open water.

These farmlands around the refuges are now irrigated by water from Upper Klamath Lake. The irrigation waters, re-collected from the fields they have served, are then pumped into Tule Lake and from there to Lower Klamath. All of the waters of the wildlife refuges established on these two bodies of water therefore represent the drainage of agricultural lands. It is important to remember this in connection with recent happenings.

In the summer of 1960 the refuge staff picked up hundreds of dead and dying birds at Tule Lake and Lower Klamath. Most of them were fish-eating species—herons, pelicans, grebes, gulls. Upon analysis, they were found to contain insecticide residues identified as toxaphene, DDD, and DDE. Fish from the lakes were also found to contain insecticides; so did samples of plankton. The refuge manager believes that pesticide residues are now building up in the waters of these refuges, being conveyed there by return irrigation flow from heavily sprayed agricultural lands.

Such poisoning of waters set aside for conservation purposes could have consequences felt by every western duck hunter and by everyone to whom the sight and sound of drifting ribbons of waterfowl across an evening sky are precious. These particular refuges occupy critical positions in the conservation of western waterfowl. They lie at a point corresponding to the narrow neck of a funnel, into which all the migratory paths composing what is known as the Pacific Flyway converge.

During the fall migration they receive many millions of ducks and geese from nesting grounds extending from the shores of Bering Sea east to Hudson Bay—fully three fourths of all the waterfowl that move south into the Pacific Coast states in autumn. In summer they provide nesting areas for waterfowl, especially for two endangered species, the redhead and the ruddy duck. If the lakes and pools of these refuges become seriously contaminated, the damage to the waterfowl populations of the Far West could be irreparable.

Water must also be thought of in terms of the chains of life it supports—from the small-as-dust green cells of the drifting plant plankton, through the minute water fleas to the fishes that strain plankton from the water and are in turn eaten by other fishes or by birds, mink, raccoons—in an endless cyclic transfer of materials from life to life. We know that the necessary minerals in the water are so passed from link to link of the food chains. Can we suppose that poisons we introduce into water will not also enter into these cycles of nature?

The answer is to be found in the amazing history of Clear Lake, California. Clear Lake lies in mountainous country some 90 miles north of San Francisco and has long been popular with anglers. The name is inappropriate, for actually it is a rather turbid lake because of the soft black ooze that covers its shallow bottom. Unfortunately for the fishermen and the resort dwellers on its shores, its waters have provided an ideal habitat for a small gnat, *Chaoborus astictopus*. Although closely related to mosquitoes, the gnat is not a bloodsucker and probably does not feed at all as an adult. However, human beings who shared its habitat found it annoying because of its sheer numbers. Efforts were made to control it but they were largely fruitless until, in the late 1940's, the chlorinated hydrocarbon insecticides offered new weapons. The chemical chosen for a fresh attack was DDD, a close relative of DDT but apparently offering fewer threats to fish life.

The new control measures undertaken in 1949 were carefully planned and few people would have supposed any harm could result. The lake was surveyed, its volume determined, and the insecticide applied in such great dilution that for every part of chemical there would be 70 million parts of water. Control of the gnats was at first good, but by 1954 the treatment had to be repeated, this time at the rate of 1 part of insecticide in 50 million parts of water. The destruction of the gnats was thought to be virtually complete.

The following winter months brought the first intimation that other life was affected: the western grebes on the lake began to die, and soon more than a hundred of them were reported dead. At Clear Lake the western grebe is a breeding bird and also a winter visitant, attracted by the abundant fish of the lake. It is a bird of spectacular appearance and beguiling habits, building its floating nests in shallow lakes of western United States and Canada. It is called the "swan grebe" with reason, for it glides with scarcely a ripple across the lake surface, the body riding low, white neck and shining black head held high. The newly hatched chick is clothed in soft gray down; in only a few hours it takes to the water and rides on the back of the father or mother, nestled under the parental wing coverts.

Following a third assault on the ever-resilient gnat population, in 1957, more grebes died. As had been true in 1954, no evidence of infectious disease could be discovered on examination of the dead birds. But when someone thought to analyze the fatty tissues of the grebes, they were found to be loaded with DDD in the extraordinary concentration of 1600 parts per million.

The maximum concentration applied to the water was $\frac{1}{50}$ part per million. How could the chemical have built up to such prodigious levels in the grebes? These birds, of course, are fish eaters. When the fish of Clear Lake also were analyzed the picture began to take form—the poison being picked up by the smallest organisms, concentrated and passed on to the larger predators. Plankton organisms were found to contain about 5 parts per million of the insecticide (about 25 times the maximum concentration ever reached in the water itself); plant-eating fishes had built up accumulations ranging from 40 to 300 parts per million; carnivorous species had stored the most of all. One, a brown bullhead, had the astounding concentration of 2500 parts per million. It was a house-that-Jack-built sequence, in which the large carnivores had eaten the smaller carnivores, that had eaten the herbivores, that had eaten the plankton, that had absorbed the poison from the water.

Even more extraordinary discoveries were made later. No trace of DDD could be found in the water shortly after the last application of the chemical. But the poison had not really left the lake; it had merely gone into the fabric of the life the lake supports. Twenty-three months after the chemical treatment had ceased, the plankton still contained as much as 5.3 parts per million. In that interval of nearly two years, successive crops of plankton had flowered and faded away, but the poison, although no longer present in the water, had somehow passed from generation to generation. And it lived on in the animal life of the lake as well. All fish, birds, and frogs examined a year after the chemical applications had ceased still contained DDD. The amount found in the flesh always exceeded by many times the original concentration in the water. Among these living carriers were fish that had hatched nine months after the last DDD application, grebes, and California gulls that had built up concentrations of more than 2000 parts per million. Meanwhile, the nesting colonies of the grebes dwindled—from more than 1000 pairs before the first insecticide treatment to about 30 pairs in 1960. And even the thirty seem to have nested in vain, for no young grebes have been observed on the lake since the last DDD application.

This whole chain of poisoning, then, seems to rest on a base of minute plants which must have been the original concentrators. But what of the opposite end of the food chain—the human being who, in probable ignorance of all this sequence of events, has rigged his fishing tackle, caught a string of fish from the waters of Clear Lake, and taken them home to fry for his supper? What could a heavy dose of DDD, or perhaps repeated doses, do to him?

Although the California Department of Public Health professed to see no hazard, nevertheless in 1959 it required that the use of DDD in the lake be stopped. In view of the scientific evidence of the vast biological potency of this chemical, the action seems a minimum safety measure. The physiological effect of DDD is probably unique among insecticides, for it destroys part of the adrenal gland—the cells of the outer layer known as the adrenal cortex, which secretes the hormone cortin. This destructive effect, known since 1948, was at first believed to be confined to dogs, because it was not revealed in such experimental animals as monkeys, rats, or rabbits: It seemed suggestive, however, that DDD produced in dogs a condition very similar to that occurring in man in the presence of Addison's disease. Recent medical research has revealed that DDD does strongly suppress the function of the human adrenal cortex. Its cell-destroying capacity is now clinically utilized in the treatment of a rare type of cancer which develops in the adrenal gland.

The Clear Lake situation brings up a question that the public needs to face: Is it wise or desirable to use substances with such strong effect on physiological processes for the control of insects, especially when the control measures involve introducing the chemical directly into a body of water? The fact that the insecticide was applied in very low concentrations is meaningless, as its explosive progress through the natural food chain in the lake demonstrates. Yet Clear Lake is typical of a large and growing number of situations where solution of an obvious and often trivial problem creates a far more serious but conveniently less tangible one. Here the problem was resolved in favor of those annoyed by gnats, and at the expense of an unstated, and probably not even clearly understood, risk to all who took food or water from the lake.

It is an extraordinary fact that the deliberate introduction of poisons into a reservoir is becoming a fairly common practice. The purpose is usually to promote recreational uses, even though the water must then be treated at some expense to make it fit for its intended use as drinking water. When sportsmen of an area want to "improve" fishing in a reservoir, they prevail on authorities to dump quantities of poison into it to kill the undesired fish, which are then replaced with hatchery fish more suited to the sportsmen's taste. The procedure has a strange, Alice-in-Wonderland quality. The reservoir was created as a public water supply, yet the community, probably unconsulted about the sportsmen's project, is forced either to drink water containing poisonous residues or to pay out tax money for treatment of the water to remove the poisons—treatments that are by no means foolproof.

As ground and surface waters are contaminated with pesticides and other chemicals, there is danger that not only poisonous but also cancer-producing substances are being introduced into public water supplies. Dr. W. C. Hueper of the National

Cancer Institute has warned that "the danger of cancer hazards from the consumption of contaminated drinking water will grow considerably within the foreseeable future." And indeed a study made in Holland in the early 1950's provides support for the view that polluted waterways may carry a cancer hazard. Cities receiving their drinking water from rivers had a higher death rate from cancer than did those whose water came from sources presumably less susceptible to pollution such as wells. Arsenic, the environmental substance most clearly established as causing cancer in man, is involved in two historic cases in which polluted water supplies caused widespread occurrence of cancer. In one case the arsenic came from the slag heaps of mining operations, in the other from rock with a high natural content of arsenic. These conditions may easily be duplicated as a result of heavy applications of arsenical insecticides. The soil in such areas becomes poisoned. Rains then carry part of the arsenic into streams, rivers, and reservoirs, as well as into the vast subterranean seas of groundwater.

Here again we are reminded that in nature nothing exists alone. To understand more clearly how the pollution of our world is happening, we must now look at another of the earth's basic resources, the soil.

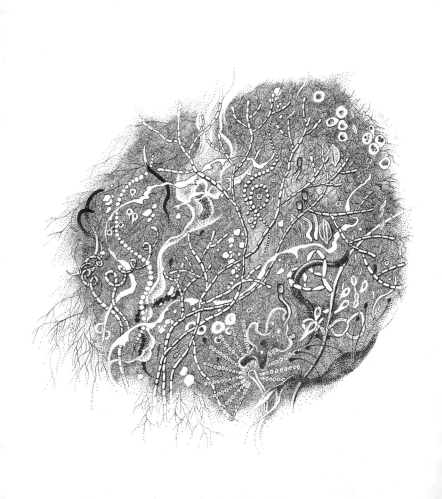

FIVE

Realms of the Soil

THE THIN LAYER of soil that forms a patchy covering over the continents controls our own existence and that of every other animal of the land. Without soil, land plants as we know them could not grow, and without plants no animals could survive.

Yet if our agriculture-based life depends on the soil, it is equally true that soil depends on life, its very origins and the maintenance of its true nature being intimately related to living plants and animals. For soil is in part a creation of life, born of a marvelous interaction of life and nonlife long eons ago. The parent materials were gathered together as volcanoes poured them out in fiery streams, as waters running over the bare rocks of the continents wore away even the hardest granite, and as the chisels of frost and ice split and shattered the rocks. Then living things began to work their creative magic and little by little these inert materials became soil. Lichens, the rocks' first covering, aided the process of disintegration by their acid secretions and made a lodging place for other life. Mosses took hold in the little pockets of simple soil—soil formed by crumbling bits of lichen, by the husks of minute insect life, by the debris of a fauna beginning its emergence from the sea.

Life not only formed the soil, but other living things of incredible abundance and diversity now exist within it; if this were not so the soil would be a dead and sterile thing. By their presence and by their activities the myriad organisms of the soil make it capable of supporting the earth's green mantle.

The soil exists in a state of constant change, taking part in cycles that have no beginning and no end. New materials are constantly being contributed as rocks disintegrate, as organic matter decays, and as nitrogen and other gases are brought down in rain from the skies. At the same time other materials are being taken away, borrowed for temporary use by living creatures. Subtle and vastly important chemical changes are constantly in progress, converting elements derived from air and water into forms suitable for use by plants. In all these changes living organisms are active agents.

There are few studies more fascinating, and at the same time more neglected, than those of the teeming populations that exist in the dark realms of the soil. We know too little of the threads that bind the soil organisms to each other and to their world, and to the world above.

Perhaps the most essential organisms in the soil are the smallest—the invisible hosts of bacteria and of threadlike fungi. Statistics of their abundance take us at once into astronomical figures. A teaspoonful of topsoil may contain billions of bacteria. In spite of their minute size, the total weight of this host of bacteria in the top foot of a single acre of fertile soil may be as much as a thousand pounds. Ray fungi, growing in long threadlike filaments, are somewhat less numerous than the bacteria, yet because they are larger their total weight in a given amount of soil may be about the same. With small green cells called algae, these make up the microscopic plant life of the soil.

Bacteria, fungi, and algae are the principal agents of decay, reducing plant and animal residues to their component minerals. The vast cyclic movements of chemical elements such as carbon and nitrogen through soil and air and living tissue could not proceed without these microplants. Without the nitrogen-fixing bacteria, for example, plants would starve for want of nitrogen, though surrounded by a sea of nitrogen-containing air. Other organisms form carbon dioxide, which, as carbonic acid, aids in dissolving rock. Still other soil microbes perform various oxidations and reductions by which minerals such as iron, manganese, and sulfur are transformed and made available to plants.

Also present in prodigious numbers are microscopic mites and primitive wingless insects called springtails. Despite their small size they play an important part in breaking down the residues of plants, aiding in the slow conversion of the litter of the forest floor to soil. The specialization of some of these minute creatures for their task is almost incredible. Several species of mites, for example, can begin life only within the fallen needles of a spruce tree. Sheltered here, they digest out the inner tissues of the needle. When the mites have completed their development only the outer layer of cells remains. The truly staggering task of dealing with the tremendous amount of plant material in the annual leaf fall belongs to some of the small insects of the soil and the forest floor. They macerate and digest the leaves, and aid in mixing the decomposed matter with the surface soil.

Besides all this horde of minute but ceaselessly toiling creatures there are of course many larger forms, for soil life runs the gamut from bacteria to mammals. Some are permanent residents of the dark subsurface layers; some hibernate or spend definite parts of their life cycles in underground chambers; some freely come and go between their burrows and the upper world. In general the effect of all this habitation of the soil is to aerate it and improve both its drainage and the penetration of water throughout the layers of plant growth.

Of all the larger inhabitants of the soil, probably none is more important than the earthworm. Over three quarters of a century ago, Charles Darwin published a book titled *The Formation of Vegetable Mould, through the Action of Worms, with Observations on Their Habits.* In it he gave the world its first understanding of the fundamental role of earthworms as geologic agents for the transport of soil—a picture of surface rocks being gradually covered by fine soil brought up from below by the worms, in annual amounts running to many tons to the acre in most favorable areas. At the same time, quantities of organic matter contained in leaves and grass (as much as 20 pounds to the square yard in six months) are drawn down into the burrows and incorporated in soil. Darwin's calculations showed that the toil of earthworms might add a layer of soil an inch to an inch and a half thick in a ten-year period. And this is by no means all they do: their burrows aerate the soil, keep it well drained, and aid the penetration of plant roots. The presence of earthworms increases the nitrifying powers of the soil bacteria and decreases putrefaction of the soil. Organic matter is broken down as it passes through the digestive tracts of the worms and the soil is enriched by their excretory products.

This soil community, then, consists of a web of interwoven lives, each in some way related to the others—the living creatures depending on the soil, but the soil in turn a vital element of the earth only so long as this community within it flourishes.

The problem that concerns us here is one that has received little consideration: What happens to these incredibly numerous and vitally necessary inhabitants of the soil when poisonous chemicals are carried down into their world, either introduced directly as soil "sterilants" or borne on the rain that has picked up a lethal contamination as it filters through the leaf canopy of forest and orchard and cropland? Is it reasonable to suppose that we can apply a broad-spectrum insecticide to kill the burrowing larval stages of a crop-destroying insect, for example, without also killing the "good" insects whose function may be the essential one of breaking down organic matter? Or can we use a nonspecific fungicide without also killing the fungi that inhabit the roots of many trees in a beneficial association that aids the tree in extracting nutrients from the soil?

The plain truth is that this critically important subject of the ecology of the soil has been largely neglected even by scientists and almost completely ignored by control men. Chemical control of insects seems to have proceeded on the assumption that the soil could and would sustain any amount of insult via the introduction of poisons without striking back. The very nature of the world of the soil has been largely ignored.

From the few studies that have been made, a picture of the impact of pesticides on the soil is slowly emerging. It is not surprising that the studies are not always in agreement, for soil types vary so enormously that what causes damage in one may be innocuous in another. Light sandy soils suffer far more heavily than humus types. Combinations of chemicals seem to do more harm than separate applications. Despite the varying results, enough solid evidence of harm is accumulating to cause apprehension on the part of many scientists.

Under some conditions, the chemical conversions and transformations that lie at the very heart of the living world are affected. Nitrification, which makes atmospheric nitrogen available to plants, is an example. The herbicide 2,4-D causes a temporary interruption of nitrification. In recent experiments in Florida, lindane, heptachlor, and BHC (benzene hexachloride) reduced nitrification after only two weeks in soil; BHC and DDT had significantly detrimental effects a year after treatment. In other experiments BHC, aldrin, lindane, heptachlor, and DDD all prevented nitrogen-fixing bacteria from forming the necessary root nodules on leguminous plants. A curious but beneficial relation between fungi and the roots of higher plants is seriously disrupted.

Sometimes the problem is one of upsetting that delicate balance of populations by which nature accomplishes far-reaching aims. Explosive increases in some kinds of soil organisms have occurred when others have been reduced by insecticides, disturbing the relation of predator to prey. Such changes could easily alter the metabolic activity of the soil and affect its productivity. They could also mean that potentially harmful organisms, formerly held in check, could escape from their natural controls and rise to pest status.

One of the most important things to remember about insecticides in soil is their long persistence, measured not in months but in years. Aldrin has been recovered after four years, both as traces and more abundantly as converted to dieldrin. Enough toxaphene remains in sandy soil ten years after its application to kill termites. Benzene hexachloride persists at least eleven years; heptachlor or a more toxic derived chemical, at least nine. Chlordane has been recovered twelve years after its application, in the amount of 15 per cent of the original quantity.

Seemingly moderate applications of insecticides over a period of years may build up fantastic quantities in soil. Since the chlorinated hydrocarbons are persistent and long-lasting, each application is merely added to the quantity remaining from the previous one. The old legend that "a pound of DDT to the acre is harmless" means nothing if spraying is repeated. Potato soils have been found to contain up to 15 pounds of DDT per acre, corn soils up to 19. A cranberry bog under study contained 34.5 pounds to the acre. Soils from apple orchards seem to reach the peak of contamination, with DDT accumulating at a rate that almost keeps pace with its rate of annual application. Even in a single season, with orchards sprayed four or more times, DDT residues may build up to peaks of 30 to 50 pounds. With repeated spraying over the years the range between trees is from 26 to 60 pounds to the acre; under trees, up to 113 pounds.

Arsenic provides a classic case of the virtually permanent poisoning of the soil. Although arsenic as a spray on growing tobacco has been largely replaced by the synthetic organic insecticides since the mid-'40's, *the arsenic content of cigarettes made from American-grown tobacco increased more than 300 per cent* between the years 1932 and 1952. Later studies have revealed increases of as much as 600 per cent. Dr. Henry S. Satterlee, an authority on arsenic toxicology, says that although organic insecticides have been largely substituted for arsenic, the tobacco plants continue to pick up the old poison, for the soils of tobacco plantations are now thoroughly impregnated with residues of a heavy and relatively insoluble poison, arsenate of lead. This will continue to release arsenic in soluble form. The soil of a large proportion of the land planted to tobacco has been subjected to "cumulative and well-nigh permanent poisoning," according to Dr. Satterlee. Tobacco grown in the eastern Mediterranean countries where arsenical insecticides are not used has shown no such increase in arsenic content.

We are therefore confronted with a second problem. We must not only be concerned with what is happening to the soil; we must wonder to what extent insecticides are absorbed from contaminated soils and introduced into plant tissues. Much depends on the type of soil, the crop, and the nature and concentration of the insecticide. Soil high in organic matter releases smaller quantities of poisons than others. Carrots absorb more insecticide than any other crop studied; if the chemical used happens to be lindane, carrots actually accumulate higher concentrations than are present in the soil. In the future it may become necessary to analyze soils for insecticides before planting certain food crops. Otherwise even unsprayed crops may take up enough insecticide merely from the soil to render them unfit for market.

This very sort of contamination has created endless problems for at least one leading manufacturer of baby foods who has been unwilling to buy any fruits or vegetables on which toxic insecticides have been used. The chemical that caused him the most trouble was benzene hexachloride (BHC), which is taken up by the roots and tubers of plants, advertising its presence by a musty taste and odor. Sweet potatoes grown on California fields where BHC had been used two years earlier contained residues and had to be rejected. In one year, in which the firm had contracted in South Carolina for its total requirements of sweet potatoes, so large a proportion of the acreage was found to be contaminated that the company was forced to buy in the open market at a considerable financial loss. Over the years a variety of fruits and vegetables, grown in various states, have had to be rejected. The most stubborn problems were concerned with peanuts. In the southern states peanuts are usually grown in rotation with cotton, on which BHC is extensively used. Peanuts grown later in this soil pick up considerable amounts of the insecticide. Actually, only a trace is enough to incorporate the telltale musty odor and taste. The chemical penetrates the nuts and cannot be removed. Processing, far from removing the mustiness, sometimes accentuates it. The only course open to a manufacturer determined to exclude BHC residues is to reject all produce treated with the chemical or grown on soils contaminated with it.

Sometimes the menace is to the crop itself—a menace that remains as long as the insecticide contamination is in the soil. Some insecticides affect sensitive plants such as beans, wheat, barley, or rye, retarding root development or depressing growth of seedlings. The experience of the hop growers in Washington and Idaho is an example. During the spring of 1955 many of these growers undertook a large-scale program to control the strawberry root weevil, whose larvae had become abundant on the roots of the hops. On the advice of agricultural experts and insecticide manufacturers, they chose heptachlor as the control agent. Within a year after the heptachlor was applied, the vines in the treated yards were wilting and dying. In the untreated fields there was no trouble; the damage stopped at the border between treated and untreated fields. The hills were replanted at great expense, but in another year the new roots, too, were found to be dead. Four years later the soil still contained heptachlor, and scientists were unable to predict how long it would remain poisonous, or to recommend any procedure for correcting the condition. The federal Department of Agriculture, which as late as March 1959 found itself in the anomalous position of declaring heptachlor to be acceptable for use on hops in the form of a soil treatment, belatedly withdrew its registration for such use. Meanwhile, the hop growers sought what redress they could in the courts.

As applications of pesticides continue and the virtually indestructible residues continue to build up in the soil, it is almost certain that we are heading for trouble. This was the consensus of a group of specialists who met at Syracuse University in 1960 to discuss the ecology of the soil. These men summed up the hazards of using "such potent and little understood tools" as chemicals and radiation: "A few false moves on the part of man may result in destruction of soil productivity and the arthropods may well take over."

Earth's Green Mantle

WATER, SOIL, and the earth's green mantle of plants make up the world that supports the animal life of the earth. Although modern man seldom remembers the fact, he could not exist without the plants that harness the sun's energy and manufacture the basic foodstuffs he depends upon for life. Our attitude toward plants is a singularly narrow one. If we see any immediate utility in a plant we foster it. If for any reason we find its presence undesirable or merely a matter of indifference, we may condemn it to destruction forthwith. Besides the various plants that are poisonous to man or his livestock, or crowd out food plants, many are marked for destruction merely because, according to our narrow view, they happen to be in the wrong place at the wrong time. Many others are destroyed merely because they happen to be associates of the unwanted plants.

The earth's vegetation is part of a web of life in which there are intimate and essential relations between plants and the earth, between plants and other plants, between plants and animals. Sometimes we have no choice but to disturb these relationships, but we should do so thoughtfully, with full awareness that what we do may have consequences remote in time and place. But no such humility marks the booming "weed killer" business of the present day, in which soaring sales and expanding uses mark the production of plant-killing chemicals.

One of the most tragic examples of our unthinking bludgeoning of the landscape is to be seen in the sagebrush lands of the West, where a vast campaign is on to destroy the sage and to substitute grasslands. If ever an enterprise needed to be illuminated with a sense of the history and meaning of the landscape, it is this. For here the natural landscape is eloquent of the interplay of forces that have created it. It is spread before us like the pages of an open book in which we can read why the land is what it is, and why we should preserve its integrity. But the pages lie unread.

The land of the sage is the land of the high western plains and the lower slopes of the mountains that rise above them, a land born of the great uplift of the Rocky Mountain system many millions of years ago. It is a place of harsh extremes of climate: of long winters when blizzards drive down from the mountains and snow lies deep on the plains, of summers whose heat is relieved by only scanty rains, with drought biting deep into the soil, and drying winds stealing moisture from leaf and stem.

As the landscape evolved, there must have been a long period of trial and error in which plants attempted the colonization of this high and windswept land. One after another must have failed. At last one group of plants evolved which combined all the qualities needed to survive. The sage—low-growing and shrubby—could hold its place on the mountain slopes and on the plains, and within its small gray leaves it could hold moisture enough to defy the thieving winds. It was no accident, but rather the result of long ages of experimentation by nature, that the great plains of the West became the land of the sage.

Along with the plants, animal life, too, was evolving in harmony with the searching requirements of the land. In time there were two as perfectly adjusted to their habitat as the sage. One was a mammal, the fleet and graceful pronghorn antelope. The other was a bird, the sage grouse—the "cock of the plains" of Lewis and Clark.

The sage and the grouse seem made for each other. The original range of the bird coincided with the range of the sage, and as the sagelands have been reduced, so the populations of grouse have dwindled. The sage is all things to these birds of the plains. The low sage of the foothill ranges shelters their nests and their young; the denser growths are loafing and roosting areas; at all times the sage provides the staple food of the grouse. Yet it is a two-way relationship. The spectacular courtship displays of the cocks help loosen the soil beneath and around the sage, aiding invasion by grasses which grow in the shelter of sagebrush.

The antelope, too, have adjusted their lives to the sage. They are primarily animals of the plains, and in winter when the first snows come those that have summered in the mountains move down to the lower elevations. There the sage provides the food that tides them over the winter. Where all other plants have shed their leaves, the sage remains evergreen, the gray-green leaves—bitter, aromatic, rich in proteins, fats, and needed minerals—clinging to the stems of the dense and shrubby plants. Though the snows pile up, the tops of the sage remain exposed, or can be reached by the sharp, pawing hoofs of the antelope. Then grouse feed on them too, finding them on bare and windswept ledges or following the antelope to feed where they have scratched away the snow.

And other life looks to the sage. Mule deer often feed on it. Sage may mean survival for winter-grazing livestock. Sheep graze many winter ranges where the big sagebrush forms almost pure stands. For half the year it is their principal forage, a plant of higher energy value than even alfalfa hay.

The bitter upland plains, the purple wastes of sage, the wild, swift antelope, and the grouse are then a natural system in perfect balance. Are? The verb must be changed—at least in those already vast and growing areas where man is attempting to improve on nature's way. In the name of progress the land management agencies have set about to satisfy the insatiable demands of the cattlemen for more grazing land. By this they mean grassland—grass without sage. So in a land which nature found suited to grass growing mixed with and under the shelter of sage, it is now proposed to eliminate the sage and create unbroken grassland. Few seem to have asked whether grasslands are a stable and desirable goal in this region. Certainly nature's own answer was otherwise. The annual precipitation in this land where the rains seldom fall is not enough to support good sod-forming grass; it favors rather the perennial bunchgrass that grows in the shelter of the sage.

Yet the program of sage eradication has been under way for a number of years. Several government agencies are active in it; industry has joined with enthusiasm to promote and encourage an enterprise which creates expanded markets not only for grass seed but for a large assortment of machines for cutting and plowing and seeding. The newest addition to the weapons is the use of chemical sprays. Now millions of acres of sagebrush lands are sprayed each year.

What are the results? The eventual effects of eliminating sage and seeding with grass are largely conjectural. Men of long experience with the ways of the land say that in this country there is better growth of grass between and under the sage than can possibly be had in pure stands, once the moisture-holding sage is gone.

But even if the program succeeds in its immediate objective, it is clear that the whole closely knit fabric of life has been ripped apart. The antelope and the grouse will disappear along with the sage. The deer will suffer, too, and the land will be poorer for the destruction of the wild things that belong to it. Even the livestock which are the intended beneficiaries will suffer; no amount of lush green grass in summer can help the sheep starving in the winter storms for lack of the sage and bitterbrush and other wild vegetation of the plains.

These are the first and obvious effects. The second is of a kind that is always associated with the shotgun approach to nature: the spraying also eliminates a great many plants that were not its intended target. Justice William O. Douglas, in his recent book *My Wilderness: East to Katahdin*, has told of an appalling example of ecological destruction wrought by the United States Forest Service in the Bridger National Forest in Wyoming. Some 10,000 acres of sagelands were sprayed by the Service, yielding to pressure of cattlemen for more grasslands. The sage was killed, as intended. But so was the green, life-giving ribbon of willows that traced its way across these plains, following the meandering streams. Moose had lived in these willow thickets, for willow is to the moose what sage is to the antelope. Beaver had lived there, too, feeding on the willows, felling them and making a strong dam across the tiny stream. Through the labor of the beavers, a lake backed up. Trout in the mountain streams seldom were more than six inches long; in the lake they thrived so prodigiously that many grew to five pounds. Waterfowl were attracted to the lake, also. Merely because of the presence of the willows and the beavers that depended on them, the region was an attractive recreational area with excellent fishing and hunting.

But with the "improvement" instituted by the Forest Service, the willows went the way of the sagebrush, killed by the same impartial spray. When Justice Douglas visited the area in 1959, the year of the spraying, he was shocked to see the shriveled and dying willows—the "vast, incredible damage." What would become of the moose? Of the beavers and the little world they had constructed? A year later he returned to read the answers in the devastated landscape. The moose were gone and so were the beaver. Their principal dam had gone out for want of attention by its skilled architects, and the lake had drained away. None of the large trout were left. None could live in the tiny creek that remained, threading its way through a bare, hot land where no shade remained. The living world was shattered.

Besides the more than four million acres of rangelands sprayed each year, tremendous areas of other types of land are also potential or actual recipients of chemical treatments for weed control. For example, an area larger than all of New England— some 50 million acres—is under management by utility corporations and much of it is routinely treated for "brush control." In the Southwest an estimated 75 million acres of mesquite lands require management by some means, and chemical spraying is the method most actively pushed. An unknown but very large acreage of timber-producing lands is now aerially sprayed in order to "weed out" the hardwoods from the more spray-resistant conifers. Treatment of agricultural lands with herbicides doubled in the decade following 1949, totaling 53 million acres in 1959. And the combined acreage of private lawns, parks, and golf courses now being treated must reach an astronomical figure.

The chemical weed killers are a bright new toy. They work in a spectacular way; they give a giddy sense of power over nature to those who wield them, and as for the long-range and less obvious effects—these are easily brushed aside as the baseless imaginings of pessimists. The "agricultural engineers" speak blithely of "chemical plowing" in a world that is urged to beat its plowshares into spray guns. The town fathers of a thousand communities lend willing ears to the chemical salesman and the eager contractors who will rid the roadsides of "brush"—for a price. It is cheaper than mowing, is the cry. So, perhaps, it appears in the neat rows of figures in the official books; but were the true costs entered, the costs not only in dollars but in the many equally valid debits we shall presently consider, the wholesale broadcasting of chemicals would be seen to be more costly in dollars as well as infinitely damaging to the long-range health of the landscape and to all the varied interests that depend on it.

Take, for instance, that commodity prized by every chamber of commerce throughout the land—the good will of vacationing tourists. There is a steadily growing chorus of outraged protest about the disfigurement of once beautiful roadsides by chemical sprays, which substitute a sere expanse of brown, withered vegetation for the beauty of fern and wildflower, of native shrubs adorned with blossom or berry. "We are making a dirty, brown, dying-looking mess along the sides of our roads," a New England woman wrote angrily to her newspaper. "This is not what the tourists expect, with all the money we are spending advertising the beautiful scenery."

In the summer of 1960 conservationists from many states converged on a peaceful Maine island to witness its presentation to the National Audubon Society by its owner, Millicent Todd Bingham. The focus that day was on the preservation of the natural landscape and of the intricate web of life whose interwoven strands lead from microbes to man. But in the background of all the conversations among the visitors to the island was indignation at the despoiling of the roads they had traveled. Once it had been a joy to follow those roads through the evergreen forests, roads lined with bayberry and sweet fern, alder and huckleberry. Now all was brown desolation. One of the conservationists wrote of that August pilgrimage to a Maine island: "I returned . . . angry at the desecration of the

Maine roadsides. Where, in previous years, the highways were bordered with wildflowers and attractive shrubs, there were only the scars of dead vegetation for mile after mile. . . . As an economic proposition, can Maine afford the loss of tourist goodwill that such sights induce?"

Maine roadsides are merely one example, though a particularly sad one for those of us who have a deep love for the beauty of that state, of the senseless destruction that is going on in the name of roadside brush control throughout the nation.

Botanists at the Connecticut Arboretum declare that the elimination of beautiful native shrubs and wildflowers has reached the proportions of a "roadside crisis." Azaleas, mountain laurel, blueberries, huckleberries, viburnums, dogwood, bayberry, sweet fern, low shadbush, winterberry, chokecherry, and wild plum are dying before the chemical barrage. So are the daisies, black-eyed Susans, Queen Anne's lace, goldenrods, and fall asters which lend grace and beauty to the landscape.

The spraying is not only improperly planned but studded with abuses such as these. In a southern New England town one contractor finished his work with some chemical remaining in his tank. He discharged this along woodland roadsides where no spraying had been authorized. As a result the community lost the blue and golden beauty of its autumn roads, where asters and goldenrod would have made a display worth traveling far to see. In another New England community a contractor changed the state specifications for town spraying without the knowledge of the highway department and sprayed roadside vegetation to a height of eight feet instead of the specified maximum of four feet, leaving a broad, disfiguring, brown swath. In a Massachusetts community the town officials purchased a weed killer from a zealous chemical salesman, unaware that it contained arsenic. One result of the subsequent roadside spraying was the death of a dozen cows from arsenic poisoning.

Trees within the Connecticut Arboretum Natural Area were seriously injured when the town of Waterford sprayed the roadsides with chemical weed killers in 1957. Even large trees not directly sprayed were affected. The leaves of the oaks began to curl and turn brown, although it was the season for spring growth. Then new shoots began to be put forth and grew with abnormal rapidity, giving a weeping appearance to the trees.

Two seasons later, large branches on these trees had died, others were without leaves, and the deformed, weeping effect of whole trees persisted.

I know well a stretch of road where nature's own landscaping has provided a border of alder, viburnum, sweet fern, and juniper with seasonally changing accents of bright flowers, or of fruits hanging in jeweled clusters in the fall. The road had no heavy load of traffic to support; there were few sharp curves or intersections where brush could obstruct the driver's vision. But the sprayers took over and the miles along that road became something to be traversed quickly, a sight to be endured with one's mind closed to thoughts of the sterile and hideous world we are letting our technicians make. But here and there authority had somehow faltered and by an unaccountable oversight there were oases of beauty in the midst of austere and regimented control—oases that made the desecration of the greater part of the road the more unbearable. In such places my spirit lifted to the sight of the drifts of white clover or the clouds of purple vetch with here and there the flaming cup of a wood lily.

Such plants are "weeds" only to those who make a business of selling and applying chemicals. In a volume of *Proceedings* of one of the weed-control conferences that are now regular institutions, I once read an extraordinary statement of a weed killer's philosophy. The author defended the killing of good plants "simply because they are in bad company." Those who complain about killing wildflowers along roadsides reminded him, he said, of antivivisectionists "to whom, if one were to judge by their actions, the life of a stray dog is more sacred than the lives of children."

To the author of this paper, many of us would unquestionably be suspect, convicted of some deep perversion of character because we prefer the sight of the vetch and the clover and the wood lily in all their delicate and transient beauty to that of roadsides scorched as by fire, the shrubs brown and brittle, the bracken that once lifted high its proud lacework now withered and drooping. We would seem deplorably weak that we can tolerate the sight of such "weeds," that we do not rejoice in their eradication, that we are not filled with exultation that man has once more triumphed over miscreant nature.

Justice Douglas tells of attending a meeting of federal field men who were discussing protests by citizens against plans for the spraying of sagebrush that I mentioned earlier in this chapter. These men considered it hilariously funny that an old lady had opposed the plan because the wildflowers would be destroyed. "Yet, was not her right to search out a banded cup or a tiger lily as inalienable as the right of stockmen to search out grass or of a lumberman to claim a tree?" asks this humane and perceptive jurist. "The esthetic values of the wilderness are as much our inheritance as the veins of copper and gold in our hills and the forests in our mountains."

There is of course more to the wish to preserve our roadside vegetation than even such esthetic considerations. In the economy of nature the natural vegetation has its essential place. Hedgerows along country roads and bordering fields provide food, cover, and nesting areas for birds and homes for many small animals. Of some 70 species of shrubs and vines that are typical roadside species in the eastern states alone, about 65 are important to wildlife as food.

Such vegetation is also the habitat of wild bees and other pollinating insects. Man is more dependent on these wild pollinators than he usually realizes. Even the farmer himself seldom understands the value of wild bees and often participates in the very measures that rob him of their services. Some agricultural crops and many wild plants are partly or wholly dependent on the services of the native pollinating insects. Several hundred species of wild bees take part in the pollination of cultivated crops—100 species visiting the flowers of alfalfa alone. Without insect pollination, most of the soil-holding and soil-enriching plants of uncultivated areas would die out, with far-reaching consequences to the ecology of the whole region. Many herbs, shrubs, and trees of forests and range depend on native insects for their reproduction; without these plants many wild animals and range stock would find little food. Now clean cultivation and the chemical destruction of hedgerows and weeds are eliminating the last sanctuaries of these pollinating insects and breaking the threads that bind life to life.

These insects, so essential to our agriculture and indeed to our landscape as we know it, deserve something better from us than the senseless destruction of their habitat. Honeybees and wild bees depend heavily on such "weeds" as goldenrod, mustard, and dandelions for pollen that serves as the food of their young. Vetch furnishes essential spring forage for bees before the alfalfa is in bloom, tiding them over this early season so that they are ready to pollinate the alfalfa. In the fall they depend on goldenrod at a season when no other food is available, to stock up for the winter. By the precise and delicate timing that is nature's own, the emergence of one species of wild bees takes place on the very day of the opening of the willow blossoms. There is no dearth of men who understand these things, but these are not the men who order the wholesale drenching of the landscape with chemicals.

And where are the men who supposedly understand the value of proper habitat for the preservation of wildlife? Too many of them are to be found defending herbicides as "harmless" to wildlife because they are thought to be less toxic than insecticides. Therefore, it is said, no harm is done. But as the herbicides rain down on forest and field, on marsh and rangeland, they are bringing about marked changes and even permanent destruction of wildlife habitat. To destroy the homes and the food of wildlife is perhaps worse in the long run than direct killing.

The irony of this all-out chemical assault on roadsides and utility rights-of-way is twofold. It is perpetuating the problem it seeks to correct, for as experience has clearly shown, the blanket application of herbicides does not permanently control roadside "brush" and the spraying has to be repeated year after year. And as a further irony, we persist in doing this despite the fact that a perfectly sound method of *selective* spraying is known, which can achieve long-term vegetational control and eliminate repeated spraying in most types of vegetation.

The object of brush control along roads and rights-of-way is not to sweep the land clear of everything but grass; it is, rather, to eliminate plants ultimately tall enough to present an obstruction to drivers' vision or interference with wires on rights-of-way. This means, in general, trees. Most shrubs are low enough to present no hazard; so, certainly, are ferns and wildflowers.

Selective spraying was developed by Dr. Frank Egler during a period of years at the American Museum of Natural History as director of a Committee for Brush Control Recommendations for Rights-of-Way. It took advantage of the inherent stability of nature, building on the fact that most communities of shrubs are strongly resistant to invasion by trees. By comparison, grasslands are easily invaded by tree seedlings. The object of selective spraying is not to produce grass on roadsides and rights-of-way but to eliminate the tall woody plants by direct treatment and to preserve all other vegetation. One treatment may be sufficient, with a possible follow-up for extremely resistant species; thereafter the shrubs assert control and the trees do not return. The best and cheapest controls for vegetation are not chemicals but other plants.

The method has been tested in research areas scattered throughout the eastern United States. Results show that once properly treated, an area becomes stabilized, *requiring no respraying for at least 20 years.* The spraying can often be done by men on foot, using knapsack sprayers, and having complete control over their material. Sometimes compressor pumps and material can be mounted on truck chassis, but there is no blanket spraying. Treatment is directed only to trees and any exceptionally tall shrubs that must be eliminated. The integrity of the environment is thereby preserved, the enormous value of the wildlife habitat remains intact, and the beauty of shrub and fern and wildflower has not been sacrificed.

Here and there the method of vegetation management by selective spraying has been adopted. For the most part, entrenched custom dies hard and blanket spraying continues to thrive, to exact its heavy annual costs from the taxpayer, and to inflict its damage on the ecological web of life. It thrives, surely, only because the facts are not known. When taxpayers understand that the bill for spraying the town roads should come due only once a generation instead of once a year, they will surely rise up and demand a change of method.

Among the many advantages of selective spraying is the fact that it minimizes the amount of chemical applied to the landscape. There is no broadcasting of material but, rather, concentrated application to the base of the trees. The potential harm to wildlife is therefore kept to a minimum.

The most widely used herbicides are 2,4-D, 2,4,5-T, and related compounds. Whether or not these are actually toxic is a matter of controversy. People spraying their lawns with 2,4-D and becoming wet with spray have occasionally developed severe neuritis and even paralysis. Although such incidents are apparently uncommon, medical authorities advise caution in use of such compounds. Other hazards, more obscure, may also attend the use of 2,4-D. It has been shown experimentally to disturb the basic physiological process of respiration in the cell, and to imitate X-rays in damaging the chromosomes. Some very recent work indicates that reproduction of birds may be adversely affected by these and certain other herbicides at levels far below those that cause death.

Apart from any directly toxic effects, curious indirect results follow the use of certain herbicides. It has been found that animals, both wild herbivores and livestock, are sometimes strangely attracted to a plant that has been sprayed, even though it is not one of their natural foods. If a highly poisonous herbicide such as arsenic has been used, this intense desire to reach the wilting vegetation inevitably has disastrous results. Fatal results may follow, also, from less toxic herbicides if the plant itself happens to be poisonous or perhaps to possess thorns or burs. Poisonous range weeds, for example, have suddenly become attractive to livestock after spraying, and the animals have died from indulging this unnatural appetite. The literature of veterinary medicine abounds in similar examples: swine eating sprayed cockleburs with consequent severe illness, lambs eating sprayed thistles, bees poisoned by pasturing on mustard sprayed after it came into bloom. Wild cherry, the leaves of which are highly poisonous, has exerted a fatal attraction for cattle once its foliage has been sprayed with 2,4-D. Apparently the wilting that follows spraying (or cutting) makes the plant attractive. Ragwort has provided other examples. Livestock ordinarily avoid this plant unless forced to turn to it in late winter and early spring by lack of other forage. However, the animals eagerly feed on it after its foliage has been sprayed with 2,4-D.

The explanation of this peculiar behavior sometimes appears to lie in the changes which the chemical brings about in the metabolism of the plant itself. There is temporarily a marked increase in sugar content, making the plant more attractive to many animals.

Another curious effect of 2,4-D has important effects for livestock, wildlife, and apparently for men as well. Experiments carried out about a decade ago showed that after treatment with this chemical there is a sharp increase in the nitrate content of corn and of sugar beets. The same effect was suspected in sorghum, sunflower, spiderwort, lambs quarters, pigweed, and smartweed. Some of these are normally ignored by cattle, but are eaten with relish after treatment with 2,4-D. A number of deaths among cattle have been traced to sprayed weeds, according to some agricultural specialists. The danger lies in the increase in nitrates, for the peculiar physiology of the ruminant at once poses a critical problem. Most such animals have a digestive system of extraordinary complexity, including a stomach divided into four chambers. The digestion of cellulose is accomplished through the action of microorganisms (rumen bacteria) in one of the chambers. When the animal feeds on vegetation containing an abnormally high level of nitrates, the microorganisms in the rumen act on the nitrates to change them into highly toxic nitrites. Thereafter a fatal chain of events ensues: the nitrites act on the blood pigment to form a chocolate-brown substance in which the oxygen is so firmly held that it cannot take part in respiration, hence oxygen is not transferred from the lungs to the tissues. Death occurs within a few hours from anoxia, or lack of oxygen. The various reports of livestock losses after grazing on certain weeds treated with 2,4-D therefore have a logical explanation. The same danger exists for wild animals belonging to the group of ruminants, such as deer, antelope, sheep, and goats.

Although various factors (such as exceptionally dry weather) can cause an increase in nitrate content, the effect of the soaring sales and applications of 2,4-D cannot be ignored. The situation was considered important enough by the University of Wisconsin Agricultural Experiment Station to justify a warning in 1957 that "plants killed by 2,4-D may contain large amounts of nitrate." The hazard extends to human beings as well as animals and may help to explain the recent mysterious increase in "silo deaths." When corn, oats, or sorghum containing large amounts of nitrates are ensiled they release poisonous nitrogen oxide gases, creating a deadly hazard to anyone entering the silo. Only a few breaths of one of these gases can cause a diffuse chemical pneumonia. In a series of such cases studied by the University of Minnesota Medical School all but one terminated fatally.

"Once again we are walking in nature like an elephant in the china cabinet." So C. J. Briejèr, a Dutch scientist of rare understanding, sums up our use of weed killers. "In my opinion too much is taken for granted. We do not know whether all weeds in crops are harmful or whether some of them are useful," says Dr. Briejèr.

Seldom is the question asked, What is the relation between the weed and the soil? Perhaps, even from our narrow standpoint of direct self-interest, the relation is a useful one. As we have seen, soil and the living things in and upon it exist in a relation of interdependence and mutual benefit. Presumably the weed is taking something from the soil; perhaps it is also contributing something to it. A practical example was provided recently by the parks in a city in Holland. The roses were doing badly. Soil samples showed heavy infestations by tiny nematode worms. Scientists of the Dutch Plant Protection Service did not recommend chemical sprays or soil treatments; instead, they suggested that marigolds be planted among the roses. This plant, which the purist would doubtless consider a weed in any rosebed, releases an excretion from its roots that kills the soil nematodes. The advice was taken; some beds were planted with marigolds, some left without as controls. The results were striking. With the aid of the marigolds the roses flourished; in the control beds they were sickly and drooping. Marigolds are now used in many places for combating nematodes.

In the same way, and perhaps quite unknown to us, other plants that we ruthlessly eradicate may be performing a function that is necessary to the health of the soil. One very useful function of natural plant communities—now pretty generally stigmatized as "weeds"—is to serve as an indicator of the condition of the soil. This useful function is of course lost where chemical weed killers have been used.

Those who find an answer to all problems in spraying also overlook a matter of great scientific importance—the need to preserve some natural plant communities. We need these as a standard against which we can measure the changes our own activities bring about. We need them as wild habitats in which original populations of insects and other organisms can be maintained, for, as will be explained in Chapter 16, the development of resistance to insecticides is changing the genetic factors of insects and perhaps other organisms. One scientist has even suggested that some sort of "zoo" should be established to preserve insects, mites, and the like, before their genetic composition is further changed.

Some experts warn of subtle but far-reaching vegetational shifts as a result of the growing use of herbicides. The chemical 2,4-D, by killing out the broad-leaved plants, allows the grasses to thrive in the reduced competition—now some of the grasses themselves have become "weeds," presenting a new problem in control and giving the cycle another turn. This strange situation is acknowledged in a recent issue of a journal devoted to crop problems: "With the widespread use of 2,4-D to control broad-leaved weeds, grass weeds in particular have increasingly become a threat to corn and soybean yields."

Ragweed, the bane of hay fever sufferers, offers an interesting example of the way efforts to control nature sometimes boomerang. Many thousands of gallons of chemicals have been discharged along roadsides in the name of ragweed control. But the unfortunate truth is that blanket spraying is resulting in more ragweed, not less. Ragweed is an annual; its seedlings require open soil to become established each year. Our best protection against this plant is therefore the maintenance of dense shrubs, ferns, and other perennial vegetation. Spraying frequently destroys this protective vegetation and creates open, barren areas which the ragweed hastens to fill. It is probable, moreover, that the pollen content of the atmosphere is not related to roadside ragweed, but to the ragweed of city lots and fallow fields.

The booming sales of chemical crabgrass killers are another example of how readily unsound methods catch on. There is a cheaper and better way to remove crabgrass than to attempt year after year to kill it out with chemicals. This is to give it competition of a kind it cannot survive, the competition of other grass. Crabgrass exists only in an unhealthy lawn. It is a symptom, not a disease in itself. By providing a fertile soil and giving the desired grasses a good start, it is possible to create an environment in which crabgrass cannot grow, for it requires open space in which it can start from seed year after year.

Instead of treating the basic condition, suburbanites—advised by nurserymen who in turn have been advised by the chemical manufacturers—continue to apply truly astonishing amounts of crabgrass killers to their lawns each year. Marketed under trade names which give no hint of their nature, many of these preparations contain such poisons as mercury, arsenic, and chlordane. Application at the recommended rates leaves tremendous amounts of these chemicals on the lawn. Users of one product, for example, apply 60 pounds of technical chlordane to the acre if they follow directions. If they use another of the many available products, they are applying 175 pounds of metallic arsenic to the acre. The toll of dead birds, as we shall see in Chapter 8, is distressing. How lethal these lawns may be for human beings is unknown.

The success of selective spraying for roadside and right-of-way vegetation, where it has been practiced, offers hope that equally sound ecological methods may be developed for other vegetation programs for farms, forests, and ranges—methods aimed not at destroying a particular species but at managing vegetation as a living community.

Other solid achievements show what can be done. Biological control has achieved some of its most spectacular successes in the area of curbing unwanted vegetation. Nature herself has met many of the problems that now beset us, and she has usually solved them in her own successful way. Where man has been intelligent enough to observe and to emulate Nature he, too, is often rewarded with success.

An outstanding example in the field of controlling unwanted plants is the handling of the Klamath-weed problem in California. Although the Klamath weed, or goatweed, is a native of Europe (where it is called St. Johnswort), it accompanied man in his westward migrations, first appearing in the United States in 1793 near Lancaster, Pennsylvania. By 1900 it had reached California in the vicinity of the Klamath River, hence the name locally given to it. By 1929 it had occupied about 100,000 acres of rangeland, and by 1952 it had invaded some two and one half million acres.

Klamath weed, quite unlike such native plants as sagebrush, has no place in the ecology of the region, and no animals or other plants require its presence. On the contrary, wherever it appeared livestock became "scabby, sore-mouthed, and unthrifty" from feeding on this toxic plant. Land values declined accordingly, for the Klamath weed was considered to hold the first mortgage.

In Europe the Klamath weed, or St. Johnswort, has never become a problem because along with the plant there have developed various species of insects; these feed on it so extensively that its abundance is severely limited. In particular, two species of beetles in southern France, pea-sized and of metallic color, have their whole beings so adapted to the presence of the weed that they feed and reproduce only upon it.

It was an event of historic importance when the first shipments of these beetles were brought to the United States in 1944, for this was the first attempt in North America to control a plant with a plant-eating insect. By 1948 both species had become so well established that no further importations were needed. Their spread was accomplished by collecting beetles from the original colonies and redistributing them at the rate of millions a year. Within small areas the beetles accomplish their own dispersion, moving on as soon as the Klamath weed dies out and locating new stands with great precision. And as the beetles thin out the weed, desirable range plants that have been crowded out are able to return.

A ten-year survey completed in 1959 showed that control of the Klamath weed had been "more effective than hoped for even by enthusiasts," with the weed reduced to a mere 1 per cent of its former abundance. This token infestation is harmless and is actually needed in order to maintain a population of beetles as protection against a future increase in the weed.

Another extraordinarily successful and economical example of weed control may be found in Australia. With the colonists' usual taste for carrying plants or animals into a new country, a Captain Arthur Phillip had brought various species of cactus into Australia about 1787, intending to use them in culturing cochineal insects for dye. Some of the cacti or prickly pears escaped from his gardens and by 1925 about 20 species could be found growing wild. Having no natural controls in this new territory, they spread prodigiously, eventually occupying about 60 million acres. At least half of this land was so densely covered as to be useless.

In 1920 Australian entomologists were sent to North and South America to study insect enemies of the prickly pears in their native habitat. After trials of several species, 3 billion eggs of an Argentine moth were released in Australia in 1930. Seven years later the last dense growth of the prickly pear had been destroyed and the once uninhabitable areas reopened to settlement and grazing. The whole operation had cost less than a penny per acre. In contrast, the unsatisfactory attempts at chemical control in earlier years had cost about £10 per acre.

Both of these examples suggest that extremely effective control of many kinds of unwanted vegetation might be achieved by paying more attention to the role of plant-eating insects. The science of range management has largely ignored this possibility, although these insects are perhaps the most selective of all grazers and their highly restricted diets could easily be turned to man's advantage.

SEVEN

Needless Havoc

A S MAN PROCEEDS toward his announced goal of the conquest of nature, he has written a depressing record of destruction, directed not only against the earth he inhabits but against the life that shares it with him. The history of the recent centuries has its black passages—the slaughter of the buffalo on the western plains, the massacre of the shorebirds by the market gunners, the near-extermination of the egrets for their plumage. Now, to these and others like them, we are adding a new chapter and a new kind of havoc—the direct killing of birds, mammals, fishes, and indeed practically every form of wildlife by chemical insecticides indiscriminately sprayed on the land.

Under the philosophy that now seems to guide our destinies, nothing must get in the way of the man with the spray gun. The incidental victims of his crusade against insects count as nothing; if robins, pheasants, raccoons, cats, or even livestock happen to inhabit the same bit of earth as the target insects and to be hit by the rain of insect-killing poisons no one must protest.

The citizen who wishes to make a fair judgment of the question of wildlife loss is today confronted with a dilemma. On the one hand conservationists and many wildlife biologists assert that the losses have been severe and in some cases even catastrophic. On the other hand the control agencies tend to deny flatly and categorically that such losses have occurred, or that they are of any importance if they have. Which view are we to accept?

The credibility of the witness is of first importance. The professional wildlife biologist on the scene is certainly best qualified to discover and interpret wildlife loss. The entomologist, whose specialty is insects, is not so qualified by training, and is not psychologically disposed to look for undesirable

side effects of his control program. Yet it is the control men in state and federal governments—and of course the chemical manufacturers—who steadfastly deny the facts reported by the biologists and declare they see little evidence of harm to wild-life. Like the priest and the Levite in the biblical story, they choose to pass by on the other side and to see nothing. Even if we charitably explain their denials as due to the shortsighted-ness of the specialist and the man with an interest this does not mean we must accept them as qualified witnesses.

The best way to form our own judgment is to look at some of the major control programs and learn, from observers famil-iar with the ways of wildlife, and unbiased in favor of chemicals, just what has happened in the wake of a rain of poison falling from the skies into the world of wildlife.

To the bird watcher, the suburbanite who derives joy from birds in his garden, the hunter, the fisherman or the explorer of wild regions, anything that destroys the wildlife of an area for even a single year has deprived him of pleasure to which he has a legitimate right. This is a valid point of view. Even if, as has sometimes happened, some of the birds and mammals and fishes are able to re-establish themselves after a single spraying, a great and real harm has been done.

But such re-establishment is unlikely to happen. Spraying tends to be repetitive, and a single exposure from which the wildlife populations might have a chance to recover is a rarity. What usually results is a poisoned environment, a lethal trap in which not only the resident populations succumb but those who come in as migrants as well. The larger the area sprayed the more serious the harm, because no oases of safety remain. Now, in a decade marked by insect-control programs in which many thousands or even millions of acres are sprayed as a unit, a decade in which private and community spraying has also surged steadily upward, a record of destruction and death of American wildlife has accumulated. Let us look at some of these programs and see what has happened.

During the fall of 1959 some 27,000 acres in southeastern Michigan, including numerous suburbs of Detroit, were heavily dusted from the air with pellets of aldrin, one of the most dan-gerous of all the chlorinated hydrocarbons. The program was conducted by the Michigan Department of Agriculture with the cooperation of the United States Department of Agriculture; its announced purpose was control of the Japanese beetle.

Little need was shown for this drastic and dangerous action. On the contrary, Walter P. Nickell, one of the best-known and best-informed naturalists in the state, who spends much of his time in the field with long periods in southern Michigan every summer, declared: "For more than thirty years, to my direct knowledge, the Japanese beetle has been present in the city of Detroit in small numbers. The numbers have not shown any appreciable increase in all this lapse of years. I have yet to see a single Japanese beetle [in 1959] other than the few caught in Government catch traps in Detroit . . . Everything is being kept so secret that I have not yet been able to obtain any information whatsoever to the effect that they have increased in numbers."

An official release by the state agency merely declared that the beetle had "put in its appearance" in the areas designated for the aerial attack upon it. Despite the lack of justification the program was launched, with the state providing the manpower and supervising the operation, the federal government providing equipment and additional men, and the communities paying for the insecticide.

The Japanese beetle, an insect accidentally imported into the United States, was discovered in New Jersey in 1916, when a few shiny beetles of a metallic green color were seen in a nursery near Riverton. The beetles, at first unrecognized, were finally identified as a common inhabitant of the main islands of Japan. Apparently they had entered the United States on nursery stock imported before restrictions were established in 1912.

From its original point of entrance the Japanese beetle has spread rather widely throughout many of the states east of the Mississippi, where conditions of temperature and rainfall are suitable for it. Each year some outward movement beyond the existing boundaries of its distribution usually takes place. In the eastern areas where the beetles have been longest established, attempts have been made to set up natural controls. Where this has been done, the beetle populations have been kept at relatively low levels, as many records attest.

Despite the record of reasonable control in eastern areas, the midwestern states now on the fringe of the beetle's range have launched an attack worthy of the most deadly enemy instead of only a moderately destructive insect, employing the most dangerous chemicals distributed in a manner that exposes large numbers of people, their domestic animals, and all wildlife to the poison intended for the beetle. As a result these Japanese beetle programs have caused shocking destruction of animal life and have exposed human beings to undeniable hazard. Sections of Michigan, Kentucky, Iowa, Indiana, Illinois, and Missouri are all experiencing a rain of chemicals in the name of beetle control.

The Michigan spraying was one of the first large-scale attacks on the Japanese beetle from the air. The choice of aldrin, one of the deadliest of all chemicals, was not determined by any peculiar suitability for Japanese beetle control, but simply by the wish to save money—aldrin was the cheapest of the compounds available. While the state in its official release to the press acknowledged that aldrin is a "poison," it implied that no harm could come to human beings in the heavily populated areas to which the chemical was applied. (The official answer to the query "What precautions should I take?" was "For you, none.") An official of the Federal Aviation Agency was later quoted in the local press to the effect that "this is a safe operation" and a representative of the Detroit Department of Parks and Recreation added his assurance that "the dust is harmless to humans and will not hurt plants or pets." One must assume that none of these officials had consulted the published and readily available reports of the United States Public Health Service, the Fish and Wildlife Service, and other evidence of the extremely poisonous nature of aldrin.

Acting under the Michigan pest control law which allows the state to spray indiscriminately without notifying or gaining permission of individual landowners, the low-lying planes began to fly over the Detroit area. The city authorities and the Federal Aviation Agency were immediately besieged by calls from worried citizens. After receiving nearly 800 calls in a single hour, the police begged radio and television stations and newspapers to "tell the watchers what they were seeing and advise them it was safe," according to the Detroit News. The Federal Aviation Agency's safety officer assured the public that "the planes are carefully supervised" and "are authorized to fly low." In a somewhat mistaken attempt to allay fears, he added that the planes had emergency valves that would allow them to dump their entire load instantaneously. This, fortunately, was not done, but as the planes went about their work the pellets of insecticide fell on beetles and humans alike, showers of "harmless" poison descending on people shopping or going to work and on children out from school for the lunch hour. Housewives swept the granules from porches and sidewalks, where they are said to have "looked like snow." As pointed out later by the Michigan Audubon Society, "In the spaces between shingles on roofs, in eaves-troughs, in the cracks in bark and twigs, the little white pellets of aldrin-and-clay, no bigger than a pin head, were lodged by the millions . . . When the snow and rain came, every puddle became a possible death potion."

Within a few days after the dusting operation, the Detroit Audubon Society began receiving calls about the birds. According to the Society's secretary, Mrs. Ann Boyes, "The first indication that the people were concerned about the spray was a call I received on Sunday morning from a woman who reported that coming home from church she saw an alarming number of dead and dying birds. The spraying there had been done on Thursday. She said there were no birds at all flying in the area, that she had found at least a dozen [dead] in her backyard and that the neighbors had found dead squirrels." All other calls received by Mrs. Boyes that day reported "a great many dead birds and no live ones . . . People who had maintained bird feeders said there were no birds at all at their feeders." Birds picked up in a dying condition showed the typical symptoms of insecticide poisoning—tremoring, loss of ability to fly, paralysis, convulsions.

Nor were birds the only forms of life immediately affected. A local veterinarian reported that his office was full of clients with dogs and cats that had suddenly sickened. Cats, who so meticulously groom their coats and lick their paws, seemed to be most affected. Their illness took the form of severe diarrhea, vomiting, and convulsions. The only advice the veterinarian could give his clients was not to let the animals out unnecessarily, or to wash the paws promptly if they did so. (But the chlorinated hydrocarbons cannot be washed even from fruits or vegetables, so little protection could be expected from this measure.)

Despite the insistence of the City–County Health Commissioner that the birds must have been killed by "some other kind of spraying" and that the outbreak of throat and chest irritations that followed the exposure to aldrin must have been due to "something else," the local Health Department received a constant stream of complaints. A prominent Detroit internist was called upon to treat four of his patients within an hour after they had been exposed while watching the planes at work. All had similar symptoms: nausea, vomiting, chills, fever, extreme fatigue, and coughing.

The Detroit experience has been repeated in many other communities as pressure has mounted to combat the Japanese beetle with chemicals. At Blue Island, Illinois, hundreds of dead and dying birds were picked up. Data collected by bird-banders here suggest that 80 per cent of the songbirds were sacrificed. In Joliet, Illinois, some 3000 acres were treated with heptachlor in 1959. According to reports from a local sportsmen's club, the bird population within the treated area was "virtually wiped out." Dead rabbits, muskrats, opossums, and fish were also found in numbers, and one of the local schools made the collection of insecticide-poisoned birds a science project.

Perhaps no community has suffered more for the sake of a beetleless world than Sheldon, in eastern Illinois, and adjacent areas in Iroquois County. In 1954 the United States Department of Agriculture and the Illinois Agriculture Department began a program to eradicate the Japanese beetle along the line of its advance into Illinois, holding out the hope, and indeed the assurance, that intensive spraying would destroy the populations of the invading insect. The first "eradication" took place that year, when dieldrin was applied to 1400 acres by air. Another 2600 acres were treated similarly in 1955, and the task was presumably considered complete. But more and more chemical treatments were called for, and by the end of 1961 some 131,000 acres had been covered. Even in the first years of the program it was apparent that heavy losses were occurring among wildlife and domestic animals. The chemical treatments were continued, nevertheless, without consultation with either the United States Fish and Wildlife Service or the

Illinois Game Management Division. (In the spring of 1960, however, officials of the federal Department of Agriculture appeared before a congressional committee in opposition to a bill that would require just such prior consultation. They declared blandly that the bill was unnecessary because cooperation and consultation were "usual." These officials were quite unable to recall situations where cooperation had not taken place "at the Washington level." In the same hearings they stated clearly their unwillingness to consult with state fish and game departments.)

Although funds for chemical control came in never-ending streams, the biologists of the Illinois Natural History Survey who attempted to measure the damage to wildlife had to operate on a financial shoestring. A mere $1100 was available for the employment of a field assistant in 1954 and no special funds were provided in 1955. Despite these crippling difficulties, the biologists assembled facts that collectively paint a picture of almost unparalleled wildlife destruction—destruction that became obvious as soon as the program got under way.

Conditions were made to order for poisoning insect-eating birds, both in the poisons used and in the events set in motion by their application. In the early programs at Sheldon, dieldrin was applied at the rate of 3 pounds to the acre. To understand its effect on birds one need only remember that in laboratory experiments on quail dieldrin has proved to be about 50 times as poisonous as DDT. The poison spread over the landscape at Sheldon was therefore roughly equivalent to 150 pounds of DDT per acre! And this was a minimum, because there seems to have been some overlapping of treatments along field borders and in corners.

As the chemical penetrated the soil the poisoned beetle grubs crawled out on the surface of the ground, where they remained for some time before they died, attractive to insect-eating birds. Dead and dying insects of various species were conspicuous for about two weeks after the treatment. The effect on the bird populations could easily have been foretold. Brown thrashers, starlings, meadowlarks, grackles, and pheasants were virtually wiped out. Robins were "almost annihilated," according to the biologists' report. Dead earthworms had been seen in numbers after a gentle rain; probably the robins had fed on the poisoned worms. For other birds, too, the once beneficial rain had been changed, through the evil power of the poison introduced into their world, into an agent of destruction. Birds seen drinking and bathing in puddles left by rain a few days after the spraying were inevitably doomed.

The birds that survived may have been rendered sterile. Although a few nests were found in the treated area, a few with eggs, none contained young birds.

Among the mammals ground squirrels were virtually annihilated; their bodies were found in attitudes characteristic of violent death by poisoning. Dead muskrats were found in the treated areas, dead rabbits in the fields. The fox squirrel had been a relatively common animal in the town; after the spraying it was gone.

It was a rare farm in the Sheldon area that was blessed by the presence of a cat after the war on beetles was begun. Ninety per cent of all the farm cats fell victims to the dieldrin during the first season of spraying. This might have been predicted because of the black record of these poisons in other places. Cats are extremely sensitive to all insecticides and especially so, it seems, to dieldrin. In western Java in the course of the antimalarial program carried out by the World Health Organization, many cats are reported to have died. In central Java so many were killed that the price of a cat more than doubled. Similarly, the World Health Organization, spraying in Venezuela, is reported to have reduced cats to the status of a rare animal.

In Sheldon it was not only the wild creatures and the domestic companions that were sacrificed in the campaign against an insect. Observations on several flocks of sheep and a herd of beef cattle are indicative of the poisoning and death that threatened livestock as well. The Natural History Survey report describes one of these episodes as follows:

> The sheep . . . were driven into a small, untreated blue-grass pasture across a gravel road from a field which had been treated with dieldrin spray on May 6. Evidently some spray had drifted across the road into the pasture, for the sheep began to show symptoms of intoxication almost at once . . . They lost interest in food and displayed extreme restlessness, following

the pasture fence around and around apparently searching for a way out . . . [They] refused to be driven, bleated almost continuously, and stood with their heads lowered; they were finally carried from the pasture . . . They displayed great desire for water. Two of the sheep were found dead in the stream passing through the pasture, and the remaining sheep were repeatedly driven out of the stream, several having to be dragged forcibly from the water. Three of the sheep eventually died; those remaining recovered to all outward appearances.

This, then, was the picture at the end of 1955. Although the chemical war went on in succeeding years, the trickle of research funds dried up completely. Requests for money for wildlife-insecticide research were included in annual budgets submitted to the Illinois legislature by the Natural History Survey, but were invariably among the first items to be eliminated. It was not until 1960 that money was somehow found to pay the expenses of one field assistant—to do work that could easily have occupied the time of four men.

The desolate picture of wildlife loss had changed little when the biologists resumed the studies broken off in 1955. In the meantime, the chemical had been changed to the even more toxic aldrin, *100 to 300 times* as toxic as DDT in tests on quail. By 1960, every species of wild mammal known to inhabit the area had suffered losses. It was even worse with the birds. In the small town of Donovan the robins had been wiped out, as had the grackles, starlings, and brown thrashers. These and many other birds were sharply reduced elsewhere. Pheasant hunters felt the effects of the beetle campaign sharply. The number of broods produced on treated lands fell off by some 50 per cent, and the number of young in a brood declined. Pheasant hunting, which had been good in these areas in former years, was virtually abandoned as unrewarding.

In spite of the enormous havoc that had been wrought in the name of eradicating the Japanese beetle, the treatment of more than 100,000 acres in Iroquois County over an eight-year period seems to have resulted in only temporary suppression of the insect, which continues its westward movement. The full extent of the toll that has been taken by this largely ineffective program may never be known, for the results measured by the Illinois biologists are a minimum figure. If the research program had been adequately financed to permit full coverage, the destruction revealed would have been even more appalling. But in the eight years of the program, only about $6000 was provided for biological field studies. Meanwhile the federal government had spent about $375,000 for control work and additional thousands had been provided by the state. The amount spent for research was therefore a small fraction of 1 per cent of the outlay for the chemical program.

These midwestern programs have been conducted in a spirit of crisis, as though the advance of the beetle presented an extreme peril justifying any means to combat it. This of course is a distortion of the facts, and if the communities that have endured these chemical drenchings had been familiar with the earlier history of the Japanese beetle in the United States they would surely have been less acquiescent.

The eastern states, which had the good fortune to sustain their beetle invasion in the days before the synthetic insecticides had been invented, have not only survived the invasion but have brought the insect under control by means that represented no threat whatever to other forms of life. There has been nothing comparable to the Detroit or Sheldon sprayings in the East. The effective methods there involved the bringing into play of natural forces of control which have the multiple advantages of permanence and environmental safety.

During the first dozen years after its entry into the United States, the beetle increased rapidly, free of the restraints that in its native land hold it in check. But by 1945 it had become a pest of only minor importance throughout much of the territory over which it had spread. Its decline was largely a consequence of the importation of parasitic insects from the Far East and of the establishment of disease organisms fatal to it.

Between 1920 and 1933, as a result of diligent searching throughout the native range of the beetle, some 34 species of predatory or parasitic insects had been imported from the Orient in an effort to establish natural control. Of these, five became well established in the eastern United States. The most effective and widely distributed is a parasitic wasp from Korea and China, *Tiphia vernalis*. The female *Tiphia*, finding a beetle grub in the soil, injects a paralyzing fluid and attaches a single egg to the undersurface of the grub. The young wasp, hatching as a larva, feeds on the paralyzed grub and destroys it. In some 25 years, colonies of *Tiphia* were introduced into 14 eastern states in a cooperative program of state and federal agencies. The wasp became widely established in this area and is generally credited by entomologists with an important role in bringing the beetle under control.

An even more important role has been played by a bacterial disease that affects beetles of the family to which the Japanese beetle belongs—the scarabaeids. It is a highly specific organism, attacking no other type of insects, harmless to earthworms, warm-blooded animals, and plants. The spores of the disease occur in soil. When ingested by a foraging beetle grub they multiply prodigiously in its blood, causing it to turn an abnormally white color, hence the popular name, "milky disease."

Milky disease was discovered in New Jersey in 1933. By 1938 it was rather widely prevalent in the older areas of Japanese beetle infestation. In 1939 a control program was launched, directed at speeding up the spread of the disease. No method had been developed for growing the disease organism in an artificial medium, but a satisfactory substitute was evolved; infected grubs are ground up, dried, and combined with chalk. In the standard mixture a gram of dust contains 100 million spores. Between 1939 and 1953 some 94,000 acres in 14 eastern states were treated in a cooperative federal-state program; other areas on federal lands were treated; and an unknown but extensive area was treated by private organizations or individuals. By 1945, milky spore disease was raging among the beetle populations of Connecticut, New York, New Jersey, Delaware, and Maryland. In some test areas infection of grubs had reached as high as 94 per cent. The distribution program was discontinued as a governmental enterprise in 1953 and production was taken over by a private laboratory, which continues to supply individuals, garden clubs, citizens' associations, and all others interested in beetle control.

The eastern areas where this program was carried out now enjoy a high degree of natural protection from the beetle. The organism remains viable in the soil for years and therefore becomes to all intents and purposes permanently established, increasing in effectiveness, and being continuously spread by natural agencies.

Why, then, with this impressive record in the East, were the same procedures not tried in Illinois and the other midwestern states where the chemical battle of the beetles is now being waged with such fury?

We are told that inoculation with milky spore disease is "too expensive"—although no one found it so in the 14 eastern states in the 1940's. And by what sort of accounting was the "too expensive" judgment reached? Certainly not by any that assessed the true costs of the total destruction wrought by such programs as the Sheldon spraying. This judgment also ignores the fact that inoculation with the spores need be done only once; the first cost is the only cost.

We are told also that milky spore disease cannot be used on the periphery of the beetle's range because it can be established only where a large grub population is *already* present in the soil. Like many other statements in support of spraying, this one needs to be questioned. The bacterium that causes milky spore disease has been found to infect at least 40 other species of beetles which collectively have quite a wide distribution and would in all probability serve to establish the disease even where the Japanese beetle population is very small or nonexistent. Furthermore, because of the long viability of the spores in soil they can be introduced even in the complete absence of grubs, as on the fringe of the present beetle infestation, there to await the advancing population.

Those who want immediate results, at whatever cost, will doubtless continue to use chemicals against the beetle. So will those who favor the modern trend to built-in obsolescence, for chemical control is self-perpetuating, needing frequent and costly repetition.

On the other hand, those who are willing to wait an extra season or two for full results will turn to milky disease; they will be rewarded with lasting control that becomes more, rather than less effective with the passage of time.

An extensive program of research is under way in the United States Department of Agriculture laboratory at Peoria, Illinois, to find a way to culture the organism of milky disease on an artificial medium. This will greatly reduce its cost and should encourage its more extensive use. After years of work, some success has now been reported. When this "breakthrough" is thoroughly established perhaps some sanity and perspective will be restored to our dealings with the Japanese beetle, which at the peak of its depredations never justified the nightmare excesses of some of these midwestern programs.

Incidents like the eastern Illinois spraying raise a question that is not only scientific but moral. The question is whether any civilization can wage relentless war on life without destroying itself, and without losing the right to be called civilized.

These insecticides are not selective poisons; they do not single out the one species of which we desire to be rid. Each of them is used for the simple reason that it is a deadly poison. It therefore poisons all life with which it comes in contact: the cat beloved of some family, the farmer's cattle, the rabbit in the field, and the horned lark out of the sky. These creatures are innocent of any harm to man. Indeed, by their very existence they and their fellows make his life more pleasant. Yet he rewards them with a death that is not only sudden but horrible. Scientific observers at Sheldon described the symptoms of a meadowlark found near death: "Although it lacked muscular coordination and could not fly or stand, it continued to beat its wings and clutch with its toes while lying on its side. Its beak was held open and breathing was labored." Even more pitiful was the mute testimony of the dead ground squirrels, which "exhibited a characteristic attitude in death. The back was bowed, and the forelegs with the toes of the feet tightly clenched were drawn close to the thorax . . . The head and neck were outstretched and the mouth often contained dirt, suggesting that the dying animal had been biting at the ground."

By acquiescing in an act that can cause such suffering to a living creature, who among us is not diminished as a human being?

EIGHT

And No Birds Sing

O VER INCREASINGLY large areas of the United States, spring now comes unheralded by the return of the birds, and the early mornings are strangely silent where once they were filled with the beauty of bird song. This sudden silencing of the song of birds, this obliteration of the color and beauty and interest they lend to our world have come about swiftly, insidiously, and unnoticed by those whose communities are as yet unaffected.

From the town of Hinsdale, Illinois, a housewife wrote in despair to one of the world's leading ornithologists, Robert Cushman Murphy, Curator Emeritus of Birds at the American Museum of Natural History.

> Here in our village the elm trees have been sprayed for several years [she wrote in 1958]. When we moved here six years ago, there was a wealth of bird life; I put up a feeder and had a steady stream of cardinals, chickadees, downies and nuthatches all winter, and the cardinals and chickadees brought their young ones in the summer.
>
> After several years of DDT spray, the town is almost devoid of robins and starlings; chickadees have not been on my shelf for two years, and this year the cardinals are gone too; the nesting population in the neighborhood seems to consist of one dove pair and perhaps one catbird family.
>
> It is hard to explain to the children that the birds have been killed off, when they have learned in school that a Federal law protects the birds from killing or capture. "Will they ever come back?" they ask, and I do not have the answer. The elms are still dying, and so are the birds. *Is* anything being done? *Can* anything be done? Can *I* do anything?

A year after the federal government had launched a massive spraying program against the fire ant, an Alabama woman wrote: "Our place has been a veritable bird sanctuary for over half a century. Last July we all remarked, 'There are more birds than ever.' Then, suddenly, in the second week of August, they all disappeared. I was accustomed to rising early to care for my favorite mare that had a young filly. There was not a sound of the song of a bird. It was eerie, terrifying. What was man doing to our perfect and beautiful world? Finally, five months later a blue jay appeared and a wren."

The autumn months to which she referred brought other somber reports from the deep South, where in Mississippi, Louisiana, and Alabama the *Field Notes* published quarterly by the National Audubon Society and the United States Fish and Wildlife Service noted the striking phenomenon of "blank spots weirdly empty of virtually *all* bird life." The *Field Notes* are a compilation of the reports of seasoned observers who have spent many years afield in their particular areas and have unparalleled knowledge of the normal bird life of the region. One such observer reported that in driving about southern Mississippi that fall she saw "no land birds at all for long distances." Another in Baton Rouge reported that the contents of her feeders had lain untouched "for weeks on end," while fruiting shrubs in her yard, that ordinarily would be stripped clean by that time, still were laden with berries. Still another reported that his picture window, "which often used to frame a scene splashed with the red of 40 or 50 cardinals and crowded with other species, seldom permitted a view of as many as a bird or two at a time." Professor Maurice Brooks of the University of West Virginia, an authority on the birds of the Appalachian region, reported that the West Virginia bird population had undergone "an incredible reduction."

One story might serve as the tragic symbol of the fate of the birds—a fate that has already overtaken some species, and that threatens all. It is the story of the robin, the bird known to everyone. To millions of Americans, the season's first robin means that the grip of winter is broken. Its coming is an event reported in newspapers and told eagerly at the breakfast table. And as the number of migrants grows and the first mists of green appear in the woodlands, thousands of people listen for the first dawn chorus of the robins throbbing in the early morning light. But now all is changed, and not even the return of the birds may be taken for granted.

The survival of the robin, and indeed of many other species as well, seems fatefully linked with the American elm, a tree that is part of the history of thousands of towns from the Atlantic to the Rockies, gracing their streets and their village squares and college campuses with majestic archways of green. Now the elms are stricken with a disease that afflicts them throughout their range, a disease so serious that many experts believe all efforts to save the elms will in the end be futile. It would be tragic to lose the elms, but it would be doubly tragic if, in vain efforts to save them, we plunge vast segments of our bird populations into the night of extinction. Yet this is precisely what is threatened.

The so-called Dutch elm disease entered the United States from Europe about 1930 in elm burl logs imported for the veneer industry. It is a fungus disease; the organism invades the water-conducting vessels of the tree, spreads by spores carried in the flow of sap, and by its poisonous secretions as well as by mechanical clogging causes the branches to wilt and the tree to die. The disease is spread from diseased to healthy trees by elm bark beetles. The galleries which the insects have tunneled out under the bark of dead trees become contaminated with spores of the invading fungus, and the spores adhere to the insect body and are carried wherever the beetle flies. Efforts to control the fungus disease of the elms have been directed largely toward control of the carrier insect. In community after community, especially throughout the strongholds of the American elm, the Midwest and New England, intensive spraying has become a routine procedure.

What this spraying could mean to bird life, and especially to the robin, was first made clear by the work of two ornithologists at Michigan State University, Professor George Wallace and one of his graduate students, John Mehner. When Mr. Mehner began work for the doctorate in 1954, he chose a research project that had to do with robin populations. This was quite by chance, for at that time no one suspected that the robins were in danger. But even as he undertook the work, events occurred that were to change its character and indeed to deprive him of his material.

Spraying for Dutch elm disease began in a small way on the university campus in 1954. The following year the city of East Lansing (where the university is located) joined in, spraying on the campus was expanded, and, with local programs for gypsy moth and mosquito control also under way, the rain of chemicals increased to a downpour.

During 1954, the year of the first light spraying, all seemed well. The following spring the migrating robins began to return to the campus as usual. Like the bluebells in Tomlinson's haunting essay "The Lost Wood," they were "expecting no evil" as they reoccupied their familiar territories. But soon it became evident that something was wrong. Dead and dying robins began to appear on the campus. Few birds were seen in their normal foraging activities or assembling in their usual roosts. Few nests were built; few young appeared. The pattern was repeated with monotonous regularity in succeeding springs. The sprayed area had become a lethal trap in which each wave of migrating robins would be eliminated in about a week. Then new arrivals would come in, only to add to the numbers of doomed birds seen on the campus in the agonized tremors that precede death.

"The campus is serving as a graveyard for most of the robins that attempt to take up residence in the spring," said Dr. Wallace. But why? At first he suspected some disease of the nervous system, but soon it became evident that "in spite of the assurances of the insecticide people that their sprays were 'harmless to birds' the robins were really dying of insecticidal poisoning; they exhibited the well-known symptoms of loss of balance, followed by tremors, convulsions, and death."

Several facts suggested that the robins were being poisoned, not so much by direct contact with the insecticides as indirectly, by eating earthworms. Campus earthworms had been fed inadvertently to crayfish in a research project and all the crayfish had promptly died. A snake kept in a laboratory cage had gone into violent tremors after being fed such worms. And earthworms are the principal food of robins in the spring.

A key piece in the jigsaw puzzle of the doomed robins was soon to be supplied by Dr. Roy Barker of the Illinois Natural History Survey at Urbana. Dr. Barker's work, published in 1958, traced the intricate cycle of events by which the robins' fate is linked to the elm trees by way of the earthworms. The trees are sprayed in the spring (usually at the rate of 2 to 5 pounds of DDT per 50-foot tree, which may be the equivalent of as much as *23 pounds per acre* where elms are numerous) and often again in July, at about half this concentration. Powerful sprayers direct a stream of poison to all parts of the tallest trees, killing directly not only the target organism, the bark beetle, but other insects, including pollinating species and predatory spiders and beetles. The poison forms a tenacious film over the leaves and bark. Rains do not wash it away. In the autumn the leaves fall to the ground, accumulate in sodden layers, and begin the slow process of becoming one with the soil. In this they are aided by the toil of the earthworms, who feed in the leaf litter, for elm leaves are among their favorite foods. In feeding on the leaves the worms also swallow the insecticide, accumulating and concentrating it in their bodies. Dr. Barker found deposits of DDT throughout the digestive tracts of the worms, their blood vessels, nerves, and body wall. Undoubtedly some of the earthworms themselves succumb, but others survive to become "biological magnifiers" of the poison. In the spring the robins return to provide another link in the cycle. As few as 11 large earthworms can transfer a lethal dose of DDT to a robin. And 11 worms form a small part of a day's rations to a bird that eats 10 to 12 earthworms in as many minutes.

Not all robins receive a lethal dose, but another consequence may lead to the extinction of their kind as surely as fatal poisoning. The shadow of sterility lies over all the bird studies and indeed lengthens to include all living things within its potential range. There are now only two or three dozen robins to be found each spring on the entire 185-acre campus of Michigan State University, compared with a conservatively estimated 370 adults in this area before spraying. In 1954 every robin nest under observation by Mehner produced young. Toward the end of June, 1957, when at least 370 young birds (the normal replacement of the adult population) would have been foraging over the campus in the years before spraying began, Mehner could find *only one young robin*. A year later Dr. Wallace was to report: "At no time during the spring or summer [of 1958] did I see a fledgling robin anywhere on the main campus, and so far I have failed to find anyone else who has seen one there."

Part of this failure to produce young is due, of course, to the fact that one or more of a pair of robins dies before the nesting cycle is completed. But Wallace has significant records which point to something more sinister—the actual destruction of the birds' capacity to reproduce. He has, for example, "records of robins and other birds building nests but laying no eggs, and others laying eggs and incubating them but not hatching them. We have one record of a robin that sat on its eggs faithfully for 21 days and they did not hatch. The normal incubation period is 13 days . . . Our analyses are showing high concentrations of DDT in the testes and ovaries of breeding birds," he told a congressional committee in 1960. "Ten males had amounts ranging from 30 to 109 parts per million in the testes, and two females had 151 and 211 parts per million respectively in the egg follicles in their ovaries."

Soon studies in other areas began to develop findings equally dismal. Professor Joseph Hickey and his students at the University of Wisconsin, after careful comparative studies of sprayed and unsprayed areas, reported the robin mortality to be at least 86 to 88 per cent. The Cranbrook Institute of Science at Bloomfield Hills, Michigan, in an effort to assess the extent of bird loss caused by the spraying of the elms, asked in 1956 that all birds thought to be victims of DDT poisoning be turned in to the institute for examination. The request had a response beyond all expectations. Within a few weeks the deep-freeze facilities of the institute were taxed to capacity, so that other specimens had to be refused. By 1959 a thousand poisoned birds from this single community had been turned in or reported. Although the robin was the chief victim (one woman calling the institute reported 12 robins lying dead on her lawn as she spoke), 63 different species were included among the specimens examined at the institute.

The robins, then, are only one part of the chain of devastation linked to the spraying of the elms, even as the elm program is only one of the multitudinous spray programs that cover our land with poisons. Heavy mortality has occurred among about 90 species of birds, including those most familiar to suburbanites and amateur naturalists. The populations of nesting birds in general have declined as much as 90 per cent in some of the sprayed towns. As we shall see, all the various types of birds are affected—ground feeders, treetop feeders, bark feeders, predators.

It is only reasonable to suppose that all birds and mammals heavily dependent on earthworms or other soil organisms for food are threatened by the robins' fate. Some 45 species of birds include earthworms in their diet. Among them is the woodcock, a species that winters in southern areas recently heavily sprayed with heptachlor. Two significant discoveries have now been made about the woodcock. Production of young birds on the New Brunswick breeding grounds is definitely reduced, and adult birds that have been analyzed contain large residues of DDT and heptachlor.

Already there are disturbing records of heavy mortality among more than 20 other species of ground-feeding birds whose food—worms, ants, grubs, or other soil organisms—has been poisoned. These include three of the thrushes whose songs are among the most exquisite of bird voices, the olive-backed, the wood, and the hermit. And the sparrows that flit through the shrubby understory of the woodlands and forage with rustling sounds amid the fallen leaves—the song sparrow and the white-throat—these, too, have been found among the victims of the elm sprays.

Mammals, also, may easily be involved in the cycle, directly or indirectly. Earthworms are important among the various foods of the raccoon, and are eaten in the spring and fall by opossums. Such subterranean tunnelers as shrews and moles capture them in some numbers, and then perhaps pass on the poison to predators such as screech owls and barn owls. Several dying screech owls were picked up in Wisconsin following heavy rains in spring, perhaps poisoned by feeding on earthworms. Hawks and owls have been found in convulsions—great horned owls, screech owls, red-shouldered hawks, sparrow hawks, marsh hawks. These may be cases of secondary poisoning, caused by eating birds or mice that have accumulated insecticides in their livers or other organs.

Nor is it only the creatures that forage on the ground or those who prey on them that are endangered by the foliar spraying of the elms. All of the treetop feeders, the birds that glean their insect food from the leaves, have disappeared from heavily sprayed areas, among them those woodland sprites the kinglets, both ruby-crowned and golden-crowned, the tiny gnatcatchers, and many of the warblers, whose migrating hordes flow through the trees in spring in a multicolored tide of life. In 1956, a late spring delayed spraying so that it coincided with the arrival of an exceptionally heavy wave of warbler migration. Nearly all species of warblers present in the area were represented in the heavy kill that followed. In Whitefish Bay, Wisconsin, at least a thousand myrtle warblers could be seen in migration during former years; in 1958, after the spraying of the elms, observers could find only two. So, with additions from other communities, the list grows, and the warblers killed by the spray include those that most charm and fascinate all who are aware of them: the black-and-white, the yellow, the magnolia, and the Cape May; the ovenbird, whose call throbs in the Maytime woods; the Blackburnian, whose wings are touched with flame; the chestnut-sided, the Canadian, and the black-throated green. These treetop feeders are affected either directly by eating poisoned insects or indirectly by a shortage of food.

The loss of food has also struck hard at the swallows that cruise the skies, straining out the aerial insects as herring strain the plankton of the sea. A Wisconsin naturalist reported: "Swallows have been hard hit. Everyone complains of how few they have compared to four or five years ago. Our sky overhead was full of them only four years ago. Now we seldom see any . . . This could be both lack of insects because of spray, or poisoned insects."

Of other birds this same observer wrote: "Another striking loss is the phoebe. Flycatchers are scarce everywhere but the early hardy common phoebe is no more. I've seen one this spring and only one last spring. Other birders in Wisconsin make the same complaint. I have had five or six pair of cardinals in the past, none now. Wrens, robins, catbirds and screech owls have nested each year in our garden. There are none now. Summer mornings are without bird song. Only pest birds, pigeons, starlings and English sparrows remain. It is tragic and I can't bear it."

The dormant sprays applied to the elms in the fall, sending the poison into every little crevice in the bark, are probably responsible for the severe reduction observed in the number of chickadees, nuthatches, titmice, woodpeckers, and brown creepers. During the winter of 1957–58, Dr. Wallace saw no chickadees or nuthatches at his home feeding station for the first time in many years. Three nuthatches he found later provided a sorry little step-by-step lesson in cause and effect: one was feeding on an elm, another was found dying of typical

DDT symptoms, the third was dead. The dying nuthatch was later found to have 226 parts per million of DDT in its tissues.

The feeding habits of all these birds not only make them especially vulnerable to insect sprays but also make their loss a deplorable one for economic as well as less tangible reasons. The summer food of the white-breasted nuthatch and the brown creeper, for example, includes the eggs, larvae, and adults of a very large number of insects injurious to trees. About three quarters of the food of the chickadee is animal, including all stages of the life cycle of many insects. The chickadee's method of feeding is described in Bent's monumental *Life Histories* of North American birds: "As the flock moves along each bird examines minutely bark, twigs, and branches, searching for tiny bits of food (spiders' eggs, cocoons, or other dormant insect life)."

Various scientific studies have established the critical role of birds in insect control in various situations. Thus, woodpeckers are the primary control of the Engelmann spruce beetle, reducing its populations from 45 to 98 per cent and are important in the control of the codling moth in apple orchards. Chickadees and other winter-resident birds can protect orchards against the cankerworm.

But what happens in nature is not allowed to happen in the modern, chemical-drenched world, where spraying destroys not only the insects but their principal enemy, the birds. When later there is a resurgence of the insect population, as almost always happens, the birds are not there to keep their numbers in check. As the Curator of Birds at the Milwaukee Public Museum, Owen J. Gromme, wrote to the Milwaukee *Journal*: "The greatest enemy of insect life is other predatory insects, birds, and some small mammals, but DDT kills indiscriminately, including nature's own safeguards or policemen . . . In the name of progress are we to become victims of our own diabolical means of insect control to provide temporary comfort, only to lose out to destroying insects later on? By what means will we control new pests, which will attack remaining tree species after the elms are gone, when nature's safeguards (the birds) have been wiped out by poison?"

Mr. Gromme reported that calls and letters about dead and dying birds had been increasing steadily during the years since spraying began in Wisconsin. Questioning always revealed that spraying or fogging had been done in the area where the birds were dying.

Mr. Gromme's experience has been shared by ornithologists and conservationists at most of the research centers of the Midwest such as the Cranbrook Institute in Michigan, the Illinois Natural History Survey, and the University of Wisconsin. A glance at the Letters-from-Readers column of newspapers almost anywhere that spraying is being done makes clear the fact that citizens are not only becoming aroused and indignant but that often they show a keener understanding of the dangers and inconsistencies of spraying than do the officials who order it done. "I am dreading the days to come soon now when many beautiful birds will be dying in our back yard," wrote a Milwaukee woman. "This is a pitiful, heartbreaking experience . . . It is, moreover, frustrating and exasperating, for it evidently does not serve the purpose this slaughter was intended to serve . . . Taking a long look, can you save trees without also saving birds? Do they not, in the economy of nature, save each other? Isn't it possible to help the balance of nature without destroying it?"

The idea that the elms, majestic shade trees though they are, are not "sacred cows" and do not justify an "open end" campaign of destruction against all other forms of life is expressed in other letters. "I have always loved our elm trees which seemed like trademarks on our landscape," wrote another Wisconsin woman. "But there are many kinds of trees . . . We must save our birds, too. Can anyone imagine anything so cheerless and dreary as a springtime without a robin's song?"

To the public the choice may easily appear to be one of stark black-or-white simplicity: Shall we have birds or shall we have elms? But it is not as simple as that, and by one of the ironies that abound throughout the field of chemical control we may very well end by having neither if we continue on our present, well-traveled road. Spraying is killing the birds but it is not saving the elms. The illusion that salvation of the elms lies at the end of a spray nozzle is a dangerous will-o'-the-wisp that is leading one community after another into a morass of heavy expenditures, without producing lasting results. Greenwich, Connecticut, sprayed regularly for ten years. Then a drought year brought conditions especially favorable to the beetle and the mortality of elms went up 1000 per cent. In Urbana, Illinois, where the University of Illinois is located, Dutch elm disease first appeared in 1951. Spraying was undertaken in 1953. By 1959, in spite of six years' spraying, the university campus had lost 86 per cent of its elms, half of them victims of Dutch elm disease.

In Toledo, Ohio, a similar experience caused the Superintendent of Forestry, Joseph A. Sweeney, to take a realistic look at the results of spraying. Spraying was begun there in 1953 and continued through 1959. Meanwhile, however, Mr. Sweeney had noticed that a city-wide infestation of the cottony maple scale was worse after the spraying recommended by "the books and the authorities" than it had been before. He decided to review the results of spraying for Dutch elm disease for himself. His findings shocked him. In the city of Toledo, he found, "the only areas under any control were the areas where we used some promptness in removing the diseased or brood trees. Where we depended on spraying the disease was out of control. In the country where nothing has been done the disease has not spread as fast as it has in the city. This indicates that spraying destroys any natural enemies.

"We are abandoning spraying for the Dutch elm disease. This has brought me into conflict with the people who back any recommendations by the United States Department of Agriculture but I have the facts and will stick with them."

It is difficult to understand why these midwestern towns, to which the elm disease spread only rather recently, have so unquestioningly embarked on ambitious and expensive spraying programs, apparently without waiting to inquire into the experience of other areas that have had longer acquaintance with the problem. New York State, for example, has certainly had the longest history of continuous experience with Dutch elm disease, for it was via the Port of New York that diseased elm wood is thought to have entered the United States about 1930. And New York State today has a most impressive record of containing and suppressing the disease. Yet it has not relied upon spraying. In fact, its agricultural extension service does not recommend spraying as a community method of control.

How, then, has New York achieved its fine record? From the early years of the battle for the elms to the present time, it has relied upon rigorous sanitation, or the prompt removal and destruction of all diseased or infected wood. In the beginning some of the results were disappointing, but this was because it was not at first understood that not only diseased trees but all elm wood in which the beetles might breed must be destroyed. Infected elm wood, after being cut and stored for firewood, will release a crop of fungus-carrying beetles unless burned before spring. It is the adult beetles, emerging from hibernation to feed in late April and May, that transmit Dutch elm disease. New York entomologists have learned by experience what kinds of beetle-breeding material have real importance in the spread of the disease. By concentrating on this dangerous material, it has been possible not only to get good results, but to keep the cost of the sanitation program within reasonable limits. By 1950 the incidence of Dutch elm disease in New York City had been reduced to $^2/_{10}$ of 1 per cent of the city's 55,000 elms. A sanitation program was launched in Westchester County in 1942. During the next 14 years the average annual loss of elms was only $^2/_{10}$ of 1 per cent a year. Buffalo, with 185,000 elms, has an excellent record of containing the disease by sanitation, with recent annual losses amounting to only $^3/_{10}$ of 1 per cent. In other words, at this rate of loss it would take about 300 years to eliminate Buffalo's elms.

What has happened in Syracuse is especially impressive. There no effective program was in operation before 1957. Between 1951 and 1956 Syracuse lost nearly 3000 elms. Then, under the direction of Howard C. Miller of the New York State University College of Forestry, an intensive drive was made to remove all diseased elm trees and all possible sources of beetle-breeding elm wood. The rate of loss is now well below 1 per cent a year.

The economy of the sanitation method is stressed by New York experts in Dutch elm disease control. "In most cases the actual expense is small compared with the probable saving," says J. G. Matthysse of the New York State College of Agriculture. "If it is a case of a dead or broken limb, the limb would have to be removed eventually, as a precaution against possible property damage or personal injury. If it is a fuel-wood pile, the wood can be used before spring, the bark can be peeled from the wood, or the wood can be stored in a dry place. In the case of dying or dead elm trees, the expense of prompt removal to prevent Dutch elm disease spread is usually no greater than would be necessary later, for most dead trees in urban regions must be removed eventually."

The situation with regard to Dutch elm disease is therefore not entirely hopeless provided informed and intelligent measures are taken. While it cannot be eradicated by any means now known, once it has become established in a community, it can be suppressed and contained within reasonable bounds by sanitation, and without the use of methods that are not only futile but involve tragic destruction of bird life. Other possibilities lie within the field of forest genetics, where experiments offer hope of developing a hybrid elm resistant to Dutch elm disease. The European elm is highly resistant, and many of them have been planted in Washington, D.C. Even during a period when a high percentage of the city's elms were affected, no cases of Dutch elm disease were found among these trees.

Replanting through an immediate tree nursery and forestry program is being urged in communities that are losing large numbers of elms. This is important, and although such programs might well include the resistant European elms, they should aim at a variety of species so that no future epidemic could deprive a community of its trees. The key to a healthy plant or animal community lies in what the British ecologist Charles Elton calls "the conservation of variety." What is happening now is in large part a result of the biological unsophistication of past generations. Even a generation ago no one knew that to fill large areas with a single species of tree was to invite disaster. And so whole towns lined their streets and dotted their parks with elms, and today the elms die and so do the birds.

Like the robin, another American bird seems to be on the verge of extinction. This is the national symbol, the eagle. Its populations have dwindled alarmingly within the past decade. The facts suggest that something is at work in the eagle's environment which has virtually destroyed its ability to reproduce. What this may be is not yet definitely known, but there is some evidence that insecticides are responsible.

The most intensively studied eagles in North America have been those nesting along a stretch of coast from Tampa to Fort Myers on the western coast of Florida. There a retired banker from Winnipeg, Charles Broley, achieved ornithological fame by banding more than 1000 young bald eagles during the years 1939–49. (Only 166 eagles had been banded in all the earlier history of birdbanding.) Mr. Broley banded eagles as young birds during the winter months before they had left their nests. Later recoveries of banded birds showed that these Florida-born eagles range northward along the coast into Canada as far as Prince Edward Island, although they had previously been considered nonmigratory. In the fall they return to the South, their migration being observed at such famous vantage points as Hawk Mountain in eastern Pennsylvania.

During the early years of his banding, Mr. Broley used to find 125 active nests a year on the stretch of coast he had chosen for his work. The number of young banded each year was about 150. In 1947 the production of young birds began to decline. Some nests contained no eggs; others contained eggs that failed to hatch. Between 1952 and 1957, about 80 per cent of the nests failed to produce young. In the last year of this period only 43 nests were occupied. Seven of them produced young (8 eaglets); 23 contained eggs that failed to hatch; 13 were used merely as feeding stations by adult eagles and contained no eggs. In 1958 Mr. Broley ranged over 100 miles of coast before finding and banding one eaglet. Adult eagles, which had been seen at 43 nests in 1957, were so scarce that he observed them at only 10 nests.

Although Mr. Broley's death in 1959 terminated this valuable series of uninterrupted observations, reports by the Florida Audubon Society, as well as from New Jersey and Pennsylvania, confirm the trend that may well make it necessary for us to find a new national emblem. The reports of Maurice Broun, curator of the Hawk Mountain Sanctuary, are especially significant. Hawk Mountain is a picturesque mountaintop in southeastern Pennsylvania, where the easternmost ridges of the Appalachians form a last barrier to the westerly winds before dropping away toward the coastal plain. Winds striking the mountains are deflected upward so that on many autumn days there is a continuous updraft on which the broad-winged hawks and eagles ride without effort, covering many miles of their southward migration in a day. At Hawk Mountain the ridges converge and so do the aerial highways. The result is that from a widespread territory to the north birds pass through this traffic bottleneck.

In his more than a score of years as custodian of the sanctuary there, Maurice Broun has observed and actually tabulated more hawks and eagles than any other American. The peak of the bald eagle migration comes in late August and early September. These are assumed to be Florida birds, returning to home territory after a summer in the North. (Later in the fall and early winter a few larger eagles drift through. These are thought to belong to a northern race, bound for an unknown wintering ground.) During the first years after the sanctuary was established, from 1935 to 1939, 40 per cent of the eagles observed were yearlings, easily identified by their uniformly dark plumage. But in recent years these immature birds have become a rarity. Between 1955 and 1959, they made up only 20 per cent of the total count, and in one year (1957) there was only one young eagle for every 32 adults.

Observations at Hawk Mountain are in line with findings elsewhere. One such report comes from Elton Fawks, an official of the Natural Resources Council of Illinois. Eagles—probably northern nesters—winter along the Mississippi and Illinois Rivers. In 1958 Mr. Fawks reported that a recent count of 59 eagles had included only one immature bird. Similar indications of the dying out of the race come from the world's only sanctuary for eagles alone, Mount Johnson Island in the Susquehanna River. The island, although only 8 miles above Conowingo Dam and about half a mile out from the Lancaster County shore, retains its primitive wildness. Since 1934 its single eagle nest has been under observation by Professor Herbert H. Beck, an ornithologist of Lancaster and custodian of the sanctuary. Between 1935 and 1947 use of the nest was regular and uniformly successful. Since 1947, although the adults have occupied the nest and there is evidence of egg laying, no young eagles have been produced.

On Mount Johnson Island as well as in Florida, then, the same situation prevails—there is some occupancy of nests by adults, some production of eggs, but few or no young birds. In seeking an explanation, only one appears to fit all the facts. This is that the reproductive capacity of the birds has been so lowered by some environmental agent that there are now almost no annual additions of young to maintain the race.

Exactly this sort of situation has been produced artificially in other birds by various experimenters, notably Dr. James

DeWitt of the United States Fish and Wildlife Service. Dr. DeWitt's now classic experiments on the effect of a series of insecticides on quail and pheasants have established the fact that exposure to DDT or related chemicals, even when doing no observable harm to the parent birds, may seriously affect reproduction. The way the effect is exerted may vary, but the end result is always the same. For example, quail into whose diet DDT was introduced throughout the breeding season survived and even produced normal numbers of fertile eggs. But few of the eggs hatched. "Many embryos appeared to develop normally during the early stages of incubation, but died during the hatching period," Dr. DeWitt said. Of those that did hatch, more than half died within 5 days. In other tests in which both pheasants and quail were the subjects, the adults produced no eggs whatever if they had been fed insecticide-contaminated diets throughout the year. And at the University of California, Dr. Robert Rudd and Dr. Richard Genelly reported similar findings. When pheasants received dieldrin in their diets, "egg production was markedly lowered and chick survival was poor." According to these authors, the delayed but lethal effect on the young birds follows from storage of dieldrin in the yolk of the egg, from which it is gradually assimilated during incubation and after hatching.

This suggestion is strongly supported by recent studies by Dr. Wallace and a graduate student, Richard F. Bernard, who found high concentrations of DDT in robins on the Michigan State University campus. They found the poison in all of the testes of male robins examined, in developing egg follicles, in the ovaries of females, in completed but unlaid eggs, in the oviducts, in unhatched eggs from deserted nests, in embryos within the eggs, and in a newly hatched, dead nestling.

These important studies establish the fact that the insecticidal poison affects a generation once removed from initial contact with it. Storage of poison in the egg, in the yolk material that nourishes the developing embryo, is a virtual death warrant and explains why so many of DeWitt's birds died in the egg or a few days after hatching.

Laboratory application of these studies to eagles presents difficulties that are nearly insuperable, but field studies are now under way in Florida, New Jersey, and elsewhere in the hope of acquiring definite evidence as to what has caused the apparent sterility of much of the eagle population. Meanwhile, the available circumstantial evidence points to insecticides. In localities where fish are abundant they make up a large part of the eagle's diet (about 65 per cent in Alaska; about 52 per cent in the Chesapeake Bay area). Almost unquestionably the eagles so long studied by Mr. Broley were predominantly fish eaters. Since 1945 this particular coastal area has been subjected to repeated sprayings with DDT dissolved in fuel oil. The principal target of the aerial spraying was the salt-marsh mosquito, which inhabits the marshes and coastal areas that are typical foraging areas for the eagles. Fishes and crabs were killed in enormous numbers. Laboratory analyses of their tissues revealed high concentrations of DDT—as much as 46 parts per million. Like the grebes of Clear Lake, which accumulated heavy concentrations of insecticide residues from eating the fish of the lake, the eagles have almost certainly been storing up the DDT in the tissues of their bodies. And like the grebes, the pheasants, the quail, and the robins, they are less and less able to produce young and to preserve the continuity of their race.

From all over the world come echoes of the peril that faces birds in our modern world. The reports differ in detail, but always repeat the theme of death to wildlife in the wake of pesticides. Such are the stories of hundreds of small birds and partridges dying in France after vine stumps were treated with an arsenic-containing herbicide, or of partridge shoots in Belgium, once famous for the numbers of their birds, denuded of partridges after the spraying of nearby farmlands.

In England the major problem seems to be a specialized one, linked with the growing practice of treating seed with insecticides before sowing. Seed treatment is not a wholly new thing, but in earlier years the chemicals principally used were fungicides. No effects on birds seem to have been noticed. Then about 1956 there was a change to dual-purpose treatment; in addition to a fungicide, dieldrin, aldrin, or heptachlor was added to combat soil insects. Thereupon the situation changed for the worse.

In the spring of 1960 a deluge of reports of dead birds reached British wildlife authorities, including the British Trust for Ornithology, the Royal Society for the Protection of Birds, and the Game Birds Association. "The place is like a battlefield," a landowner in Norfolk wrote. "My keeper has found innumerable corpses, including masses of small birds—Chaffinches, Greenfinches, Linnets, Hedge Sparrows, also House Sparrows . . . the destruction of wild life is quite piti-ful." A gamekeeper wrote: "My Partridges have been wiped out with the dressed corn, also some Pheasants and all other birds, hundreds of birds have been killed . . . As a lifelong gamekeeper it has been a distressing experience for me. It is bad to see pairs of Partridges that have died together."

In a joint report, the British Trust for Ornithology and the Royal Society for the Protection of Birds described some 67 kills of birds—a far from complete listing of the destruction that took place in the spring of 1960. Of these 67, 59 were caused by seed dressings, 8 by toxic sprays.

A new wave of poisoning set in the following year. The death of 600 birds on a single estate in Norfolk was reported to the House of Lords, and 100 pheasants died on a farm in North Essex. It soon became evident that more counties were involved than in 1960 (34 compared with 23). Lincolnshire, heavily agricultural, seemed to have suffered most, with reports of 10,000 birds dead. But destruction involved all of agricultural England, from Angus in the north to Cornwall in the south, from Anglesey in the west to Norfolk in the east.

In the spring of 1961 concern reached such a peak that a special committee of the House of Commons made an investigation of the matter, taking testimony from farmers, landowners, and representatives of the Ministry of Agriculture and of various governmental and nongovernmental agencies concerned with wildlife.

"Pigeons are suddenly dropping out of the sky dead," said one witness. "You can drive a hundred or two hundred miles outside London and not see a single kestrel," reported another. "There has been no parallel in the present century, or at any time so far as I am aware, [this is] the biggest risk to wildlife and game that ever occurred in the country," officials of the Nature Conservancy testified.

Facilities for chemical analysis of the victims were most inadequate to the task, with only two chemists in the country able to make the tests (one the government chemist, the other in the employ of the Royal Society for the Protection of Birds). Witnesses described huge bonfires on which the bodies of the birds were burned. But efforts were made to have carcasses collected for examination, and of the birds analyzed, all but one contained pesticide residues. The single exception was a snipe, which is not a seed-eating bird.

Along with the birds, foxes also may have been affected, probably indirectly by eating poisoned mice or birds. England, plagued by rabbits, sorely needs the fox as a predator. But between November 1959 and April 1960 at least 1300 foxes died. Deaths were heaviest in the same counties from which sparrow hawks, kestrels, and other birds of prey virtually disappeared, suggesting that the poison was spreading through the food chain, reaching out from the seed eaters to the furred and feathered carnivores. The actions of the moribund foxes were those of animals poisoned by chlorinated hydrocarbon insecticides. They were seen wandering in circles, dazed and half blind, before dying in convulsions.

The hearings convinced the committee that the threat to wildlife was "most alarming"; it accordingly recommended to the House of Commons that "the Minister of Agriculture and the Secretary of State for Scotland should secure the immediate prohibition for the use as seed dressings of compounds containing dieldrin, aldrin, or heptachlor, or chemicals of comparable toxicity." The committee also recommended more adequate controls to ensure that chemicals were adequately tested under field as well as laboratory conditions before being put on the market. This, it is worth emphasizing, is one of the great blank spots in pesticide research everywhere. Manufacturers' tests on the common laboratory animals—rats, dogs, guinea pigs—include no wild species, no birds as a rule, no fishes, and are conducted under controlled and artificial conditions. Their application to wildlife in the field is anything but precise.

England is by no means alone in its problem of protecting birds from treated seeds. Here in the United States the problem has been most troublesome in the rice-growing areas of California and the South. For a number of years California rice growers have been treating seed with DDT as protection against tadpole shrimp and scavenger beetles which sometimes

damage seedling rice. California sportsmen have enjoyed excellent hunting because of the concentrations of waterfowl and
pheasants in the rice fields. But for the past decade persistent
reports of bird losses, especially among pheasants, ducks, and
blackbirds, have come from the rice-growing counties. "Pheasant sickness" became a well-known phenomenon: birds "seek
water, become paralyzed, and are found on the ditch banks and
rice checks quivering," according to one observer. The "sickness" comes in the spring, at the time the rice fields are seeded.
The concentration of DDT used is many times the amount
that will kill an adult pheasant.

The passage of a few years and the development of even more
poisonous insecticides served to increase the hazard from treated
seed. Aldrin, which is 100 times as toxic as DDT to pheasants,
is now widely used as a seed coating. In the rice fields of eastern
Texas, this practice has seriously reduced the populations of the
fulvous tree duck, a tawny-colored, gooselike duck of the Gulf
Coast. Indeed, there is some reason to think that the rice growers, having found a way to reduce the populations of blackbirds,
are using the insecticide for a dual purpose, with disastrous effects on several bird species of the rice fields.

As the habit of killing grows—the resort to "eradicating"
any creature that may annoy or inconvenience us—birds are
more and more finding themselves a direct target of poisons
rather than an incidental one. There is a growing trend toward
aerial applications of such deadly poisons as parathion to "control" concentrations of birds distasteful to farmers. The Fish
and Wildlife Service has found it necessary to express serious
concern over this trend, pointing out that "parathion treated
areas constitute a potential hazard to humans, domestic animals, and wildlife." In southern Indiana, for example, a group
of farmers went together in the summer of 1959 to engage
a spray plane to treat an area of river bottomland with parathion. The area was a favored roosting site for thousands of
blackbirds that were feeding in nearby cornfields. The problem
could have been solved easily by a slight change in agricultural
practice—a shift to a variety of corn with deep-set ears not
accessible to the birds—but the farmers had been persuaded of
the merits of killing by poison, and so they sent in the planes
on their mission of death.

The results probably gratified the farmers, for the casualty list included some 65,000 red-winged blackbirds and starlings. What other wildlife deaths may have gone unnoticed and unrecorded is not known. Parathion is not a specific for blackbirds: it is a universal killer. But such rabbits or raccoons or opossums as may have roamed those bottomlands and perhaps never visited the farmers' cornfields were doomed by a judge and jury who neither knew of their existence nor cared.

And what of human beings? In California orchards sprayed with this same parathion, workers handling foliage that had been treated *a month* earlier collapsed and went into shock, and escaped death only through skilled medical attention. Does Indiana still raise any boys who roam through woods or fields and might even explore the margins of a river? If so, who guarded the poisoned area to keep out any who might wander in, in misguided search for unspoiled nature? Who kept vigilant watch to tell the innocent stroller that the fields he was about to enter were deadly—all their vegetation coated with a lethal film? Yet at so fearful a risk the farmers, with none to hinder them, waged their needless war on blackbirds.

In each of these situations, one turns away to ponder the question: Who has made the decision that sets in motion these chains of poisonings, this ever-widening wave of death that spreads out, like ripples when a pebble is dropped into a still pond? Who has placed in one pan of the scales the leaves that might have been eaten by the beetles and in the other the pitiful heaps of many-hued feathers, the lifeless remains of the birds that fell before the unselective bludgeon of insecticidal poisons? Who has decided—who has the *right* to decide—for the countless legions of people who were not consulted that the supreme value is a world without insects, even though it be also a sterile world ungraced by the curving wing of a bird in flight? The decision is that of the authoritarian temporarily entrusted with power; he has made it during a moment of inattention by millions to whom beauty and the ordered world of nature still have a meaning that is deep and imperative.

NINE

Rivers of Death

FROM THE GREEN DEPTHS of the offshore Atlantic many paths lead back to the coast. They are paths followed by fish; although unseen and intangible, they are linked with the outflow of waters from the coastal rivers. For thousands upon thousands of years the salmon have known and followed these threads of fresh water that lead them back to the rivers, each returning to the tributary in which it spent the first months or years of life. So, in the summer and fall of 1953, the salmon of the river called Miramichi on the coast of New Brunswick moved in from their feeding grounds in the far Atlantic and ascended their native river. In the upper reaches of the Miramichi, in streams that gather together a network of shadowed brooks, the salmon deposited their eggs that autumn in beds of gravel over which the stream water flowed swift and cold. Such places, the watersheds of the great coniferous forests of spruce and balsam, of hemlock and pine, provide the kind of spawning grounds that salmon must have in order to survive.

These events repeated a pattern that was age-old, a pattern

that had made the Miramichi one of the finest salmon streams in North America. But that year the pattern was to be broken.

During the fall and winter the salmon eggs, large and thick-shelled, lay in shallow gravel-filled troughs, or redds, which the mother fish had dug in the stream bottom. In the cold of winter they developed slowly, as was their way, and only when spring at last brought thawing and release to the forest streams did the young hatch. At first they hid among the pebbles of the stream bed—tiny fish about half an inch long. They took no food, living on the large yolk sac. Not until it was absorbed would they begin to search the stream for small insects.

With the newly hatched salmon in the Miramichi that spring of 1954 were young of previous hatchings, salmon a year or two old, young fish in brilliant coats marked with bars and bright red spots. These young fed voraciously, seeking out the strange and varied insect life of the stream.

As the summer approached, all this was changed. That year the watershed of the Northwest Miramichi was included in a vast spraying program which the Canadian Government had embarked upon the previous year—a program designed to save the forests from the spruce budworm. The budworm is a native insect that attacks several kinds of evergreens. In eastern Canada it seems to become extraordinarily abundant about every 35 years. The early 1950's had seen such an upsurge in the budworm populations. To combat it, spraying with DDT was begun, first in a small way, then at a suddenly accelerated rate in 1953. Millions of acres of forests were sprayed instead of thousands as before, in an effort to save the balsams, which are the mainstay of the pulp and paper industry.

So in 1954, in the month of June, the planes visited the forests of the Northwest Miramichi and white clouds of settling mist marked the crisscross pattern of their flight. The spray—one-half pound of DDT to the acre in a solution of oil—filtered down through the balsam forests and some of it finally reached the ground and the flowing streams. The pilots, their thoughts only on their assigned task, made no effort to avoid the streams or to shut off the spray nozzles while flying over them; but because spray drifts so far in even the slightest stirrings of air, perhaps the result would have been little different if they had.

Soon after the spraying had ended there were unmistakable signs that all was not well. Within two days dead and dying fish, including many young salmon, were found along the banks of the stream. Brook trout also appeared among the dead fish, and along the roads and in the woods birds were dying. All the life of the stream was stilled. Before the spraying there had been a rich assortment of the water life that forms the food of salmon and trout—caddis fly larvae, living in loosely fitting protective cases of leaves, stems or gravel cemented together with saliva, stonefly nymphs clinging to rocks in the swirling currents, and the wormlike larvae of blackflies edging the stones under riffles or where the stream spills over steeply slanting rocks. But now the stream insects were dead, killed by the DDT, and there was nothing for a young salmon to eat.

Amid such a picture of death and destruction, the young salmon themselves could hardly have been expected to escape, and they did not. By August not one of the young salmon that had emerged from the gravel beds that spring remained. A whole year's spawning had come to nothing. The older young, those hatched a year or more earlier, fared only slightly better. For every six young of the 1953 hatch that had foraged in the stream as the planes approached, only one remained. Young salmon of the 1952 hatch, almost ready to go to sea, lost a third of their numbers.

All these facts are known because the Fisheries Research Board of Canada had been conducting a salmon study on the Northwest Miramichi since 1950. Each year it had made a census of the fish living in this stream. The records of the biologists covered the number of adult salmon ascending to spawn, the number of young of each age group present in the stream, and the normal population not only of salmon but of other species of fish inhabiting the stream. With this complete record of prespraying conditions, it was possible to measure the damage done by the spraying with an accuracy that has seldom been matched elsewhere.

The survey showed more than the loss of young fish; it revealed a serious change in the streams themselves. Repeated sprayings have now completely altered the stream environment, and the aquatic insects that are the food of salmon and trout have been killed. A great deal of time is required, even

after a single spraying, for most of these insects to build up sufficient numbers to support a normal salmon population—time
measured in years rather than months.

The smaller species, such as midges and blackflies, become
re-established rather quickly. These are suitable food for the
smallest salmon, the fry only a few months old. But there is
no such rapid recovery of the larger aquatic insects, on which
salmon in their second and third years depend. These are the
larval stages of caddis flies, stoneflies, and mayflies. Even in the
second year after DDT enters a stream, a foraging salmon parr
would have trouble finding anything more than an occasional
small stonefly. There would be no large stoneflies, no mayflies,
no caddis flies. In an effort to supply this natural food, the
Canadians have attempted to transplant caddis fly larvae and
other insects to the barren reaches of the Miramichi. But of
course such transplants would be wiped out by any repeated
spraying.

The budworm populations, instead of dwindling as expected, have proved refractory, and from 1955 to 1957 spraying
was repeated in various parts of New Brunswick and Quebec,
some places being sprayed as many as three times. By 1957,
nearly 15 million acres had been sprayed. Although spraying
was then tentatively suspended, a sudden resurgence of budworms led to its resumption in 1960 and 1961. Indeed there
is no evidence anywhere that chemical spraying for budworm
control is more than a stopgap measure (aimed at saving the
trees from death through defoliation over several successive
years), and so its unfortunate side effects will continue to be
felt as spraying is continued. In an effort to minimize the destruction of fish, the Canadian forestry officials have reduced
the concentration of DDT from the ½ pound previously used
to ¼ pound to the acre, on the recommendation of the Fisheries Research Board. (In the United States the standard and
highly lethal pound-to-the-acre still prevails.) Now, after several years in which to observe the effects of spraying, the Canadians find a mixed situation, but one that affords very little
comfort to devotees of salmon fishing, provided spraying is
continued.

A very unusual combination of circumstances has so far saved
the runs of the Northwest Miramichi from the destruction that

was anticipated—a constellation of happenings that might not occur again in a century. It is important to understand what has happened there, and the reasons for it.

In 1954, as we have seen, the watershed of this branch of the Miramichi was heavily sprayed. Thereafter, except for a narrow band sprayed in 1956, the whole upper watershed of this branch was excluded from the spraying program. In the fall of 1954 a tropical storm played its part in the fortunes of the Miramichi salmon. Hurricane Edna, a violent storm to the very end of its northward path, brought torrential rains to the New England and Canadian coasts. The resulting freshets carried streams of fresh water far out to sea and drew in unusual numbers of salmon. As a result, the gravel beds of the streams which the salmon seek out for spawning received an unusual abundance of eggs. The young salmon hatching in the Northwest Miramichi in the spring of 1955 found circumstances practically ideal for their survival. While the DDT had killed off all stream insects the year before, the smallest of the insects—the midges and blackflies—had returned in numbers. These are the normal food of baby salmon. The salmon fry of that year not only found abundant food but they had few competitors for it. This was because of the grim fact that the older young salmon had been killed off by the spraying in 1954. Accordingly, the fry of 1955 grew very fast and survived in exceptional numbers. They completed their stream growth rapidly and went to sea early. Many of them returned in 1959 to give large runs of grilse to the native stream.

If the runs in the Northwest Miramichi are still in relatively good condition this is because spraying was done in one year only. The results of repeated spraying are clearly seen in other streams of the watershed, where alarming declines in the salmon populations are occurring.

In all sprayed streams, young salmon of every size are scarce. The youngest are often "practically wiped out," the biologists report. In the main Southwest Miramichi, which was sprayed in 1956 and 1957, the 1959 catch was the lowest in a decade. Fishermen remarked on the extreme scarcity of grilse—the youngest group of returning fish. At the sampling trap in the estuary of the Miramichi the count of grilse was only a fourth as large in 1959 as the year before. In 1959 the whole Miramichi watershed produced only about 600,000 smolt (young salmon descending to the sea). This was less than a third of the runs of the three preceding years.

Against such a background, the future of the salmon fisheries in New Brunswick may well depend on finding a substitute for drenching forests with DDT.

The eastern Canadian situation is not unique, except perhaps in the extent of forest spraying and the wealth of facts that have been collected. Maine, too, has its forests of spruce and balsam, and its problem of controlling forest insects. Maine, too, has its salmon runs—a remnant of the magnificent runs of former days, but a remnant hard won by the work of biologists and conservationists to save some habitat for salmon in streams burdened with industrial pollution and choked with logs. Although spraying has been tried as a weapon against the ubiquitous budworm, the areas affected have been relatively small and have not, as yet, included important spawning streams for salmon. But what happened to stream fish in an area observed by the Maine Department of Inland Fisheries and Game is perhaps a portent of things to come.

"Immediately after the 1958 spraying," the Department reported, "moribund suckers were observed in large numbers in Big Goddard Brook. These fish exhibited the typical symptoms of DDT poisoning; they swam erratically, gasped at the surface, and exhibited tremors and spasms. In the first five days after spraying, 668 dead suckers were collected from two blocking nets. Minnows and suckers were also killed in large numbers in Little Goddard, Carry, Alder, and Blake Brooks. Fish were often seen floating passively downstream in a weakened and moribund condition. In several instances, blind and dying trout were found floating passively downstream more than a week after spraying."

(The fact that DDT may cause blindness in fish is confirmed by various studies. A Canadian biologist who observed spraying on northern Vancouver Island in 1957 reported that cutthroat trout fingerlings could be picked out of the streams by hand, for the fish were moving sluggishly and made no attempt to escape. On examination, they were found to have an opaque white film covering the eye, indicating that vision had been impaired or destroyed. Laboratory studies by the Canadian Department of Fisheries showed that almost all fish [Coho salmon] not actually killed by exposure to low concentrations of DDT [3 parts per million] showed symptoms of blindness, with marked opacity of the lens.)

Wherever there are great forests, modern methods of insect control threaten the fishes inhabiting the streams in the shelter of the trees. One of the best-known examples of fish destruction in the United States took place in 1955, as a result of spraying in and near Yellowstone National Park. By the fall of that year, so many dead fish had been found in the Yellowstone River that sportsmen and Montana fish-and-game administrators became alarmed. About 90 miles of the river were affected. In one 300-yard length of shoreline, 600 dead fish were counted, including brown trout, whitefish, and suckers. Stream insects, the natural food of trout, had disappeared.

Forest Service officials declared they had acted on advice that 1 pound of DDT to the acre was "safe." But the results of the spraying should have been enough to convince anyone that the advice had been far from sound. A cooperative study was begun in 1956 by the Montana Fish and Game Department and two federal agencies, the Fish and Wildlife Service and the Forest Service. Spraying in Montana that year covered 900,000 acres; 800,000 acres were also treated in 1957. The biologists therefore had no trouble finding areas for their study.

Always, the pattern of death assumed a characteristic shape: the smell of DDT over the forests, an oil film on the water surface, dead trout along the shoreline. All fish analyzed, whether taken alive or dead, had stored DDT in their tissues. As in eastern Canada, one of the most serious effects of spraying was the severe reduction of food organisms. On many study areas aquatic insects and other stream-bottom fauna were reduced to a tenth of their normal populations. Once destroyed, populations of these insects, so essential to the survival of trout, take a long time to rebuild. Even by the end of the second summer after spraying, only meager quantities of aquatic insects had reestablished themselves, and on one stream—formerly rich in bottom fauna—scarcely any could be found. In this particular stream, game fish had been reduced by 80 per cent.

The fish do not necessarily die immediately. In fact, delayed mortality may be more extensive than the immediate kill and, as the Montana biologists discovered, it may go unreported because it occurs after the fishing season. Many deaths occurred in the study streams among autumn spawning fish, including brown trout, brook trout, and whitefish. This is not surprising, because in time of physiological stress the organism, be it fish or man, draws on stored fat for energy. This exposes it to the full lethal effect of the DDT stored in the tissues.

It was therefore more than clear that spraying at the rate of a pound of DDT to the acre posed a serious threat to the fishes in forest streams. Moreover, control of the budworm had not been achieved and many areas were scheduled for respraying. The Montana Fish and Game Department registered strong opposition to further spraying, saying it was "not willing to compromise the sport fishery resource for programs of questionable necessity and doubtful success." The Department declared, however, that it would continue to cooperate with the Forest Service "in determining ways to minimize adverse effects."

But can such cooperation actually succeed in saving the fish? An experience in British Columbia speaks volumes on this point. There an outbreak of the black-headed budworm had been raging for several years. Forestry officials, fearing that another season's defoliation might result in severe loss of trees, decided to carry out control operations in 1957. There were many consultations with the Game Department, whose officials were concerned about the salmon runs. The Forest Biology Division agreed to modify the spraying program in every possible way short of destroying its effectiveness, in order to reduce risks to the fish.

Despite these precautions, and despite the fact that a sincere effort was apparently made, *in at least four major streams almost 100 per cent of the salmon were killed.*

In one of the rivers, the young of a run of 40,000 adult Coho salmon were almost completely annihilated. So were the young stages of several thousand steelhead trout and other species of trout. The Coho salmon has a three-year life cycle and the runs are composed almost entirely of fish of a single age group. Like other species of salmon, the Coho has a strong homing instinct, returning to its natal stream. There will be no repopulation from other streams. This means, then, that every third year the run of salmon into this river will be almost nonexistent, until such time as careful management, by artificial propagation or other means, has been able to rebuild this commercially important run.

There are ways to solve this problem—to preserve the forests and to save the fishes, too. To assume that we must resign ourselves to turning our waterways into rivers of death is to follow the counsel of despair and defeatism. We must make wider use of alternative methods that are now known, and we must devote our ingenuity and resources to developing others. There are cases on record where natural parasitism has kept the budworm under control more effectively than spraying. Such natural control needs to be utilized to the fullest extent. There are possibilities of using less toxic sprays or, better still, of introducing microorganisms that will cause disease among the budworms without affecting the whole web of forest life. We shall see later what some of these alternative methods are and what they promise. Meanwhile, it is important to realize that chemical spraying of forest insects is neither the only way nor the best way.

The pesticide threat to fishes may be divided into three parts. One, as we have seen, relates to the fishes of running streams in northern forests and to the single problem of forest spraying. It is confined almost entirely to the effects of DDT. Another is vast, sprawling, and diffuse, for it concerns the many different kinds of fishes—bass, sunfish, crappies, suckers, and others—that inhabit many kinds of waters, still or flowing, in many parts of the country. It also concerns almost the whole gamut of insecticides now in agricultural use, although a few principal offenders like endrin, toxaphene, dieldrin, and heptachlor can easily be picked out. Still another problem must now be considered largely in terms of what we may logically suppose will happen in the future, because the studies that will disclose the facts are only beginning to be made. This has to do with the fishes of salt marshes, bays, and estuaries.

It was inevitable that serious destruction of fishes would follow the widespread use of the new organic pesticides. Fishes are almost fantastically sensitive to the chlorinated hydrocarbons that make up the bulk of modern insecticides. And when millions of tons of poisonous chemicals are applied to the surface of the land, it is inevitable that some of them will find their way into the ceaseless cycle of waters moving between land and sea.

Reports of fish kills, some of disastrous proportions, have now become so common that the United States Public Health Service has set up an office to collect such reports from the states as an index of water pollution.

This is a problem that concerns a great many people. Some 25 million Americans look to fishing as a major source of recreation and another 15 million are at least casual anglers. These people spend three billion dollars annually for licenses, tackle, boats, camping equipment, gasoline, and lodgings. Anything that deprives them of their sport will also reach out and affect a large number of economic interests. The commercial fisheries represent such an interest, and even more importantly, an essential source of food. Inland and coastal fisheries (excluding the offshore catch) yield an estimated three billion pounds a year. Yet, as we shall see, the invasion of streams, ponds, rivers, and bays by pesticides is now a threat to both recreational and commercial fishing.

Examples of the destruction of fish by agricultural crop sprayings and dustings are everywhere to be found. In California, for example, the loss of some 60,000 game fish, mostly bluegill and other sunfish, followed an attempt to control the rice-leaf miner with dieldrin. In Louisiana 30 or more instances of heavy fish mortality occurred in one year alone (1960) because of the use of endrin in the sugarcane fields. In Pennsylvania fish have been killed in numbers by endrin, used in orchards to combat mice. The use of chlordane for grasshopper control on the high western plains has been followed by the death of many stream fish.

Probably no other agricultural program has been carried out on so large a scale as the dusting and spraying of millions of acres of land in southern United States to control the fire ant. Heptachlor, the chemical chiefly used, is only slightly less toxic to fish than DDT. Dieldrin, another fire ant poison, has a well-documented history of extreme hazard to all aquatic life. Only endrin and toxaphene represent a greater danger to fish.

All areas within the fire ant control area, whether treated with heptachlor or dieldrin, reported disastrous effects on aquatic life. A few excerpts will give the flavor of the reports from biologists who studied the damage: From Texas, "Heavy loss of aquatic life despite efforts to protect canals," "Dead fish . . . were present in all treated water," "Fish kill was heavy and continued for over 3 weeks." From Alabama, "Most adult fish were killed [in Wilcox County] within a few days after treatment," "The fish in temporary waters and small tributary streams appeared to have been completely eradicated."

In Louisiana, farmers complained of loss in farm ponds. Along one canal more than 500 dead fish were seen floating or lying on the bank on a stretch of less than a quarter of a mile. In another parish 150 dead sunfish could be found for every 4 that remained alive. Five other species appeared to have been wiped out completely.

In Florida, fish from ponds in a treated area were found to contain residues of heptachlor and a derived chemical, heptachlor epoxide. Included among these fish were sunfish and bass, which of course are favorites of anglers and commonly find their way to the dinner table. Yet the chemicals they contained are among those the Food and Drug Administration considers too dangerous for human consumption, even in minute quantities.

So extensive were the reported kills of fish, frogs, and other life of the waters that the American Society of Ichthyologists and Herpetologists, a venerable scientific organization devoted to the study of fishes, reptiles, and amphibians, passed a resolution in 1958 calling on the Department of Agriculture and the associated state agencies to cease "aerial distribution of heptachlor, dieldrin, and equivalent poisons—before irreparable harm is done." The Society called attention to the great variety of species of fish and other forms of life inhabiting the southeastern part of the United States, including species that occur nowhere else in the world. "Many of these animals," the Society warned, "occupy only small areas and therefore might readily be completely exterminated."

Fishes of the southern states have also suffered heavily from insecticides used against cotton insects. The summer of 1950 was a season of disaster in the cotton-growing country of northern Alabama. Before that year, only limited use had been made of organic insecticides for the control of the boll weevil. But in 1950 there were many weevils because of a series of mild winters, and so an estimated 80 to 95 per cent of the farmers, on the urging of the county agents, turned to the use of insecticides. The chemical most popular with the farmers was toxaphene, one of the most destructive to fishes.

Rains were frequent and heavy that summer. They washed the chemicals into the streams, and as this happened the farmers applied more. An average acre of cotton that year received 63 pounds of toxaphene. Some farmers used as much as 200 pounds per acre; one, in an extraordinary excess of zeal, applied more than a quarter of a ton to the acre.

The results could easily have been foreseen. What happened in Flint Creek, flowing through 50 miles of Alabama cotton country before emptying into Wheeler Reservoir, was typical of the region. On August 1, torrents of rain descended on the Flint Creek watershed. In trickles, in rivulets, and finally in floods the water poured off the land into the streams. The water level rose six inches in Flint Creek. By the next morning it was obvious that a great deal more than rain had been carried into the stream. Fish swam about in aimless circles near the surface. Sometimes one would throw itself out of the water onto the bank. They could easily be caught; one farmer picked up several and took them to a spring-fed pool. There, in the pure water, these few recovered. But in the stream dead fish floated down all day. This was but the prelude to more, for each rain washed more of the insecticide into the river, killing more fish. The rain of August 10 resulted in such a heavy fish kill throughout the river that few remained to become victims of the next surge of poison into the stream, which occurred on August 15. But evidence of the deadly presence of the chemicals was obtained by placing test goldfish in cages in the river; they were dead within a day.

The doomed fish of Flint Creek included large numbers of white crappies, a favorite among anglers. Dead bass and sunfish were also found, occurring abundantly in Wheeler Reservoir, into which the creek flows. All the rough-fish population of these waters was destroyed also—the carp, buffalo, drum, gizzard shad, and catfish. None showed signs of disease—only the erratic movements of the dying and a strange deep wine color of the gills.

In the warm enclosed waters of farm ponds, conditions are very likely to be lethal for fish when insecticides are applied in the vicinity. As many examples show, the poison is carried in by rains and runoff from surrounding lands. Sometimes the ponds receive not only contaminated runoff but also a direct dose as crop-dusting pilots neglect to shut off the duster in passing over a pond. Even without such complications, normal agricultural use subjects fish to far heavier concentrations of chemicals than would be required to kill them. In other words, a marked reduction in the poundages used would hardly alter the lethal situation, for applications of over 0.1 pound per acre to the pond itself are generally considered hazardous. And the poison, once introduced, is hard to get rid of. One pond that had been treated with DDT to remove unwanted shiners remained so poisonous through repeated drainings and flushings that it killed 94 per cent of the sunfish with which it was later stocked. Apparently the chemical remained in the mud of the pond bottom.

Conditions are evidently no better now than when the modern insecticides first came into use. The Oklahoma Wildlife Conservation Department stated in 1961 that reports of fish losses in farm ponds and small lakes had been coming in at the rate of at least one a week, and that such reports were increasing. The conditions usually responsible for these losses in Oklahoma were those made familiar by repetition over the years: the application of insecticides to crops, a heavy rain, and poison washed into the ponds.

In some parts of the world the cultivation of fish in ponds provides an indispensable source of food. In such places the use of insecticides without regard for the effects on fish creates immediate problems. In Rhodesia, for example, the young of an important food fish, the Kafue bream, are killed by exposure to only 0.04 parts per million of DDT in shallow pools. Even smaller doses of many other insecticides would be lethal. The shallow waters in which these fish live are favorable mosquito-breeding places. The problem of controlling mosquitoes and at the same time conserving a fish important in the Central African diet has obviously not been solved satisfactorily.

Milkfish farming in the Philippines, China, Vietnam, Thailand, Indonesia, and India faces a similar problem. The milkfish is cultivated in shallow ponds along the coasts of these countries. Schools of young suddenly appear in the coastal waters (from no one knows where) and are scooped up and placed in impoundments, where they complete their growth. So important is this fish as a source of animal protein for the rice-eating millions of Southeast Asia and India that the Pacific Science Congress has recommended an international effort to search for the now unknown spawning grounds, in order to develop the farming of these fish on a massive scale. Yet spraying has been permitted to cause heavy losses in existing impoundments. In the Philippines aerial spraying for mosquito control has cost pond owners dearly. In one such pond containing 120,000 milkfish, more than half the fish died after a spray plane had passed over, in spite of desperate efforts by the owner to dilute the poison by flooding the pond.

One of the most spectacular fish kills of recent years occurred in the Colorado River below Austin, Texas, in 1961. Shortly after daylight on Sunday morning, January 15, dead fish appeared in the new Town Lake in Austin and in the river for a distance of about 5 miles below the lake. None had been seen the day before. On Monday there were reports of dead fish 50 miles downstream. By this time it was clear that a wave of some poisonous substance was moving down in the river water. By January 21, fish were being killed 100 miles downstream near La Grange, and a week later the chemicals were doing their lethal work 200 miles below Austin. During the last week of January the locks on the Intracoastal Waterway were closed to exclude the toxic waters from Matagorda Bay and divert them into the Gulf of Mexico.

Meanwhile, investigators in Austin noticed an odor associated with the insecticides chlordane and toxaphene. It was especially strong in the discharge from one of the storm sewers. This sewer had in the past been associated with trouble from industrial wastes, and when officers of the Texas Game and Fish Commission followed it back from the lake, they noticed an odor like that of benzene hexachloride at all openings as far back as a feeder line from a chemical plant. Among the major products of this plant were DDT, benzene hexachloride, chlordane, and toxaphene, as well as smaller quantities of other insecticides. The manager of the plant admitted that quantities of powdered insecticide had been washed into the storm sewer recently and, more significantly, he acknowledged that such disposal of insecticide spillage and residues had been common practice for the past 10 years.

On searching further, the fishery officers found other plants where rains or ordinary clean-up waters would carry insecticides into the sewer. The fact that provided the final link in the chain, however, was the discovery that a few days before the water in lake and river became lethal to fish the entire storm-sewer system had been flushed out with several million gallons of water under high pressure to clear it of debris. This flushing had undoubtedly released insecticides lodged in the accumulation of gravel, sand, and rubble and carried them into the lake and thence to the river, where chemical tests later established their presence.

As the lethal mass drifted down the Colorado it carried death before it. For 140 miles downstream from the lake the kill of fish must have been almost complete, for when seines were used later in an effort to discover whether any fish had escaped they came up empty. Dead fish of 27 species were observed, totaling about 1000 pounds to a mile of riverbank. There were channel cats, the chief game fish of the river. There were blue and flathead catfish, bullheads, four species of sunfish, shiners, dace, stone rollers, largemouth bass, carp, mullet, suckers. There were eels, gar, carp, river carpsuckers, gizzard shad, and buffalo. Among them were some of the patriarchs of the river, fish that by their size must have been of great age—many flathead catfish weighing over 25 pounds, some of 60 pounds reportedly picked up by local residents along the river, and a giant blue catfish officially recorded as weighing 84 pounds.

The Game and Fish Commission predicted that even without further pollution the pattern of the fish population of the river would be altered for years. Some species—those existing at the limits of their natural range—might never be able to re-establish themselves, and the others could do so only with the aid of extensive stocking operations by the state.

This much of the Austin fish disaster is known, but there was almost certainly a sequel. The toxic river water was still possessed of its death-dealing power after passing more than 200 miles downstream. It was regarded as too dangerous to be admitted to the waters of Matagorda Bay, with its oyster beds and shrimp fisheries, and so the whole toxic outflow was diverted to the waters of the open Gulf. What were its effects there? And what of the outflow of scores of other rivers, carrying contaminants perhaps equally lethal?

At present our answers to these questions are for the most part only conjectures, but there is growing concern about the role of pesticide pollution in estuaries, salt marshes, bays, and other coastal waters. Not only do these areas receive the contaminated discharge of rivers but all too commonly they are sprayed directly in efforts to control mosquitoes or other insects.

Nowhere has the effect of pesticides on the life of salt marshes, estuaries, and all quiet inlets from the sea been more graphically demonstrated than on the eastern coast of Florida, in the Indian River country. There, in the spring of 1955, some 2000 acres of salt marsh in St. Lucie County were treated with dieldrin in an attempt to eliminate the larvae of the sandfly. The concentration used was one pound of active ingredient to the acre. The effect on the life of the waters was catastrophic. Scientists from the Entomology Research Center of the State Board of Health surveyed the carnage after the spraying and reported that the fish kill was "substantially complete." Everywhere dead fishes littered the shores. From the air sharks could be seen moving in, attracted by the helpless and dying fishes in the water. No species was spared. Among the dead were mullets, snook, mojarras, gambusia.

> The minimum immediate over-all kill throughout the marshes, exclusive of the Indian River shoreline, was 20–30 tons of fishes, or about 1,175,000 fishes, of at least 30 species [reported R. W. Harrington, Jr., and W. L. Bidlingmayer of the survey team].
>
> Mollusks seemed to be unharmed by dieldrin. Crustaceans were virtually exterminated throughout the area. The entire aquatic crab population was apparently destroyed and the fiddler crabs, all but annihilated, survived temporarily only in patches of marsh evidently missed by the pellets.
>
> The larger game and food fishes succumbed most rapidly . . . Crabs set upon and destroyed the moribund fishes,

but the next day were dead themselves. Snails continued to devour fish carcasses. After two weeks, no trace remained of the litter of dead fishes.

The same melancholy picture was painted by the late Dr. Herbert R. Mills from his observations in Tampa Bay on the opposite coast of Florida, where the National Audubon Society operates a sanctuary for seabirds in the area including Whiskey Stump Key. The sanctuary ironically became a poor refuge after the local health authorities undertook a campaign to wipe out the salt-marsh mosquitoes. Again fishes and crabs were the principal victims. The fiddler crab, that small and picturesque crustacean whose hordes move over mud flats or sand flats like grazing cattle, has no defense against the sprayers. After successive sprayings during the summer and fall months (some areas were sprayed as many as 16 times), the state of the fiddler crabs was summed up by Dr. Mills: "A progressive scarcity of fiddlers had by this time become apparent. Where there should have been in the neighborhood of 100,000 fiddlers under the tide and weather conditions of the day [October 12] there were not over 100 which could be seen anywhere on the beach, and these were all dead or sick, quivering, twitching, stumbling, scarcely able to crawl; although in neighboring unsprayed areas fiddlers were plentiful."

The place of the fiddler crab in the ecology of the world it inhabits is a necessary one, not easily filled. It is an important source of food for many animals. Coastal raccoons feed on them. So do marsh-inhabiting birds like the clapper rail, shorebirds, and even visiting seabirds. In one New Jersey salt marsh sprayed with DDT, the normal population of laughing gulls was decreased by 85 per cent for several weeks, presumably because the birds could not find sufficient food after the spraying. The marsh fiddlers are important in other ways as well, being useful scavengers and aerating the mud of the marshes by their extensive burrowings. They also furnish quantities of bait for fishermen.

The fiddler crab is not the only creature of tidal marsh and estuary to be threatened by pesticides; others of more obvious importance to man are endangered. The famous blue crab of the Chesapeake Bay and other Atlantic Coast areas is

an example. These crabs are so highly susceptible to insecticides that every spraying of creeks, ditches, and ponds in tidal
marshes kills most of the crabs living there. Not only do the
local crabs die, but others moving into a sprayed area from the
sea succumb to the lingering poison. And sometimes poisoning may be indirect, as in the marshes near Indian River, where
scavenger crabs attacked the dying fishes, but soon themselves
succumbed to the poison. Less is known about the hazard to
the lobster. However, it belongs to the same group of arthropods as the blue crab, has essentially the same physiology, and
would presumably suffer the same effects. This would be true
also of the stone crab and other crustaceans which have direct
economic importance as human food.

The inshore waters—the bays, the sounds, the river estuaries, the tidal marshes—form an ecological unit of the utmost
importance. They are linked so intimately and indispensably
with the lives of many fishes, mollusks, and crustaceans that
were they no longer habitable these seafoods would disappear
from our tables.

Even among fishes that range widely in coastal waters, many
depend upon protected inshore areas to serve as nursery and
feeding grounds for their young. Baby tarpon are abundant in
all that labyrinth of mangrove-lined streams and canals bordering the lower third of the western coast of Florida. On the
Atlantic Coast the sea trout, croaker, spot, and drum spawn
on sandy shoals off the inlets between the islands or "banks"
that lie like a protective chain off much of the coast south of
New York. The young fish hatch and are carried through the
inlets by the tides. In the bays and sounds—Currituck, Pam-
lico, Bogue, and many others—they find abundant food and
grow rapidly. Without these nursery areas of warm, protected,
food-rich waters the populations of these and many other species could not be maintained. Yet we are allowing pesticides to
enter them via the rivers and by direct spraying over bordering
marshlands. And the early stages of these fishes, even more
than the adults, are especially susceptible to direct chemical
poisoning.

Shrimp, too, depend on inshore feeding grounds for their
young. One abundant and widely ranging species supports the
entire commercial fishery of the southern Atlantic and Gulf

states. Although spawning occurs at sea, the young come into the estuaries and bays when a few weeks old to undergo successive molts and changes of form. There they remain from May or June until fall, feeding on the bottom detritus. In the entire period of their inshore life, the welfare of the shrimp populations and of the industry they support depends upon favorable conditions in the estuaries.

Do pesticides represent a threat to the shrimp fisheries and to the supply for the markets? The answer may be contained in recent laboratory experiments carried out by the Bureau of Commercial Fisheries. The insecticide tolerance of young commercial shrimp just past larval life was found to be exceedingly low—measured in parts per *billion* instead of the more commonly used standard of parts per million. For example, half the shrimp in one experiment were killed by dieldrin at a concentration of only 15 parts per billion. Other chemicals were even more toxic. Endrin, always one of the most deadly of the pesticides, killed half the shrimp at a concentration of only *half of one part per billion*.

The threat to oysters and clams is multiple. Again, the young stages are most vulnerable. These shellfish inhabit the bottoms of bays and sounds and tidal rivers from New England to Texas and sheltered areas of the Pacific Coast. Although sedentary in adult life, they discharge their spawn into the sea, where the young are free-living for a period of several weeks. On a summer day a fine-meshed tow net drawn behind a boat will collect, along with the other drifting plant and animal life that make up the plankton, the infinitely small, fragile-as-glass larvae of oysters and clams. No larger than grains of dust, these transparent larvae swim about in the surface waters, feeding on the microscopic plant life of the plankton. If the crop of minute sea vegetation fails, the young shellfish will starve. Yet pesticides may well destroy substantial quantities of plankton. Some of the herbicides in common use on lawns, cultivated fields, and roadsides and even in coastal marshes are extraordinarily toxic to the plant plankton which the larval mollusks use as food—some at only a few parts per billion.

The delicate larvae themselves are killed by very small quantities of many of the common insecticides. Even exposures to less than lethal quantities may in the end cause death of the larvae, for inevitably the growth rate is retarded. This prolongs the period the larvae must spend in the hazardous world of the plankton and so decreases the chance they will live to adulthood.

For adult mollusks there is apparently less danger of direct poisoning, at least by some of the pesticides. This is not necessarily reassuring, however. Oysters and clams may concentrate these poisons in their digestive organs and other tissues. Both types of shellfish are normally eaten whole and sometimes raw. Dr. Philip Butler of the Bureau of Commercial Fisheries has pointed out an ominous parallel in that we may find ourselves in the same situation as the robins. The robins, he reminds us, did not die as a direct result of the spraying of DDT. They died because they had eaten earthworms that had already concentrated the pesticides in their tissues.

Although the sudden death of thousands of fish or crustaceans in some stream or pond as the direct and visible effect of insect control is dramatic and alarming, these unseen and as yet largely unknown and unmeasurable effects of pesticides reaching estuaries indirectly in streams and rivers may in the end be more disastrous. The whole situation is beset with questions for which there are at present no satisfactory answers. We know that pesticides contained in runoff from farms and forests are now being carried to the sea in the waters of many and perhaps all of the major rivers. But we do not know the identity of all the chemicals or their total quantity, and we do not presently have any dependable tests for identifying them in highly diluted state once they have reached the sea. Although we know that the chemicals have almost certainly undergone change during the long period of transit, we do not know whether the altered chemical is more toxic than the original or less. Another almost unexplored area is the question of interactions between chemicals, a question that becomes especially urgent when they enter the marine environment where so many different minerals are subjected to mixing and transport. All of these questions urgently require the precise answers that only extensive research can provide, yet funds for such purposes are pitifully small.

The fisheries of fresh and salt water are a resource of great importance, involving the interests and the welfare of a very large number of people. That they are now seriously threatened by the chemicals entering our waters can no longer be doubted. If we would divert to constructive research even a small fraction of the money spent each year on the development of ever more toxic sprays, we could find ways to use less dangerous materials and to keep poisons out of our waterways. When will the public become sufficiently aware of the facts to demand such action?

TEN

Indiscriminately from the Skies

FROM SMALL BEGINNINGS over farmlands and forests the scope of aerial spraying has widened and its volume has increased so that it has become what a British ecologist recently called "an amazing rain of death" upon the surface of the earth. Our attitude toward poisons has undergone a subtle change. Once they were kept in containers marked with skull and crossbones; the infrequent occasions of their use were marked with utmost care that they should come in contact with the target and with nothing else. With the development of the new organic insecticides and the abundance of surplus planes after the Second World War, all this was forgotten. Although today's poisons are more dangerous than any known before, they have amazingly become something to be showered down indiscriminately from the skies. Not only the target insect or plant, but anything—human or nonhuman—within range of the chemical fallout may know the sinister touch of the poison. Not only forests and cultivated fields are sprayed, but towns and cities as well.

A good many people now have misgivings about the aerial distribution of lethal chemicals over millions of acres, and two mass-spraying campaigns undertaken in the late 1950's have done much to increase these doubts. These were the campaigns against the gypsy moth in the northeastern states and the fire ant in the South. Neither is a native insect but both have been in this country for many years without creating a situation calling for desperate measures. Yet drastic action was suddenly taken against them, under the end-justifies-the-means philosophy that has too long directed the control divisions of our Department of Agriculture.

The gypsy moth program shows what a vast amount of damage can be done when reckless large-scale treatment is substituted for local and moderate control. The campaign against the fire ant is a prime example of a campaign based on gross exaggeration of the need for control, blunderingly launched without scientific knowledge of the dosage of poison required to destroy the target or of its effects on other life. Neither program has achieved its goal.

The gypsy moth, a native of Europe, has been in the United States for nearly a hundred years. In 1869 a French scientist, Leopold Trouvelot, accidentally allowed a few of these moths to escape from his laboratory in Medford, Massachusetts, where he was attempting to cross them with silkworms. Little by little the gypsy moth has spread throughout New England. The primary agent of its progressive spread is the wind; the larval, or caterpillar, stage is extremely light and can be carried to considerable heights and over great distances. Another means is the shipment of plants carrying the egg masses, the form in which the species exists over winter. The gypsy moth, which in its larval stage attacks the foliage of oak trees and a few other hardwoods for a few weeks each spring, now occurs in all the New England states. It also occurs sporadically in New Jersey, where it was introduced in 1911 on a shipment of spruce trees from Holland, and in Michigan, where its method of entry is not known. The New England hurricane of 1938 carried it into Pennsylvania and New York, but the Adirondacks have generally served as a barrier to its westward advance, being forested with species not attractive to it.

The task of confining the gypsy moth to the northeastern corner of the country has been accomplished by a variety of methods, and in the nearly one hundred years since its arrival on this continent the fear that it would invade the great hard-wood forests of the southern Appalachians has not been jus-tified. Thirteen parasites and predators were imported from abroad and successfully established in New England. The Ag-riculture Department itself has credited these importations with appreciably reducing the frequency and destructiveness of gypsy moth outbreaks. This natural control, plus quarantine measures and local spraying, achieved what the Department in 1955 described as "outstanding restriction of distribution and damage."

Yet only a year after expressing satisfaction with the state of affairs, its Plant Pest Control Division embarked on a program calling for the blanket spraying of several million acres a year with the announced intention of eventually "eradicating" the gypsy moth. ("Eradication" means the complete and final ex-tinction or extermination of a species throughout its range. Yet as successive programs have failed, the Department has found it necessary to speak of second or third "eradications" of the same species in the same area.)

The Department's all-out chemical war on the gypsy moth began on an ambitious scale. In 1956 nearly a million acres were sprayed in the states of Pennsylvania, New Jersey, Michi-gan, and New York. Many complaints of damage were made by people in the sprayed areas. Conservationists became increas-ingly disturbed as the pattern of spraying huge areas began to establish itself. When plans were announced for spraying 3 million acres in 1957 opposition became even stronger. State and federal agriculture officials characteristically shrugged off individual complaints as unimportant.

The Long Island area included within the gypsy moth spray-ing in 1957 consisted chiefly of heavily populated towns and suburbs and of some coastal areas with bordering salt marsh. Nassau County, Long Island, is the most densely settled county in New York apart from New York City itself. In what seems the height of absurdity, the "threat of infestation of the New York City metropolitan area" has been cited as an important justification of the program. The gypsy moth is a forest insect, certainly not an inhabitant of cities. Nor does it live in mead-ows, cultivated fields, gardens, or marshes. Nevertheless, the planes hired by the United States Department of Agriculture and the New York Department of Agriculture and Markets in 1957 showered down the prescribed DDT-in-fuel-oil with impartiality. They sprayed truck gardens and dairy farms, fish ponds and salt marshes. They sprayed the quarter-acre lots of suburbia, drenching a housewife making a desperate effort to cover her garden before the roaring plane reached her, and showering insecticide over children at play and commuters at railway stations. At Setauket a fine quarter horse drank from a trough in a field which the planes had sprayed; ten hours later it was dead. Automobiles were spotted with the oily mixture; flowers and shrubs were ruined. Birds, fish, crabs, and useful insects were killed.

A group of Long Island citizens led by the world-famous ornithologist Robert Cushman Murphy had sought a court injunction to prevent the 1957 spraying. Denied a preliminary injunction, the protesting citizens had to suffer the prescribed drenching with DDT, but thereafter persisted in efforts to obtain a permanent injunction. But because the act had already been performed the courts held that the petition for an injunction was "moot." The case was carried all the way to the Supreme Court, which declined to hear it. Justice William O. Douglas, strongly dissenting from the decision not to review the case, held that "the alarms that many experts and responsible officials have raised about the perils of DDT underline the public importance of this case."

The suit brought by the Long Island citizens at least served to focus public attention on the growing trend to mass application of insecticides, and on the power and inclination of the control agencies to disregard supposedly inviolate property rights of private citizens.

The contamination of milk and of farm produce in the course of the gypsy moth spraying came as an unpleasant surprise to many people. What happened on the 200-acre Waller farm in northern Westchester County, New York, was revealing. Mrs. Waller had specifically requested Agriculture officials not to spray her property, because it would be impossible to avoid the pastures in spraying the woodlands. She offered to have the land checked for gypsy moths and to have any infestation destroyed by spot spraying. Although she was assured that no farms would be sprayed, her property received two direct sprayings and, in addition, was twice subjected to drifting spray. Milk samples taken from the Wallers' purebred Guernsey cows 48 hours later contained DDT in the amount of 14 parts per million. Forage samples from fields where the cows had grazed were of course contaminated also. Although the county Health Department was notified, no instructions were given that the milk should not be marketed. This situation is unfortunately typical of the lack of consumer protection that is all too common. Although the Food and Drug Administration permits no residues of pesticides in milk, its restrictions are not only inadequately policed but they apply solely to interstate shipments. State and county officials are under no compulsion to follow the federal pesticides tolerances unless local laws happen to conform—and they seldom do.

Truck gardeners also suffered. Some leaf crops were so burned and spotted as to be unmarketable. Others carried heavy residues; a sample of peas analyzed at Cornell University's Agricultural Experiment Station contained 14 to 20 parts per million of DDT. The legal maximum is 7 parts per million. Growers therefore had to sustain heavy losses or find themselves in the position of selling produce carrying illegal residues. Some of them sought and collected damages.

As the aerial spraying of DDT increased, so did the number of suits filed in the courts. Among them were suits brought by beekeepers in several areas of New York State. Even before the 1957 spraying, the beekeepers had suffered heavily from use of DDT in orchards. "Up to 1953 I had regarded as gospel everything that emanated from the U.S. Department of Agriculture and the agricultural colleges," one of them remarked bitterly. But in May of that year this man lost 800 colonies after the state had sprayed a large area. So widespread and heavy was the loss that 14 other beekeepers joined him in suing the state for a quarter of a million dollars in damages. Another beekeeper, whose 400 colonies were incidental targets of the 1957 spray, reported that 100 per cent of the field force of bees (the workers out gathering nectar and pollen for the hives) had been killed in forested areas and up to 50 per cent in farming areas sprayed less intensively. "It is a very distressful thing," he wrote, "to walk into a yard in May and not hear a bee buzz."

The gypsy moth programs were marked by many acts of ir-responsibility. Because the spray planes were paid by the gallon rather than by the acre there was no effort to be conservative, and many properties were sprayed not once but several times. Contracts for aerial spraying were in at least one case awarded to an out-of-state firm with no local address, which had not complied with the legal requirement of registering with state officials for the purpose of establishing legal responsibility. In this exceedingly slippery situation, citizens who suffered direct financial loss from damage to apple orchards or bees discov-ered that there was no one to sue.

After the disastrous 1957 spraying the program was abruptly and drastically curtailed, with vague statements about "evalu-ating" previous work and testing alternative insecticides. In-stead of the 3½ million acres sprayed in 1957, the treated areas fell to ½ million in 1958 and to about 100,000 acres in 1959, 1960, and 1961. During this interval, the control agencies must have found news from Long Island disquieting. The gypsy moth had reappeared there in numbers. The expensive spray-ing operation that had cost the Department dearly in public confidence and good will—the operation that was intended to wipe out the gypsy moth for ever—had in reality accomplished nothing at all.

Meanwhile, the Department's Plant Pest Control men had temporarily forgotten gypsy moths, for they had been busy launching an even more ambitious program in the South. The word "eradication" still came easily from the Department's mimeograph machines; this time the press releases were prom-ising the eradication of the fire ant.

The fire ant, an insect named for its fiery sting, seems to have entered the United States from South America by way of the port of Mobile, Alabama, where it was discovered shortly after the end of the First World War. By 1928 it had spread into the suburbs of Mobile and thereafter continued an invasion that has now carried it into most of the southern states.

During most of the forty-odd years since its arrival in the United States the fire ant seems to have attracted little atten-tion. The states where it was most abundant considered it a nuisance, chiefly because it builds large nests or mounds a foot or more high. These may hamper the operation of farm machinery. But only two states listed it among their 20 most important insect pests, and these placed it near the bottom of the list. No official or private concern seems to have been felt about the fire ant as a menace to crops or livestock.

With the development of chemicals of broad lethal powers, there came a sudden change in the official attitude toward the fire ant. In 1957 the United States Department of Agriculture launched one of the most remarkable publicity campaigns in its history. The fire ant suddenly became the target of a barrage of government releases, motion pictures, and government-inspired stories portraying it as a despoiler of southern agriculture and a killer of birds, livestock, and man. A mighty campaign was announced, in which the federal government in cooperation with the afflicted states would ultimately treat some 20,000,000 acres in nine southern states.

"United States pesticide makers appear to have tapped a sales bonanza in the increasing numbers of broad-scale pest elimination programs conducted by the U.S. Department of Agriculture," cheerfully reported one trade journal in 1958, as the fire ant program got under way.

Never has any pesticide program been so thoroughly and deservedly damned by practically everyone except the beneficiaries of this "sales bonanza." It is an outstanding example of an ill-conceived, badly executed, and thoroughly detrimental experiment in the mass control of insects, an experiment so expensive in dollars, in destruction of animal life, and in loss of public confidence in the Agriculture Department that it is incomprehensible that any funds should still be devoted to it.

Congressional support of the project was initially won by representations that were later discredited. The fire ant was pictured as a serious threat to southern agriculture through destruction of crops and to wildlife because of attacks on the young of ground-nesting birds. Its sting was said to make it a serious menace to human health.

Just how sound were these claims? The statements made by Department witnesses seeking appropriations were not in accord with those contained in key publications of the Agriculture Department. The 1957 bulletin *Insecticide Recommendations . . . for the Control of Insects Attacking Crops and*

Livestock did not so much as mention the fire ant—an extraordinary omission if the Department believes its own propaganda. Moreover, its encyclopedic *Yearbook* for 1952, which was devoted to insects, contained only one short paragraph on the fire ant out of its half-million words of text.

Against the Department's undocumented claim that the fire ant destroys crops and attacks livestock is the careful study of the Agricultural Experiment Station in the state that has had the most intimate experience with this insect, Alabama. According to Alabama scientists, "damage to plants in general is rare." Dr. F. S. Arant, an entomologist at the Alabama Polytechnic Institute and in 1961 president of the Entomological Society of America, states that his department "has not received a single report of damage to plants by ants in the past five years . . . No damage to livestock has been observed." These men, who have actually observed the ants in the field and in the laboratory, say that the fire ants feed chiefly on a variety of other insects, many of them considered harmful to man's interests. Fire ants have been observed picking larvae of the boll weevil off cotton. Their mound-building activities serve a useful purpose in aerating and draining the soil. The Alabama studies have been substantiated by investigations at the Mississippi State University, and are far more impressive than the Agriculture Department's evidence, apparently based either on conversations with farmers, who may easily mistake one ant for another, or on old research. Some entomologists believe that the ant's food habits have changed as it has become more abundant, so that observations made several decades ago have little value now.

The claim that the ant is a menace to health and life also bears considerable modification. The Agriculture Department sponsored a propaganda movie (to gain support for its program) in which horror scenes were built around the fire ant's sting. Admittedly this is painful and one is well advised to avoid being stung, just as one ordinarily avoids the sting of wasp or bee. Severe reactions may occasionally occur in sensitive individuals, and medical literature records one death possibly, though not definitely, attributable to fire ant venom. In contrast to this, the Office of Vital Statistics records 33 deaths in 1959 alone from the sting of bees and wasps. Yet no one seems to have proposed "eradicating" these insects. Again, local evidence is most convincing. Although the fire ant has inhabited Alabama for 40 years and is most heavily concentrated there, the Alabama State Health Officer declares that "there has never been recorded in Alabama a human death resulting from the bites of imported fire ants," and considers the medical cases resulting from the bites of fire ants "incidental." Ant mounds on lawns or playgrounds may create a situation where children are likely to be stung, but this is hardly an excuse for drenching millions of acres with poisons. These situations can easily be handled by individual treatment of the mounds.

Damage to game birds was also alleged, without supporting evidence. Certainly a man well qualified to speak on this issue is the leader of the Wildlife Research Unit at Auburn, Alabama, Dr. Maurice F. Baker, who has had many years' experience in the area. But Dr. Baker's opinion is directly opposite to the claims of the Agriculture Department. He declares: "In south Alabama and northwest Florida we are able to have excellent hunting and bobwhite populations coexistent with heavy populations of the imported fire ant . . . in the almost 40 years that south Alabama has had the fire ant, game populations have shown a steady and very substantial increase. Certainly, if the imported fire ant were a serious menace to wildlife, these conditions could not exist."

What would happen to wildlife as a result of the insecticide used against the ants was another matter. The chemicals to be used were dieldrin and heptachlor, both relatively new. There was little experience of field use for either, and no one knew what their effects would be on wild birds, fishes, or mammals when applied on a massive scale. It was known, however, that both poisons were many times more toxic than DDT, which had been used by that time for approximately a decade, and had killed some birds and many fish even at a rate of 1 pound per acre. And the dosage of dieldrin and heptachlor was heavier— 2 pounds to the acre under most conditions, or 3 pounds of dieldrin if the white-fringed beetle was also to be controlled. In terms of their effects on birds, the prescribed use of heptachlor would be equivalent to 20 pounds of DDT to the acre, that of dieldrin to 120 pounds!

Urgent protests were made by most of the state conservation departments, by national conservation agencies, and by ecologists and even by some entomologists, calling upon the then Secretary of Agriculture, Ezra Benson, to delay the program at least until some research had been done to determine the effects of heptachlor and dieldrin on wild and domestic animals and to find the minimum amount that would control the ants. The protests were ignored and the program was launched in 1958. A million acres were treated the first year. It was clear that any research would be in the nature of a post mortem.

As the program continued, facts began to accumulate from studies made by biologists of state and federal wildlife agencies and several universities. The studies revealed losses running all the way up to complete destruction of wildlife on some of the treated areas. Poultry, livestock, and pets were also killed. The Agriculture Department brushed away all evidence of damage as exaggerated and misleading.

The facts, however, continue to accumulate. In Hardin County, Texas, for example, opossums, armadillos, and an abundant raccoon population virtually disappeared after the chemical was laid down. Even the second autumn after treatment these animals were scarce. The few raccoons then found in the area carried residues of the chemical in their tissues.

Dead birds found in the treated areas had absorbed or swallowed the poisons used against the fire ants, a fact clearly shown by chemical analysis of their tissues. (The only bird surviving in any numbers was the house sparrow, which in other areas too has given some evidence that it may be relatively immune.) On a tract in Alabama treated in 1959 half of the birds were killed. Species that live on the ground or frequent low vegetation suffered 100 per cent mortality. Even a year after treatment, a spring die-off of songbirds occurred and much good nesting territory lay silent and unoccupied. In Texas, dead blackbirds, dickcissels, and meadowlarks were found at the nests, and many nests were deserted. When specimens of dead birds from Texas, Louisiana, Alabama, Georgia, and Florida were sent to the Fish and Wildlife Service for analysis, more than 90 per cent were found to contain residues of dieldrin or a form of heptachlor, in amounts up to 38 parts per million.

Woodcocks, which winter in Louisiana but breed in the North, now carry the taint of the fire ant poisons in their bodies. The source of this contamination is clear. Woodcocks feed heavily on earthworms, which they probe for with their long bills. Surviving worms in Louisiana were found to have as much as 20 parts per million of heptachlor in their tissues 6 to 10 months after treatment of the area. A year later they had up to 10 parts per million. The consequences of the sublethal poisoning of the woodcock are now seen in a marked decline in the proportion of young birds to adults, first observed in the season after fire ant treatments began.

Some of the most upsetting news for southern sportsmen concerned the bobwhite quail. This bird, a ground nester and forager, was all but eliminated on treated areas. In Alabama, for example, biologists of the Alabama Cooperative Wildlife Research Unit conducted a preliminary census of the quail population in a 3600-acre area that was scheduled for treatment. Thirteen resident coveys—121 quail—ranged over the area. Two weeks after treatment only dead quail could be found. All specimens sent to the Fish and Wildlife Service for analysis were found to contain insecticides in amounts sufficient to cause their death. The Alabama findings were duplicated in Texas, where a 2500-acre area treated with heptachlor lost all of its quail. Along with the quail went 90 per cent of the songbirds. Again, analysis revealed the presence of heptachlor in the tissues of dead birds.

In addition to quail, wild turkeys were seriously reduced by the fire ant program. Although 80 turkeys had been counted on an area in Wilcox County, Alabama, before heptachlor was applied, none could be found the summer after treatment— none, that is, except a clutch of unhatched eggs and one dead poult. The wild turkeys may have suffered the same fate as their domestic brethren, for turkeys on farms in the area treated with chemicals also produced few young. Few eggs hatched and almost no young survived. This did not happen on nearby untreated areas.

The fate of the turkeys was by no means unique. One of the most widely known and respected wildlife biologists in the country, Dr. Clarence Cottam, called on some of the farmers whose property had been treated. Besides remarking that "all the little tree birds" seemed to have disappeared after the land had been treated, most of these people reported losses of livestock, poultry, and household pets. One man was "irate against the control workers," Dr. Cottam reported, "as he said he buried or otherwise disposed of 19 carcasses of his cows that had been killed by the poison and he knew of three or four additional cows that died as a result of the same treatment. Calves died that had been given only milk since birth."

The people Dr. Cottam interviewed were puzzled by what had happened in the months following the treatment of their land. One woman told him she had set several hens after the surrounding land had been covered with poison, "and for reasons she did not understand very few young were hatched or survived." Another farmer "raises hogs and for fully nine months after the broadcast of poisons, he could raise no young pigs. The litters were born dead or they died after birth." A similar report came from another, who said that out of 37 litters that might have numbered as many as 250 young, only 31 little pigs survived. This man had also been quite unable to raise chickens since the land was poisoned.

The Department of Agriculture has consistently denied livestock losses related to the fire ant program. However, a veterinarian in Bainbridge, Georgia, Dr. Otis L. Poitevint, who was called upon to treat many of the affected animals, has summarized his reasons for attributing the deaths to the insecticide as follows. Within a period of two weeks to several months after the fire ant poison was applied, cattle, goats, horses, chickens, and birds and other wildlife began to suffer an often fatal disease of the nervous system. It affected only animals that had access to contaminated food or water. Stabled animals were not affected. The condition was seen only in areas treated for fire ants. Laboratory tests for disease were negative. The symptoms observed by Dr. Poitevint and other veterinarians were those described in authoritative texts as indicating poisoning by dieldrin or heptachlor.

Dr. Poitevint also described an interesting case of a two-month-old calf that showed symptoms of poisoning by heptachlor. The animal was subjected to exhaustive laboratory tests. The only significant finding was the discovery of 79 parts per million of heptachlor in its fat. But it was five months since the poison had been applied. Did the calf get it directly from grazing or indirectly from its mother's milk or even before birth? "If from the milk," asked Dr. Poitevint, "why were not special precautions taken to protect our children who drank milk from local dairies?"

Dr. Poitevint's report brings up a significant problem about the contamination of milk. The area included in the fire ant program is predominantly fields and croplands. What about the dairy cattle that graze on these lands? In treated fields the grasses will inevitably carry residues of heptachlor in one of its forms, and if the residues are eaten by the cows the poison will appear in the milk. This direct transmission into milk had been demonstrated experimentally for heptachlor in 1955, long before the control program was undertaken, and was later reported for dieldrin, also used in the fire ant program.

The Department of Agriculture's annual publications now list heptachlor and dieldrin among the chemicals that make forage plants unsuitable for feeding to dairy animals or animals being finished for slaughter, yet the control divisions of the Department promote programs that spread heptachlor and dieldrin over substantial areas of grazing land in the South. Who is safeguarding the consumer to see that no residues of dieldrin or heptachlor are appearing in milk? The United States Department of Agriculture would doubtless answer that it has advised farmers to keep milk cows out of treated pastures for 30 to 90 days. Given the small size of many of the farms and the large-scale nature of the program—much of the chemical applied by planes—it is extremely doubtful that this recommendation was followed or could be. Nor is the prescribed period adequate in view of the persistent nature of the residues.

The Food and Drug Administration, although frowning on the presence of any pesticide residues in milk, has little authority in this situation. In most of the states included in the fire ant program the dairy industry is small and its products do not cross state lines. Protection of the milk supply endangered by a federal program is therefore left to the states themselves. Inquiries addressed to the health officers or other appropriate officials of Alabama, Louisiana, and Texas in 1959 revealed that no tests had been made and that it simply was not known whether the milk was contaminated with pesticides or not.

Meanwhile, after rather than before the control program was launched, some research into the peculiar nature of heptachlor was done. Perhaps it would be more accurate to say that someone looked up the research already published, since the basic fact that brought about belated action by the federal government had been discovered several years before, and should have influenced the initial handling of the program. This is the fact that heptachlor, after a short period in the tissues of animals or plants or in the soil, assumes a considerably more toxic form known as heptachlor epoxide. The epoxide is popularly described as "an oxidation product" produced by weathering. The fact that this transformation could occur had been known since 1952, when the Food and Drug Administration discovered that female rats, fed 30 parts per million of heptachlor, had stored 165 parts per million of the more poisonous epoxide only 2 weeks later.

These facts were allowed to come out of the obscurity of biological literature in 1959, when the Food and Drug Administration took action which had the effect of banning any residues of heptachlor or its epoxide on food. This ruling put at least a temporary damper on the program; although the Agriculture Department continued to press for its annual appropriations for fire ant control, local agricultural agents became increasingly reluctant to advise farmers to use chemicals which would probably result in their crops being legally unmarketable.

In short, the Department of Agriculture embarked on its program without even elementary investigation of what was already known about the chemical to be used—or if it investigated, it ignored the findings. It must also have failed to do preliminary research to discover the minimum amount of the chemical that would accomplish its purpose. After three years of heavy dosages, it abruptly reduced the rate of application of heptachlor from 2 pounds to 1¼ pounds per acre in 1959; later on to ½ pound per acre, applied in two treatments of ¼ pound each, 3 to 6 months apart. An official of the Department explained that "an aggressive methods improvement program" showed the lower rate to be effective. Had this information been acquired before the program was launched, a vast amount of damage could have been avoided and the taxpayers could have been saved a great deal of money.

In 1959, perhaps in an attempt to offset the growing dissatisfaction with the program, the Agriculture Department offered the chemicals free to Texas landowners who would sign a release absolving federal, state, and local governments of responsibility for damage. In the same year the State of Alabama, alarmed and angry at the damage done by the chemicals, refused to appropriate any further funds for the project. One of its officials characterized the whole program as "ill advised, hastily conceived, poorly planned, and a glaring example of riding roughshod over the responsibilities of other public and private agencies." Despite the lack of state funds, federal money continued to trickle into Alabama, and in 1961 the legislature was again persuaded to make a small appropriation. Meanwhile, farmers in Louisiana showed growing reluctance to sign up for the project as it became evident that use of chemicals against the fire ant was causing an upsurge of insects destructive to sugarcane. Moreover, the program was obviously accomplishing nothing. Its dismal state was tersely summarized in the spring of 1962 by the director of entomology research at Louisiana State University Agricultural Experiment Station, Dr. L. D. Newsom: "The imported fire ant 'eradication' program which has been conducted by state and federal agencies is thus far a failure. There are more infested acres in Louisiana now than when the program began."

A swing to more sane and conservative methods seems to have begun. Florida, reporting that "there are more fire ants in Florida now than there were when the program started," announced it was abandoning any idea of a broad eradication program and would instead concentrate on local control.

Effective and inexpensive methods of local control have been known for years. The mound-building habit of the fire ant makes the chemical treatment of individual mounds a simple matter. Cost of such treatment is about one dollar per acre. For situations where mounds are numerous and mechanized methods are desirable, a cultivator which first levels and then applies chemical directly to the mounds has been developed by Mississippi's Agricultural Experiment Station. The method gives 90 to 95 per cent control of the ants. Its cost is only $.23 per acre. The Agriculture Department's mass control program, on the other hand, cost about $3.50 per acre—the most expensive, the most damaging, and the least effective program of all.

ELEVEN

Beyond the Dreams of the Borgias

THE CONTAMINATION of our world is not alone a matter of mass spraying. Indeed, for most of us this is of less importance than the innumerable small-scale exposures to which we are subjected day by day, year after year. Like the constant dripping of water that in turn wears away the hardest stone, this birth-to-death contact with dangerous chemicals may in the end prove disastrous. Each of these recurrent exposures, no matter how slight, contributes to the progressive buildup of chemicals in our bodies and so to cumulative poisoning. Probably no person is immune to contact with this spreading contamination unless he lives in the most isolated situation imaginable. Lulled by the soft sell and the hidden persuader, the average citizen is seldom aware of the deadly materials with which he is surrounding himself; indeed, he may not realize he is using them at all.

So thoroughly has the age of poisons become established that anyone may walk into a store and, without questions being asked, buy substances of far greater death-dealing power than the medicinal drug for which he may be required to sign a "poison book" in the pharmacy next door. A few minutes' research in any supermarket is enough to alarm the most stout-hearted customer—provided, that is, he has even a rudimentary knowledge of the chemicals presented for his choice.

If a huge skull and crossbones were suspended above the insecticide department the customer might at least enter it with the respect normally accorded death-dealing materials. But instead the display is homey and cheerful, and, with the pickles and olives across the aisle and the bath and laundry soaps adjoining, the rows upon rows of insecticides are displayed. Within easy reach of a child's exploring hand are chemicals in *glass* containers. If dropped to the floor by a child or careless adult everyone nearby could be splashed with the same chemical that has sent spraymen using it into convulsions. These hazards of course follow the purchaser right into his home. A can of a mothproofing material containing DDD, for example, carries in very fine print the warning that its contents are under pressure and that it may burst if exposed to heat or open flame. A common insecticide for household use, including assorted uses in the kitchen, is chlordane. Yet the Food and Drug Administration's chief pharmacologist has declared the hazard of living in a house sprayed with chlordane to be "very great." Other household preparations contain the even more toxic dieldrin.

Use of poisons in the kitchen is made both attractive and easy. Kitchen shelf paper, white or tinted to match one's color scheme, may be impregnated with insecticide, not merely on one but on both sides. Manufacturers offer us do-it-yourself booklets on how to kill bugs. With push-button ease, one may send a fog of dieldrin into the most inaccessible nooks and crannies of cabinets, corners, and baseboards.

If we are troubled by mosquitoes, chiggers, or other insect pests on our persons we have a choice of innumerable lotions, creams, and sprays for application to clothing or skin. Although we are warned that some of these will dissolve varnish, paint, and synthetic fabrics, we are presumably to infer that the human skin is impervious to chemicals. To make certain that we shall at all times be prepared to repel insects, an exclusive New York store advertises a pocket-sized insecticide dispenser, suitable for the purse or for beach, golf, or fishing gear.

We can polish our floors with a wax guaranteed to kill any insect that walks over it. We can hang strips impregnated with the chemical lindane in our closets and garment bags or place them in our bureau drawers for a half year's freedom from worry over moth damage. The advertisements contain no suggestion that lindane is dangerous. Neither do the ads for an electronic device that dispenses lindane fumes—we are told that it is safe and odorless. Yet the truth of the matter is that the American Medical Association considers lindane vaporizers so dangerous that it conducted an extended campaign against them in its *Journal*.

The Department of Agriculture, in a *Home and Garden Bulletin*, advises us to spray our clothing with oil solutions of DDT, dieldrin, chlordane, or any of several other moth killers. If excessive spraying results in a white deposit of insecticide on the fabric, this may be removed by brushing, the Department says, omitting to caution us to be careful where and how the brushing is done. All these matters attended to, we may round out our day with insecticides by going to sleep under a moth-proof blanket impregnated with dieldrin.

Gardening is now firmly linked with the super poisons. Every hardware store, garden-supply shop, and supermarket has rows of insecticides for every conceivable horticultural situation. Those who fail to make wide use of this array of lethal sprays and dusts are by implication remiss, for almost every newspaper's garden page and the majority of the gardening magazines take their use for granted.

So extensively are even the rapidly lethal organic phosphorus insecticides applied to lawns and ornamental plants that in 1960 the Florida State Board of Health found it necessary to forbid the commercial use of pesticides in residential areas by anyone who had not first obtained a permit and met certain requirements. A number of deaths from parathion had occurred in Florida before this regulation was adopted.

Little is done, however, to warn the gardener or homeowner that he is handling extremely dangerous materials. On the contrary, a constant stream of new gadgets make it easier to use poisons on lawn and garden—and increase the gardener's contact with them. One may get a jar-type attachment for the garden hose, for example, by which such extremely dangerous chemicals as chlordane or dieldrin are applied as one waters the lawn. Such a device is not only a hazard to the person using the hose; it is also a public menace. The *New York Times* found it necessary to issue a warning on its garden page to the effect that unless special protective devices were installed poisons might get into the water supply by back siphonage. Considering the number of such devices that are in use, and the scarcity of warnings such as this, do we need to wonder why our public waters are contaminated?

As an example of what may happen to the gardener himself, we might look at the case of a physician—an enthusiastic spare-time gardener—who began using DDT and then malathion on his shrubs and lawn, making regular weekly applications. Sometimes he applied the chemicals with a hand spray, sometimes with an attachment to his hose. In doing so, his skin and clothing were often soaked with spray. After about a year of this sort of thing, he suddenly collapsed and was hospitalized. Examination of a biopsy specimen of fat showed an accumulation of 23 parts per million of DDT. There was extensive nerve damage, which his physicians regarded as permanent. As time went on he lost weight, suffered extreme fatigue, and experienced a peculiar muscular weakness, a characteristic effect of malathion. All of these persisting effects were severe enough to make it difficult for the physician to carry on his practice.

Besides the once innocuous garden hose, power mowers also have been fitted with devices for the dissemination of pesticides, attachments that will dispense a cloud of vapor as the homeowner goes about the task of mowing his lawn. So to the potentially dangerous fumes from gasoline are added the finely divided particles of whatever insecticide the probably unsuspecting suburbanite has chosen to distribute, raising the level of air pollution above his own grounds to something few cities could equal.

Yet little is said about the hazards of the fad of gardening by poisons, or of insecticides used in the home; warnings on labels are printed so inconspicuously in small type that few take the trouble to read or follow them. An industrial firm recently undertook to find out just *how* few. Its survey indicated that fewer than fifteen people out of a hundred of those using insecticide aerosols and sprays are even aware of the warnings on the containers.

The mores of suburbia now dictate that crabgrass must go at whatever cost. Sacks containing chemicals designed to rid the lawn of such despised vegetation have become almost a status symbol. These weed-killing chemicals are sold under brand names that never suggest their identity or nature. To learn that they contain chlordane or dieldrin one must read exceedingly fine print placed on the least conspicuous part of the sack. The descriptive literature that may be picked up in any hardware- or garden-supply store seldom if ever reveals the true hazard involved in handling or applying the material. Instead, the typical illustration portrays a happy family scene, father and son smilingly preparing to apply the chemical to the lawn, small children tumbling over the grass with a dog.

The question of chemical residues on the food we eat is a hotly debated issue. The existence of such residues is either played down by the industry as unimportant or is flatly denied. Simultaneously, there is a strong tendency to brand as fanatics or cultists all who are so perverse as to demand that their food be free of insect poisons. In all this cloud of controversy, what are the actual facts?

It has been medically established that, as common sense would tell us, persons who lived and died before the dawn of the DDT era (about 1942) contained no trace of DDT or any similar material in their tissues. As mentioned in Chapter 3, samples of body fat collected from the general population between 1954 and 1956 averaged from 5.3 to 7.4 parts per million of DDT. There is some evidence that the average level has risen since then to a consistently higher figure, and individuals with occupational or other special exposures to insecticides of course store even more.

Among the general population with no known gross exposures to insecticides it may be assumed that much of the DDT stored in fat deposits has entered the body in food. To test this assumption, a scientific team from the United States Public Health Service sampled restaurant and institutional meals.

Every meal sampled contained DDT. From this the investigators concluded, reasonably enough, that "few if any foods can
be relied upon to be entirely free of DDT."

The quantities in such meals may be enormous. In a separate
Public Health Service study, analysis of prison meals disclosed
such items as stewed dried fruit containing 69.6 parts per million and bread containing 100.9 parts per million of DDT!

In the diet of the average home, meats and any products
derived from animal fats contain the heaviest residues of chlorinated hydrocarbons. This is because these chemicals are soluble
in fat. Residues on fruits and vegetables tend to be somewhat
less. These are little affected by washing—the only remedy is
to remove and discard all outside leaves of such vegetables as
lettuce or cabbage, to peel fruit and to use no skins or outer
covering whatever. Cooking does not destroy residues.

Milk is one of the few foods in which no pesticide residues
are permitted by Food and Drug Administration regulations.
In actual fact, however, residues turn up whenever a check is
made. They are heaviest in butter and other manufactured
dairy products. A check of 461 samples of such products in
1960 showed that a third contained residues, a situation which
the Food and Drug Administration characterized as "far from
encouraging."

To find a diet free from DDT and related chemicals, it seems
one must go to a remote and primitive land, still lacking the
amenities of civilization. Such a land appears to exist, at least
marginally, on the far Arctic shores of Alaska—although even
there one may see the approaching shadow. When scientists
investigated the native diet of the Eskimos in this region it
was found to be free from insecticides. The fresh and dried
fish; the fat, oil, or meat from beaver, beluga, caribou, moose,
oogruk, polar bear, and walrus; cranberries, salmonberries, and
wild rhubarb all had so far escaped contamination. There was
only one exception—two white owls from Point Hope carried
small amounts of DDT, perhaps acquired in the course of some
migratory journey.

When some of the Eskimos themselves were checked by
analysis of fat samples, small residues of DDT were found (0
to 1.9 parts per million). The reason for this was clear. The
fat samples were taken from people who had left their native

villages to enter the United States Public Health Service Hospital in Anchorage for surgery. There the ways of civilization prevailed, and the meals in this hospital were found to contain as much DDT as those in the most populous city. For their brief stay in civilization the Eskimos were rewarded with a taint of poison.

The fact that every meal we eat carries its load of chlorinated hydrocarbons is the inevitable consequence of the almost universal spraying or dusting of agricultural crops with these poisons. If the farmer scrupulously follows the instructions on the labels, his use of agricultural chemicals will produce no residues larger than are permitted by the Food and Drug Administration. Leaving aside for the moment the question whether these legal residues are as "safe" as they are represented to be, there remains the well-known fact that farmers very frequently exceed the prescribed dosages, use the chemical too close to the time of harvest, use several insecticides where one would do, and in other ways display the common human failure to read the fine print.

Even the chemical industry recognizes the frequent misuse of insecticides and the need for education of farmers. One of its leading trade journals recently declared that "many users do not seem to understand that they may exceed insecticide tolerances if they use higher dosages than recommended. And haphazard use of insecticides on many crops may be based on farmers' whims."

The files of the Food and Drug Administration contain records of a disturbing number of such violations. A few examples will serve to illustrate the disregard of directions: a lettuce farmer who applied not one but eight different insecticides to his crop within a short time of harvest, a shipper who had used the deadly parathion on celery in an amount five times the recommended maximum, growers using endrin—most toxic of all the chlorinated hydrocarbons—on lettuce although no residue was allowable, spinach sprayed with DDT a week before harvest.

There are also cases of chance or accidental contamination. Large lots of green coffee in burlap bags have become contaminated while being transported by vessels also carrying a cargo of insecticides. Packaged foods in warehouses are subjected to repeated aerosol treatments with DDT, lindane, and other insecticides, which may penetrate the packaging materials and occur in measurable quantities on the contained foods. The longer the food remains in storage, the greater the danger of contamination.

To the question "But doesn't the government protect us from such things?" the answer is, "Only to a limited extent." The activities of the Food and Drug Administration in the field of consumer protection against pesticides are severely limited by two facts. The first is that it has jurisdiction only over foods shipped in interstate commerce; foods grown and marketed within a state are entirely outside its sphere of authority, no matter what the violation. The second and critically limiting fact is the small number of inspectors on its staff—fewer than 600 men for all its varied work. According to a Food and Drug official, only an infinitesimal part of the crop products moving in interstate commerce—far less than 1 per cent—can be checked with existing facilities, and this is not enough to have statistical significance. As for food produced and sold within a state, the situation is even worse, for most states have woefully inadequate laws in this field.

The system by which the Food and Drug Administration establishes maximum permissible limits of contamination, called "tolerances," has obvious defects. Under the conditions prevailing it provides mere paper security and promotes a completely unjustified impression that safe limits have been established and are being adhered to. As to the safety of allowing a sprinkling of poisons on our food—a little on this, a little on that—many people contend, with highly persuasive reasons, that no poison is safe or desirable on food. In setting a tolerance level the Food and Drug Administration reviews tests of the poison on laboratory animals and then establishes a maximum level of contamination that is much less than required to produce symptoms in the test animal. This system, which is supposed to ensure safety, ignores a number of important facts. A laboratory animal, living under controlled and highly artificial conditions, consuming a given amount of a specific chemical, is very different from a human being whose exposures to pesticides are not only multiple but for the most part unknown, unmeasurable, and uncontrollable. Even if 7 parts per million of DDT on the lettuce in his luncheon salad were "safe," the meal includes other foods, each with allowable residues, and the pesticides on his food are, as we have seen, only a part, and possibly a small part, of his total exposure. This piling up of chemicals from many different sources creates a total exposure that cannot be measured. It is meaningless, therefore, to talk about the "safety" of any specific amount of residue.

And there are other defects. Tolerances have sometimes been established against the better judgment of Food and Drug Administration scientists, as in the case cited on page 195 ff., or they have been established on the basis of inadequate knowledge of the chemical concerned. Better information has led to later reduction or withdrawal of the tolerance, but only after the public has been exposed to admittedly dangerous levels of the chemical for months or years. This happened when heptachlor was given a tolerance that later had to be revoked. For some chemicals no practical field method of analysis exists before a chemical is registered for use. Inspectors are therefore frustrated in their search for residues. This difficulty greatly hampered the work on the "cranberry chemical," aminotriazole. Analytical methods are lacking, too, for certain fungicides in common use for the treatment of seeds—seeds which if unused at the end of the planting season, may very well find their way into human food.

In effect, then, to establish tolerances is to authorize contamination of public food supplies with poisonous chemicals in order that the farmer and the processor may enjoy the benefit of cheaper production—then to penalize the consumer by taxing him to maintain a policing agency to make certain that he shall not get a lethal dose. But to do the policing job properly would cost money beyond any legislator's courage to appropriate, given the present volume and toxicity of agricultural chemicals. So in the end the luckless consumer pays his taxes but gets his poisons regardless.

What is the solution? The first necessity is the elimination of tolerances on the chlorinated hydrocarbons, the organic phosphorus group, and other highly toxic chemicals. It will immediately be objected that this will place an intolerable burden on the farmer. But if, as is now the presumable goal, it is possible to use chemicals in such a way that they leave a residue of only 7 parts per million (the tolerance for DDT), or of 1 part per million (the tolerance for parathion), or even of only 0.1 part per million as is required for dieldrin on a great variety of fruits and vegetables, then why is it not possible, with only a little more care, to prevent the occurrence of any residues at all? This, in fact, is what is required for some chemicals such as heptachlor, endrin, and dieldrin on certain crops. If it is considered practical in these instances, why not for all?

But this is not a complete or final solution, for a zero tolerance on paper is of little value. At present, as we have seen, more than 99 per cent of the interstate food shipments slip by without inspection. A vigilant and aggressive Food and Drug Administration, with a greatly increased force of inspectors, is another urgent need.

This system, however—deliberately poisoning our food, then policing the result—is too reminiscent of Lewis Carroll's White Knight who thought of "a plan to dye one's whiskers green, and always use so large a fan that they could not be seen." The ultimate answer is to use less toxic chemicals so that the public hazard from their misuse is greatly reduced. Such chemicals already exist: the pyrethrins, rotenone, ryania, and others derived from plant substances. Synthetic substitutes for the pyrethrins have recently been developed so that an otherwise critical shortage can be averted. Public education as to the nature of the chemicals offered for sale is sadly needed. The average purchaser is completely bewildered by the array of available insecticides, fungicides, and weed killers, and has no way of knowing which are the deadly ones, which reasonably safe.

In addition to making this change to less dangerous agricultural pesticides, we should diligently explore the possibilities of non-chemical methods. Agricultural use of insect diseases, caused by a bacterium highly specific for certain types of insects, is already being tried in California, and more extended tests of this method are under way. A great many other possibilities exist for effective insect control by methods that will leave no residues on foods (see Chapter 17). Until a large-scale conversion to these methods has been made, we shall have little relief from a situation that, by any common-sense standards, is intolerable. As matters stand now, we are in little better position than the guests of the Borgias.

TWELVE

The Human Price

A S THE TIDE of chemicals born of the Industrial Age has arisen to engulf our environment, a drastic change has come about in the nature of the most serious public health problems. Only yesterday mankind lived in fear of the scourges of smallpox, cholera, and plague that once swept nations before them. Now our major concern is no longer with the disease organisms that once were omnipresent; sanitation, better living conditions, and new drugs have given us a high degree of control over infectious disease. Today we are concerned with a different kind of hazard that lurks in our environment—a hazard we ourselves have introduced into our world as our modern way of life has evolved.

The new environmental health problems are multiple—created by radiation in all its forms, born of the never-ending stream of chemicals of which pesticides are a part, chemicals now pervading the world in which we live, acting upon us directly and indirectly, separately and collectively. Their presence casts a shadow that is no less ominous because it is formless and obscure, no less frightening because it is simply impossible to

165

predict the effects of lifetime exposure to chemical and physical agents that are not part of the biological experience of man.

"We all live under the haunting fear that something may corrupt the environment to the point where man joins the dinosaurs as an obsolete form of life," says Dr. David Price of the United States Public Health Service. "And what makes these thoughts all the more disturbing is the knowledge that our fate could perhaps be sealed twenty or more years before the development of symptoms."

Where do pesticides fit into the picture of environmental disease? We have seen that they now contaminate soil, water, and food, that they have the power to make our streams fishless and our gardens and woodlands silent and birdless. Man, however much he may like to pretend the contrary, is part of nature. Can he escape a pollution that is now so thoroughly distributed throughout our world?

We know that even single exposures to these chemicals, if the amount is large enough, can precipitate acute poisoning. But this is not the major problem. The sudden illness or death of farmers, spraymen, pilots, and others exposed to appreciable quantities of pesticides is tragic and should not occur. For the population as a whole, we must be more concerned with the delayed effects of absorbing small amounts of the pesticides that invisibly contaminate our world.

Responsible public health officials have pointed out that the biological effects of chemicals are cumulative over long periods of time, and that the hazard to the individual may depend on the sum of the exposures received throughout his lifetime. For these very reasons the danger is easily ignored. It is human nature to shrug off what may seem to us a vague threat of future disaster. "Men are naturally most impressed by diseases which have obvious manifestations," says a wise physician, Dr. René Dubos, "yet some of their worst enemies creep on them unobtrusively."

For each of us, as for the robin in Michigan or the salmon in the Miramichi, this is a problem of ecology, of interrelationships, of interdependence. We poison the caddis flies in a stream and the salmon runs dwindle and die. We poison the gnats in a lake and the poison travels from link to link of the food chain and soon the birds of the lake margins become its victims. We spray our elms and the following springs are silent of robin song, not because we sprayed the robins directly but because the poison traveled, step by step, through the now familiar elm leaf–earthworm–robin cycle. These are matters of record, observable, part of the visible world around us. They reflect the web of life—or death—that scientists know as ecology.

But there is also an ecology of the world within our bodies. In this unseen world minute causes produce mighty effects; the effect, moreover, is often seemingly unrelated to the cause, appearing in a part of the body remote from the area where the original injury was sustained. "A change at one point, in one molecule even, may reverberate throughout the entire system to initiate changes in seemingly unrelated organs and tissues," says a recent summary of the present status of medical research. When one is concerned with the mysterious and wonderful functioning of the human body, cause and effect are seldom simple and easily demonstrated relationships. They may be widely separated both in space and time. To discover the agent of disease and death depends on a patient piecing together of many seemingly distinct and unrelated facts developed through a vast amount of research in widely separated fields.

We are accustomed to look for the gross and immediate effect and to ignore all else. Unless this appears promptly and in such obvious form that it cannot be ignored, we deny the existence of hazard. Even research men suffer from the handicap of inadequate methods of detecting the beginnings of injury. The lack of sufficiently delicate methods to detect injury before symptoms appear is one of the great unsolved problems in medicine.

"But," someone will object, "I have used dieldrin sprays on the lawn many times but I have never had convulsions like the World Health Organization spraymen—so it hasn't harmed me." It is not that simple. Despite the absence of sudden and dramatic symptoms, one who handles such materials is unquestionably storing up toxic materials in his body. Storage of the chlorinated hydrocarbons, as we have seen, is cumulative, beginning with the smallest intake. The toxic materials become lodged in all the fatty tissues of the body. When these reserves of fat are drawn upon the poison may then strike quickly. A

New Zealand medical journal recently provided an example. A man under treatment for obesity suddenly developed symptoms of poisoning. On examination his fat was found to contain stored dieldrin, which had been metabolized as he lost weight. The same thing could happen with loss of weight in illness.

The results of storage, on the other hand, could be even less obvious. Several years ago the *Journal* of the American Medical Association warned strongly of the hazards of insecticide storage in adipose tissue, pointing out that drugs or chemicals that are cumulative require greater caution than those having no tendency to be stored in the tissues. The adipose tissue, we are warned, is not merely a place for the deposition of fat (which makes up about 18 per cent of the body weight), but has many important functions with which the stored poisons may interfere. Furthermore, fats are very widely distributed in the organs and tissues of the whole body, even being constituents of cell membranes. It is important to remember, therefore, that the fat-soluble insecticides become stored in individual cells, where they are in position to interfere with the most vital and necessary functions of oxidation and energy production. This important aspect of the problem will be taken up in the next chapter.

One of the most significant facts about the chlorinated hydrocarbon insecticides is their effect on the liver. Of all organs in the body the liver is most extraordinary. In its versatility and in the indispensable nature of its functions it has no equal. It presides over so many vital activities that even the slightest damage to it is fraught with serious consequences. Not only does it provide bile for the digestion of fats, but because of its location and the special circulatory pathways that converge upon it the liver receives blood directly from the digestive tract and is deeply involved in the metabolism of all the principal foodstuffs. It stores sugar in the form of glycogen and releases it as glucose in carefully measured quantities to keep the blood sugar at a normal level. It builds body proteins, including some essential elements of blood plasma concerned with blood-clotting. It maintains cholesterol at its proper level in the blood plasma, and inactivates the male and female hormones when they reach excessive levels. It is a storehouse of many vitamins, some of which in turn contribute to its own proper functioning.

Without a normally functioning liver the body would be disarmed—defenseless against the great variety of poisons that continually invade it. Some of these are normal by-products of metabolism, which the liver swiftly and efficiently makes harmless by withdrawing their nitrogen. But poisons that have no normal place in the body may also be detoxified. The "harmless" insecticides malathion and methoxychlor are less poisonous than their relatives only because a liver enzyme deals with them, altering their molecules in such a way that their capacity for harm is lessened. In similar ways the liver deals with the majority of the toxic materials to which we are exposed.

Our line of defense against invading poisons or poisons from within is now weakened and crumbling. A liver damaged by pesticides is not only incapable of protecting us from poisons, the whole wide range of its activities may be interfered with. Not only are the consequences far-reaching, but because of their variety and the fact that they may not immediately appear they may not be attributed to their true cause.

In connection with the nearly universal use of insecticides that are liver poisons, it is interesting to note the sharp rise in hepatitis that began during the 1950's and is continuing a fluctuating climb. Cirrhosis also is said to be increasing. While it is admittedly difficult, in dealing with human beings rather than laboratory animals, to "prove" that cause A produces effect B, plain common sense suggests that the relation between a soaring rate of liver disease and the prevalence of liver poisons in the environment is no coincidence. Whether or not the chlorinated hydrocarbons are the primary cause, it seems hardly sensible under the circumstances to expose ourselves to poisons that have a proven ability to damage the liver and so presumably to make it less resistant to disease.

Both major types of insecticides, the chlorinated hydrocarbons and the organic phosphates, directly affect the nervous system, although in somewhat different ways. This has been made clear by an infinite number of experiments on animals and by observations on human subjects as well. As for DDT, the first of the new organic insecticides to be widely used, its action is primarily on the central nervous system of man; the cerebellum and the higher motor cortex are thought to be the areas chiefly affected. Abnormal sensations as of prickling, burning, or itching, as well as tremors or even convulsions may follow exposure to appreciable amounts, according to a standard textbook of toxicology.

Our first knowledge of the symptoms of acute poisoning by DDT was furnished by several British investigators, who deliberately exposed themselves in order to learn the consequences. Two scientists at the British Royal Navy Physiological Laboratory invited absorption of DDT through the skin by direct contact with walls covered with a water-soluble paint containing 2 per cent DDT, overlaid with a thin film of oil. The direct effect on the nervous system is apparent in their eloquent description of their symptoms: "The tiredness, heaviness, and aching of limbs were very real things, and the mental state was also most distressing . . . [there was] extreme irritability . . . great distaste for work of any sort . . . a feeling of mental incompetence in tackling the simplest mental task. The joint pains were quite violent at times."

Another British experimenter who applied DDT in acetone solution to his skin reported heaviness and aching of limbs, muscular weakness, and "spasms of extreme nervous tension." He took a holiday and improved, but on return to work his condition deteriorated. He then spent three weeks in bed, made miserable by constant aching in limbs, insomnia, nervous tension, and feelings of acute anxiety. On occasion tremors shook his whole body—tremors of the sort now made all too familiar by the sight of birds poisoned by DDT. The experimenter lost 10 weeks from his work, and at the end of a year, when his case was reported in a British medical journal, recovery was not complete.

(Despite this evidence, several American investigators conducting an experiment with DDT on volunteer subjects dismissed the complaint of headache and "pain in every bone" as "obviously of psychoneurotic origin.")

There are now many cases on record in which both the symptoms and the whole course of the illness point to insecticides as the cause. Typically, such a victim has had a known exposure to one of the insecticides, his symptoms have subsided under treatment which included the exclusion of all insecticides from his environment, and most significantly *have returned with each renewed contact* with the offending chemicals. This sort of evidence—and no more—forms the basis of a vast amount of medical therapy in many other disorders. There is no reason why it should not serve as a warning that it is no longer sensible to take the "calculated risk" of saturating our environment with pesticides.

Why does not everyone handling and using insecticides develop the same symptoms? Here the matter of individual sensitivity enters in. There is some evidence that women are more susceptible than men, the very young more than adults, those who lead sedentary, indoor lives more than those leading a rugged life of work or exercise in the open. Beyond these differences are others that are no less real because they are intangible. What makes one person allergic to dust or pollen, sensitive to a poison, or susceptible to an infection whereas another is not is a medical mystery for which there is at present no explanation. The problem nevertheless exists and it affects significant numbers of the population. Some physicians estimate that a third or more of their patients show signs of some form of sensitivity, and that the number is growing. And unfortunately, sensitivity may suddenly develop in a person previously insensitive. In fact, some medical men believe that intermittent exposures to chemicals may produce just such sensitivity. If this is true, it may explain why some studies on men subjected to continuous occupational exposure find little evidence of toxic effects. By their constant contact with the chemicals these men keep themselves desensitized—as an allergist keeps his patients desensitized by repeated small injections of the allergen.

The whole problem of pesticide poisoning is enormously complicated by the fact that a human being, unlike a laboratory animal living under rigidly controlled conditions, is never exposed to one chemical alone. Between the major groups of insecticides, and between them and other chemicals, there are interactions that have serious potentials. Whether released into soil or water or a man's blood, these unrelated chemicals do not remain segregated; there are mysterious and unseen changes by which one alters the power of another for harm.

There is interaction even between the two major groups of insecticides usually thought to be completely distinct in their action. The power of the organic phosphates, those poisoners of the nerve-protective enzyme cholinesterase, may become greater if the body has first been exposed to a chlorinated hydrocarbon which injures the liver. This is because, when liver function is disturbed, the cholinesterase level drops below normal. The added depressive effect of the organic phosphate may then be enough to precipitate acute symptoms. And as we have seen, pairs of the organic phosphates themselves may interact in such a way as to increase their toxicity a hundredfold. Or the organic phosphates may interact with various drugs, or with synthetic materials, food additives—who can say what else of the infinite number of man-made substances that now pervade our world?

The effect of a chemical of supposedly innocuous nature can be drastically changed by the action of another; one of the best examples is a close relative of DDT called methoxychlor. (Actually, methoxychlor may not be as free from dangerous qualities as it is generally said to be, for recent work on experimental animals shows a direct action on the uterus and a blocking effect on some of the powerful pituitary hormones—reminding us again that these are chemicals with enormous biologic effect. Other work shows that methoxychlor has a potential ability to damage the kidneys.) Because it is not stored to any great extent when given alone, we are told that methoxychlor is a safe chemical. But this is not necessarily true. If the liver has been damaged by another agent, methoxychlor is stored in the body at *100 times* its normal rate, and will then imitate the effects of DDT with long-lasting effects on the nervous system. Yet the liver damage that brings this about might be so slight as to pass unnoticed. It might have been the result of any of a number of commonplace situations—using another insecticide, using a cleaning fluid containing carbon tetrachloride, or taking one of the so-called tranquilizing drugs, a number (but not all) of which are chlorinated hydrocarbons and possess power to damage the liver.

Damage to the nervous system is not confined to acute poisoning; there may also be delayed effects from exposure. Long-lasting damage to brain or nerves has been reported for methoxychlor and others. Dieldrin, besides its immediate consequences, can have long delayed effects ranging from "loss of memory, insomnia, and nightmares to mania." Lindane, according to medical findings, is stored in significant amounts in the brain and functioning liver tissue and may induce "profound and long lasting effects on the central nervous system." Yet this chemical, a form of benzene hexachloride, is much used in vaporizers, devices that pour a stream of volatilized insecticide vapor into homes, offices, restaurants.

The organic phosphates, usually considered only in relation to their more violent manifestations in acute poisoning, also have the power to produce lasting physical damage to nerve tissues and, according to recent findings, to induce mental disorders. Various cases of delayed paralysis have followed use of one or another of these insecticides. A bizarre happening in the United States during the prohibition era about 1930 was an omen of things to come. It was caused not by an insecticide but by a substance belonging chemically to the same group as the organic phosphate insecticides. During that period some medicinal substances were being pressed into service as substitutes for liquor, being exempt from the prohibition law. One of these was Jamaica ginger. But the *United States Pharmacopeia* product was expensive, and bootleggers conceived the idea of making a substitute Jamaica ginger. They succeeded so well that their spurious product responded to the appropriate chemical tests and deceived the government chemists. To give their false ginger the necessary tang they had introduced a chemical known as triorthocresyl phosphate. This chemical, like parathion and its relatives, destroys the protective enzyme cholinesterase. As a consequence of drinking the bootleggers' product some 15,000 people developed a permanently crippling type of paralysis of the leg muscles, a condition now called "ginger paralysis." The paralysis was accompanied by destruction of the nerve sheaths and by degeneration of the cells of the anterior horns of the spinal cord.

About two decades later various other organic phosphates came into use as insecticides, as we have seen, and soon cases reminiscent of the ginger paralysis episode began to occur. One was a greenhouse worker in Germany who became paralyzed several months after experiencing mild symptoms of poisoning on a few occasions after using parathion. Then a group of three chemical plant workers developed acute poisoning from exposure to other insecticides of this group. They recovered under treatment, but ten days later two of them developed muscular weakness in the legs. This persisted for 10 months in one; the other, a young woman chemist, was more severely affected, with paralysis in both legs and some involvement of the hands and arms. Two years later when her case was reported in a medical journal she was still unable to walk.

The insecticide responsible for these cases has been withdrawn from the market, but some of those now in use may be capable of like harm. Malathion (beloved of gardeners) has induced severe muscular weakness in experiments on chickens. This was attended (as in ginger paralysis) by destruction of the sheaths of the sciatic and spinal nerves.

All these consequences of organic phosphate poisoning, if survived, may be a prelude to worse. In view of the severe damage they inflict upon the nervous system, it was perhaps inevitable that these insecticides would eventually be linked with mental disease. That link has recently been supplied by investigators at the University of Melbourne and Prince Henry's Hospital in Melbourne, who reported on 16 cases of mental disease. All had a history of prolonged exposure to organic phosphorus insecticides. Three were scientists checking the efficacy of sprays; 8 worked in greenhouses; 5 were farm workers. Their symptoms ranged from impairment of memory to schizophrenic and depressive reactions. All had normal medical histories before the chemicals they were using boomeranged and struck them down.

Echoes of this sort of thing are to be found, as we have seen, widely scattered throughout medical literature, sometimes involving the chlorinated hydrocarbons, sometimes the organic phosphates. Confusion, delusions, loss of memory, mania—a heavy price to pay for the temporary destruction of a few insects, but a price that will continue to be exacted as long as we insist upon using chemicals that strike directly at the nervous system.

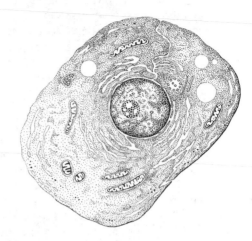

THIRTEEN

Through a Narrow Window

THE BIOLOGIST George Wald once compared his work on an exceedingly specialized subject, the visual pigments of the eye, to "a very narrow window through which at a distance one can see only a crack of light. As one comes closer the view grows wider and wider, until finally through this same narrow window one is looking at the universe."

So it is that only when we bring our focus to bear, first on the individual cells of the body, then on the minute structures within the cells, and finally on the ultimate reactions of molecules within these structures—only when we do this can we comprehend the most serious and far-reaching effects of the haphazard introduction of foreign chemicals into our internal environment. Medical research has only rather recently turned to the functioning of the individual cell in producing the energy that is the indispensable quality of life. The extraordinary energy-producing mechanism of the body is basic not only to health but to life; it transcends in importance even the most vital organs, for without the smooth and effective functioning of energy-yielding oxidation none of the body's functions can be performed. Yet the nature of many of the chemicals used against insects, rodents, and weeds is such that they may strike directly at this system, disrupting its beautifully functioning mechanism.

The research that led to our present understanding of cellular oxidation is one of the most impressive accomplishments in all biology and biochemistry. The roster of contributors to this work includes many Nobel Prize winners. Step by step it has been going on for a quarter of a century, drawing on even earlier work for some of its foundation stones. Even yet it is not complete in all details. And only within the past decade have all the varied pieces of research come to form a whole so that biological oxidation could become part of the common knowledge of biologists. Even more important is the fact that medical men who received their basic training before 1950 have had little opportunity to realize the critical importance of the process and the hazards of disrupting it.

The ultimate work of energy production is accomplished not in any specialized organ but in every cell of the body. A living cell, like a flame, burns fuel to produce the energy on which life depends. The analogy is more poetic than precise, for the cell accomplishes its "burning" with only the moderate heat of the body's normal temperature. Yet all these billions of gently burning little fires spark the energy of life. Should they cease to burn, "no heart could beat, no plant could grow upward defying gravity, no amoeba could swim, no sensation could speed along a nerve, no thought could flash in the human brain," said the chemist Eugene Rabinowitch.

The transformation of matter into energy in the cell is an ever-flowing process, one of nature's cycles of renewal, like a wheel endlessly turning. Grain by grain, molecule by molecule, carbohydrate fuel in the form of glucose is fed into this wheel; in its cyclic passage the fuel molecule undergoes fragmentation and a series of minute chemical changes. The changes are made in orderly fashion, step by step, each step directed and controlled by an enzyme of so specialized a function that it does this one thing and nothing else. At each step energy is produced, waste products (carbon dioxide and water) are given off, and the altered molecule of fuel is passed on to the next stage. When the turning wheel comes full cycle the fuel molecule has been stripped down to a form in which it is ready to combine with a new molecule coming in and to start the cycle anew.

This process by which the cell functions as a chemical factory is one of the wonders of the living world. The fact that all the functioning parts are of infinitesimal size adds to the miracle. With few exceptions cells themselves are minute, seen only with the aid of a microscope. Yet the greater part of the work of oxidation is performed in a theater far smaller, in tiny granules within the cell called mitochondria. Although known for more than 60 years, these were formerly dismissed as cellular elements of unknown and probably unimportant function. Only in the 1950's did their study become an exciting and fruitful field of research; suddenly they began to engage so much attention that 1000 papers on this subject alone appeared within a five-year period.

Again one stands in awe at the marvelous ingenuity and patience by which the mystery of the mitochondria has been solved. Imagine a particle so small that you can barely see it even though a microscope has enlarged it for you 300 times. Then imagine the skill required to isolate this particle, to take it apart and analyze its components and determine their highly complex functioning. Yet this has been done with the aid of the electron microscope and the techniques of the biochemist.

It is now known that the mitochondria are tiny packets of enzymes, a varied assortment including all the enzymes necessary for the oxidative cycle, arranged in precise and orderly array on walls and partitions. The mitochondria are the "powerhouses" in which most of the energy-producing reactions occur. After the first, preliminary steps of oxidation have been performed in the cytoplasm the fuel molecule is taken into the mitochondria. It is here that oxidation is completed; it is here that enormous amounts of energy are released.

The endlessly turning wheels of oxidation within the mitochondria would turn to little purpose if it were not for this all-important result. The energy produced at each stage of the oxidative cycle is in a form familiarly spoken of by the biochemists as ATP (adenosine triphosphate), a molecule containing three phosphate groups. The role of ATP in furnishing energy comes from the fact that it can transfer one of its phosphate groups to other substances, along with the energy of its bonds of electrons shuttling back and forth at high speed. Thus, in a muscle cell, energy to contract is gained when a terminal phosphate group is transferred to the contracting muscle. So another cycle takes place—a cycle within a cycle: a molecule of ATP gives up one of its phosphate groups and retains only two, becoming a diphosphate molecule, ADP. But as the wheel turns further another phosphate group is coupled on and the potent ATP is restored. The analogy of the storage battery has been used: ATP represents the charged, ADP the discharged battery.

ATP is the universal currency of energy—found in all organisms from microbes to man. It furnishes mechanical energy to muscle cells; electrical energy to nerve cells. The sperm cell, the fertilized egg ready for the enormous burst of activity that will transform it into a frog or a bird or a human infant, the cell that must create a hormone, all are supplied with ATP. Some of the energy of ATP is used in the mitochondrion but most of it is immediately dispatched into the cell to provide power for other activities. The location of the mitochondria within certain cells is eloquent of their function, since they are placed so that energy can be delivered precisely where it is needed. In muscle cells they cluster around contracting fibers; in nerve cells they are found at the junction with another cell, supplying energy for the transfer of impulses; in sperm cells they are concentrated at the point where the propellant tail is joined to the head.

The charging of the battery, in which ADP and a free phosphate group are combined to restore ATP, is coupled to the oxidative process; the close linking is known as coupled phosphorylation. If the combination becomes uncoupled, the means is lost for providing usable energy. Respiration continues but no energy is produced. The cell has become like a racing engine, generating heat but yielding no power. Then the muscle cannot contract, nor can the impulse race along the nerve pathways. Then the sperm cannot move to its destination; the fertilized egg cannot carry to completion its complex divisions and elaborations. The consequences of uncoupling could indeed be disastrous for any organism from embryo to adult: in time it could lead to the death of the tissue or even of the organism.

How can uncoupling be brought about? Radiation is an uncoupler, and the death of cells exposed to radiation is thought by some to be brought about in this way. Unfortunately, a good many chemicals also have the power to separate oxidation from energy production, and the insecticides and weed killers are well represented on the list. The phenols, as we have seen, have a strong effect on metabolism, causing a potentially fatal rise in temperature; this is brought about by the "racing engine" effect of uncoupling. The dinitrophenols and pentachlorophenols are examples of this group that have widespread use as herbicides. Another uncoupler among the herbicides is 2,4-D. Of the chlorinated hydrocarbons, DDT is a proven uncoupler and further study will probably reveal others among this group.

But uncoupling is not the only way to extinguish the little fires in some or all of the body's billions of cells. We have seen that each step in oxidation is directed and expedited by a specific enzyme. When any of these enzymes—even a single one of them—is destroyed or weakened, the cycle of oxidation within the cell comes to a halt. It makes no difference which enzyme is affected. Oxidation progresses in a cycle like a turning wheel. If we thrust a crowbar between the spokes of a wheel it makes no difference where we do it, the wheel stops turning. In the same way, if we destroy an enzyme that functions at any point in the cycle, oxidation ceases. There is then no further energy production, so the end effect is very similar to uncoupling.

The crowbar to wreck the wheels of oxidation can be supplied by any of a number of chemicals commonly used as pesticides. DDT, methoxychlor, malathion, phenothiazine, and various dinitro compounds are among the numerous pesticides that have been found to inhibit one or more of the enzymes concerned in the cycle of oxidation. They thus appear as agents potentially capable of blocking the whole process of energy production and depriving the cells of utilizable oxygen. This is an injury with most disastrous consequences, only a few of which can be mentioned here.

Merely by systematically withholding oxygen, experimenters have caused normal cells to turn into cancer cells, as we shall see in the following chapter. Some hint of other drastic consequences of depriving a cell of oxygen can be seen in animal experiments on developing embryos. With insufficient oxygen the orderly processes by which the tissues unfold and the organs develop are disrupted; malformations and other abnormalities then occur. Presumably the human embryo deprived of oxygen may also develop congenital deformities.

There are signs that an increase in such disasters is being noticed, even though few look far enough to find all of the causes. In one of the more unpleasant portents of the times, the Office of Vital Statistics in 1961 initiated a national tabulation of malformations at birth, with the explanatory comment that the resulting statistics would provide needed facts on the incidence of congenital malformations and the circumstances under which they occur. Such studies will no doubt be directed largely toward measuring the effects of radiation, but it must not be overlooked that many chemicals are the partners of radiation, producing precisely the same effects. Some of the defects and malformations in tomorrow's children, grimly anticipated by the Office of Vital Statistics, will almost certainly be caused by these chemicals that permeate our outer and inner worlds.

It may well be that some of the findings about diminished reproduction are also linked with interference with biological oxidation, and consequent depletion of the all-important storage batteries of ATP. The egg, even before fertilization, needs to be generously supplied with ATP, ready and waiting for the enormous effort, the vast expenditure of energy that will be required once the sperm has entered and fertilization has occurred. Whether the sperm cell will reach and penetrate the egg depends upon its own supply of ATP, generated in the mitochondria thickly clustered in the neck of the cell. Once fertilization is accomplished and cell division has begun, the supply of energy in the form of ATP will largely determine whether the development of the embryo will proceed to completion. Embryologists studying some of their most convenient subjects, the eggs of frogs and of sea urchins, have found that if the ATP content is reduced below a certain critical level the egg simply stops dividing and soon dies.

It is not an impossible step from the embryology laboratory to the apple tree where a robin's nest holds its complement of blue-green eggs; but the eggs lie cold, the fires of life that flickered for a few days now extinguished. Or to the top of a tall Florida pine where a vast pile of twigs and sticks in ordered disorder holds three large white eggs, cold and lifeless. Why did the robins and the eaglets not hatch? Did the eggs of the birds, like those of the laboratory frogs, stop developing simply because they lacked enough of the common currency of energy—the ATP molecules—to complete their development? And was the lack of ATP brought about because in the body of the parent birds and in the eggs there were stored enough insecticides to stop the little turning wheels of oxidation on which the supply of energy depends?

It is no longer necessary to guess about the storage of insecticides in the eggs of birds, which obviously lend themselves to this kind of observation more readily than the mammalian ovum. Large residues of DDT and other hydrocarbons have been found whenever looked for in the eggs of birds subjected to these chemicals, either experimentally or in the wild. And the concentrations have been heavy. Pheasant eggs in a California experiment contained up to 349 parts per million of DDT. In Michigan, eggs taken from the oviducts of robins dead of DDT poisoning showed concentrations up to 200 parts per million. Other eggs were taken from nests left unattended as parent robins were stricken with poison; these too contained DDT. Chickens poisoned by aldrin used on a neighboring farm have passed on the chemical to their eggs; hens experimentally fed DDT laid eggs containing as much as 65 parts per million.

Knowing that DDT and other (perhaps all) chlorinated hydrocarbons stop the energy-producing cycle by inactivating a specific enzyme or uncoupling the energy-producing mechanism, it is hard to see how any egg so loaded with residues could complete the complex process of development: the infinite number of cell divisions, the elaboration of tissues and organs, the synthesis of vital substances that in the end produce a living creature. All this requires vast amounts of energy—the little packets of ATP which the turning of the metabolic wheel alone can produce.

There is no reason to suppose these disastrous events are confined to birds. ATP is the universal currency of energy, and the metabolic cycles that produce it turn to the same purpose in birds and bacteria, in men and mice. The fact of insecticide storage in the germ cells of any species should therefore disturb us, suggesting comparable effects in human beings.

And there are indications that these chemicals lodge in tissues concerned with the manufacture of germ cells as well as in the cells themselves. Accumulations of insecticides have been discovered in the sex organs of a variety of birds and mammals—in pheasants, mice, and guinea pigs under controlled conditions, in robins in an area sprayed for elm disease, and in deer roaming western forests sprayed for spruce budworm. In one of the robins the concentration of DDT in the testes was heavier than in any other part of the body. Pheasants also accumulated extraordinary amounts in the testes, up to 1500 parts per million.

Probably as an effect of such storage in the sex organs, atrophy of the testes has been observed in experimental mammals. Young rats exposed to methoxychlor had extraordinarily small testes. When young roosters were fed DDT, the testes made only 18 per cent of their normal growth; combs and wattles, dependent for their development upon the testicular hormone, were only a third the normal size.

The spermatozoa themselves may well be affected by loss of ATP. Experiments show that the motility of bull sperm is decreased by dinitrophenol, which interferes with the energy-coupling mechanism with inevitable loss of energy. The same effect would probably be found with other chemicals were the matter investigated. Some indication of the possible effect on human beings is seen in medical reports of oligospermia, or reduced production of spermatozoa, among aviation crop dusters applying DDT.

For mankind as a whole, a possession infinitely more valuable than individual life is our genetic heritage, our link with past and future. Shaped through long eons of evolution, our genes not only make us what we are, but hold in their minute beings the future—be it one of promise or threat. Yet genetic deterioration through man-made agents is the menace of our time, "the last and greatest danger to our civilization."

Again the parallel between chemicals and radiation is exact and inescapable.

The living cell assaulted by radiation suffers a variety of injuries: its ability to divide normally may be destroyed, it may suffer changes in chromosome structure, or the genes, carriers of hereditary material, may undergo those sudden changes known as mutations, which cause them to produce new characteristics in succeeding generations. If especially susceptible the cell may be killed outright, or finally, after the passage of time measured in years, it may become malignant.

All these consequences of radiation have been duplicated in laboratory studies by a large group of chemicals known as radiomimetic or radiation-imitating. Many chemicals used as pesticides—herbicides as well as insecticides—belong to this group of substances that have the ability to damage the chromosomes, interfere with normal cell division, or cause mutations. These injuries to the genetic material are of a kind that may lead to disease in the individual exposed or they may make their effects felt in future generations.

Only a few decades ago, no one knew these effects of either radiation or chemicals. In those days the atom had not been split and few of the chemicals that were to duplicate radiation had as yet been conceived in the test tubes of chemists. Then in 1927, a professor of zoology in a Texas university, Dr. H. J. Muller, found that by exposing an organism to X-radiation, he could produce mutations in succeeding generations. With Muller's discovery a vast new field of scientific and medical knowledge was opened up. Muller later received the Nobel Prize in Medicine for his achievement, and in a world that soon gained unhappy familiarity with the gray rains of fallout, even the nonscientist now knows the potential results of radiation.

Although far less noticed, a companion discovery was made by Charlotte Auerbach and William Robson at the University of Edinburgh in the early 1940's. Working with mustard gas, they found that this chemical produces permanent chromosome abnormalities that cannot be distinguished from those induced by radiation. Tested on the fruit fly, the same organism Muller had used in his original work with X-rays, mustard gas also produced mutations. Thus the first chemical mutagen was discovered.

Mustard gas as a mutagen has now been joined by a long list of other chemicals known to alter genetic material in plants and animals. To understand how chemicals can alter the course of heredity, we must first watch the basic drama of life as it is played on the stage of the living cell.

The cells composing the tissues and organs of the body must have the power to increase in number if the body is to grow and if the stream of life is to be kept flowing from generation to generation. This is accomplished by the process of mitosis, or nuclear division. In a cell that is about to divide, changes of the utmost importance occur, first within the nucleus, but eventually involving the entire cell. Within the nucleus, the chromosomes mysteriously move and divide, ranging themselves in age-old patterns that will serve to distribute the determiners of heredity, the genes, to the daughter cells. First they assume the form of elongated threads, on which the genes are aligned, like beads on a string. Then each chromosome divides lengthwise (the genes dividing also). When the cell divides into two, half of each goes to each of the daughter cells. In this way each new cell will contain a complete set of chromosomes, and all the genetic information encoded within them. In this way the integrity of the race and of the species is preserved; in this way like begets like.

A special kind of cell division occurs in the formation of the germ cells. Because the chromosome number for a given species is constant, the egg and the sperm, which are to unite to form a new individual, must carry to their union only half the species number. This is accomplished with extraordinary precision by a change in the behavior of the chromosomes that occurs at one of the divisions producing those cells. At this time the chromosomes do not split, but one whole chromosome of each pair goes into each daughter cell.

In this elemental drama all life is revealed as one. The events of the process of cell division are common to all earthly life; neither man nor amoeba, the giant sequoia nor the simple yeast cell can long exist without carrying on this process of cell division. Anything that disturbs mitosis is therefore a grave threat to the welfare of the organism affected and to its descendants.

"The major features of cellular organization, including, for instance, mitosis, must be much older than 500 million years—more nearly 1000 million," wrote George Gaylord Simpson and his colleagues Pittendrigh and Tiffany in their broadly encompassing book entitled *Life*. "In this sense the world of life, while surely fragile and complex, is incredibly durable through time—more durable than mountains. This durability is wholly

dependent on the almost incredible accuracy with which the inherited information is copied from generation to generation."

But in all the thousand million years envisioned by these
authors no threat has struck so directly and so forcefully at
that "incredible accuracy" as the mid-20th century threat of
man-made radiation and man-made and man-disseminated
chemicals. Sir Macfarlane Burnet, a distinguished Australian
physician and a Nobel Prize winner, considers it "one of the
most significant medical features" of our time that, "as a
by-product of more and more powerful therapeutic procedures
and the production of chemical substances outside of biological
experiences, the normal protective barriers that kept mutagenic
agents from the internal organs have been more and more frequently penetrated."

The study of human chromosomes is in its infancy, and so
it has only recently become possible to study the effect of environmental factors upon them. It was not until 1956 that new
techniques made it possible to determine accurately the number of chromosomes in the human cell—46—and to observe
them in such detail that the presence or absence of whole chromosomes or even parts of chromosomes could be detected.
The whole concept of genetic damage by something in the
environment is also relatively new, and is little understood except by the geneticists, whose advice is too seldom sought. The
hazard from radiation in its various forms is now reasonably
well understood—although still denied in surprising places. Dr.
Muller has frequently had occasion to deplore the "resistance
to the acceptance of genetic principles on the part of so many,
not only of governmental appointees in the policy-making positions, but also of so many of the medical profession." The fact
that chemicals may play a role similar to radiation has scarcely
dawned on the public mind, nor on the minds of most medical
or scientific workers. For this reason the role of chemicals in
general use (rather than in laboratory experiments) has not yet
been assessed. It is extremely important that this be done.

Sir Macfarlane is not alone in his estimate of the potential
danger. Dr. Peter Alexander, an outstanding British authority,
has said that the radiomimetic chemicals "may well represent a
greater danger" than radiation. Dr. Muller, with the perspective gained by decades of distinguished work in genetics, warns

that various chemicals (including groups represented by pesticides) "can raise the mutation frequency as much as radiation. . . . As yet far too little is known of the extent to which our genes, under modern conditions of exposure to unusual chemicals, are being subjected to such mutagenic influences."

The widespread neglect of the problem of chemical mutagens is perhaps due to the fact that those first discovered were of scientific interest only. Nitrogen mustard, after all, is not sprayed upon whole populations from the air; its use is in the hands of experimental biologists or of physicians who use it in cancer therapy. (A case of chromosome damage in a patient receiving such therapy has recently been reported.) But insecticides and weed killers *are* brought into intimate contact with large numbers of people.

Despite the scant attention that has been given to the matter, it is possible to assemble specific information on a number of these pesticides, showing that they disturb the cell's vital processes in ways ranging from slight chromosome damage to gene mutation, and with consequences extending to the ultimate disaster of malignancy.

Mosquitoes exposed to DDT for several generations turned into strange creatures called gynandromorphs—part male and part female.

Plants treated with various phenols suffered profound destruction of chromosomes, changes in genes, a striking number of mutations, "irreversible hereditary changes." Mutations also occurred in fruit flies, the classic subject of genetics experiments, when subjected to phenol; these flies developed mutations so damaging as to be fatal on exposure to one of the common herbicides or to urethane. Urethane belongs to the group of chemicals called carbamates, from which an increasing number of insecticides and other agricultural chemicals are drawn. Two of the carbamates are actually used to prevent sprouting of potatoes in storage—precisely because of their proven effect in stopping cell division. One of these, maleic hydrazide, is rated a powerful mutagen.

Plants treated with benzene hexachloride (BHC) or lindane became monstrously deformed with tumorlike swellings on their roots. Their cells grew in size, being swollen with chromosomes which doubled in number. The doubling continued in future divisions until further cell division became mechanically impossible.

The herbicide 2,4-D has also produced tumorlike swellings in treated plants. Chromosomes become short, thick, clumped together. Cell division is seriously retarded. The general effect is said to parallel closely that produced by X-rays.

These are but a few illustrations; many more could be cited. As yet there has been no comprehensive study aimed at testing the mutagenic effects of pesticides as such. The facts cited above are by-products of research in cell physiology or genetics. What is urgently needed is a direct attack on the problem.

Some scientists who are willing to concede the potent effect of environmental radiation on man nevertheless question whether mutagenic chemicals can, as a practical proposition, have the same effect. They cite the great penetrating power of radiation, but doubt that chemicals could reach the germ cells. Once again we are hampered by the fact that there has been little direct investigation of the problem in man. However, the finding of large residues of DDT in the gonads and germ cells of birds and mammals is strong evidence that the chlorinated hydrocarbons, at least, not only become widely distributed throughout the body but come into contact with genetic materials. Professor David E. Davis at Pennsylvania State University has recently discovered that a potent chemical which prevents cells from dividing and has had limited use in cancer therapy can also be used to cause sterility in birds. Sublethal levels of the chemical halt cell division in the gonads. Professor Davis has had some success in field trials. Obviously, then, there is little basis for the hope or belief that the gonads of any organism are shielded from chemicals in the environment.

Recent medical findings in the field of chromosome abnormalities are of extreme interest and significance. In 1959 several British and French research teams found their independent studies pointing to a common conclusion—that some of humanity's ills are caused by a disturbance of the normal chromosome number. In certain diseases and abnormalities studied by these investigators the number differed from the normal. To illustrate: it is now known that all typical mongoloids have one extra chromosome. Occasionally this is attached to another so that the chromosome number remains the normal 46. As a rule, however, the extra is a separate chromosome, making the number 47. In such individuals, the original cause of the defect must have occurred in the generation preceding its appearance.

A different mechanism seems to operate in a number of patients, both in America and Great Britain, who are suffering from a chronic form of leukemia. These have been found to have a consistent chromosome abnormality in some of the blood cells. The abnormality consists of the loss of part of a chromosome. In these patients the skin cells have a normal complement of chromosomes. This indicates that the chromosome defect did not occur in the germ cells that gave rise to these individuals, but represents damage to particular cells (in this case, the precursors of blood cells) that occurred during the life of the individual. The loss of part of a chromosome has perhaps deprived these cells of their "instructions" for normal behavior.

The list of defects linked to chromosome disturbances has grown with surprising speed since the opening of this territory, hitherto beyond the boundaries of medical research. One, known only as Klinefelter's syndrome, involves a duplication of one of the sex chromosomes. The resulting individual is a male, but because he carries two of the X chromosomes (becoming XXY instead of XY, the normal male complement) he is somewhat abnormal. Excessive height and mental defects often accompany the sterility caused by this condition. In contrast, an individual who receives only one sex chromosome (becoming XO instead of either XX or XY) is actually female but lacks many of the secondary sexual characteristics. The condition is accompanied by various physical (and sometimes mental) defects, for of course the X chromosome carries genes for a variety of characteristics. This is known as Turner's syndrome. Both conditions had been described in medical literature long before the cause was known.

An immense amount of work on the subject of chromosome abnormalities is being done by workers in many countries. A group at the University of Wisconsin, headed by Dr. Klaus Patau, has been concentrating on a variety of congenital abnormalities, usually including mental retardation, that seem to result from the duplication of only part of a chromosome, as if somewhere in the formation of one of the germ cells a chromosome had broken and the pieces had not been properly redistributed. Such a mishap is likely to interfere with the normal development of the embryo.

According to present knowledge, the occurrence of an entire extra body chromosome is usually lethal, preventing survival of the embryo. Only three such conditions are known to be viable; one of them, of course, is mongolism. The presence of an extra attached fragment, on the other hand, although seriously damaging is not necessarily fatal, and according to the Wisconsin investigators this situation may well account for a substantial part of the so far unexplained cases in which a child is born with multiple defects, usually including mental retardation.

This is so new a field of study that as yet scientists have been more concerned with identifying the chromosome abnormalities associated with disease and defective development than with speculating about the causes. It would be foolish to assume that any single agent is responsible for damaging the chromosomes or causing their erratic behavior during cell division. But can we afford to ignore the fact that we are now filling the environment with chemicals that have the power to strike directly at the chromosomes, affecting them in the precise ways that could cause such conditions? Is this not too high a price to pay for a sproutless potato or a mosquitoless patio?

We can, if we wish, reduce this threat to our genetic heritage, a possession that has come down to us through some two billion years of evolution and selection of living protoplasm, a possession that is ours for the moment only, until we must pass it on to generations to come. We are doing little now to preserve its integrity. Although chemical manufacturers are required by law to test their materials for toxicity, they are not required to make the tests that would reliably demonstrate genetic effect, and they do not do so.

FOURTEEN

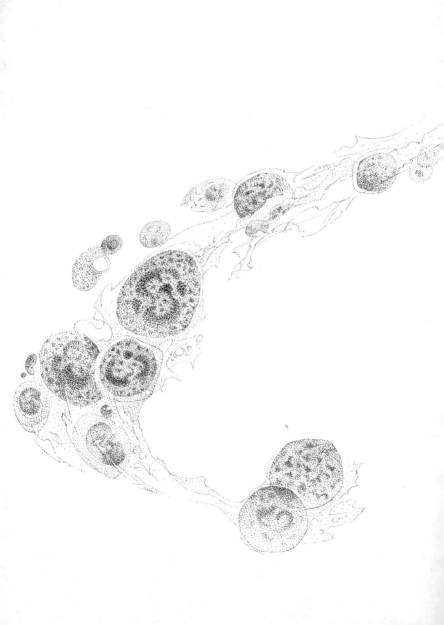

One in Every Four

THE BATTLE of living things against cancer began so long ago that its origin is lost in time. But it must have begun in a natural environment, in which whatever life inhabited the earth was subjected, for good or ill, to influences that had their origin in sun and storm and the ancient nature of the earth. Some of the elements of this environment created hazards to which life had to adjust or perish. The ultraviolet radiation in sunlight could cause malignancy. So could radiations from certain rocks, or arsenic washed out of soil or rocks to contaminate food or water supplies.

The environment contained these hostile elements even before there was life; yet life arose, and over the millions of years it came to exist in infinite numbers and endless variety. Over the eons of unhurried time that is nature's, life reached an adjustment with destructive forces as selection weeded out the less adaptable and only the most resistant survived. These natural cancer-causing agents are still a factor in producing malignancy; however, they are few in number and they belong to that ancient array of forces to which life has been accustomed from the beginning.

With the advent of man the situation began to change, for man, alone of all forms of life, can *create* cancer-producing substances, which in medical terminology are called carcinogens. A few man-made carcinogens have been part of the environment for centuries. An example is soot, containing aromatic hydrocarbons. With the dawn of the industrial era the world became a place of continuous, ever-accelerating change. Instead of the natural environment there was rapidly substituted an artificial one composed of new chemical and physical agents, many of them possessing powerful capacities for inducing biologic change. Against these carcinogens which his own activities had created man had no protection, for even as his biological heritage has evolved slowly, so it adapts slowly to new conditions. As a result these powerful substances could easily penetrate the inadequate defenses of the body.

The history of cancer is long, but our recognition of the agents that produce it has been slow to mature. The first awareness that external or environmental agents could produce malignant change dawned in the mind of a London physician nearly two centuries ago. In 1775 Sir Percival Pott declared that the scrotal cancer so common among chimney sweeps must be caused by the soot that accumulated on their bodies. He could not furnish the "proof" we would demand today, but modern research methods have now isolated the deadly chemical in soot and proved the correctness of his perception.

For a century or more after Pott's discovery there seems to have been little further realization that certain of the chemicals in the human environment could cause cancer by repeated skin contact, inhalation, or swallowing. True, it had been noticed that skin cancer was prevalent among workers exposed to arsenic fumes in copper smelters and tin foundries in Cornwall and Wales. And it was realized that workers in the cobalt mines in Saxony and in the uranium mines at Joachimsthal in Bohemia were subject to a disease of the lungs, later identified as cancer. But these were phenomena of the pre-industrial era, before the flowering of the industries whose products were to pervade the environment of almost every living thing.

The first recognition of malignancies traceable to the age of industry came during the last quarter of the 19th century. About the time that Pasteur was demonstrating the microbial origin of many infectious diseases, others were discovering the chemical origin of cancer—skin cancers among workers in the new lignite industry in Saxony and in the Scottish shale industry, along with other cancers caused by occupational exposure to tar and pitch. By the end of the 19th century a half-dozen sources of industrial carcinogens were known; the 20th century was to create countless new cancer-causing chemicals and to bring the general population into intimate contact with them. In the less than two centuries intervening since the work of Pott, the environmental situation has been vastly changed. No longer are exposures to dangerous chemicals occupational alone; they have entered the environment of everyone—even of children as yet unborn. It is hardly surprising, therefore, that we are now aware of an alarming increase in malignant disease.

The increase itself is no mere matter of subjective impressions. The monthly report of the Office of Vital Statistics for July 1959 states that malignant growths, including those of the lymphatic and blood-forming tissues, accounted for 15 per cent of the deaths in 1958 compared with only 4 per cent in 1900. Judging by the present incidence of the disease, the American Cancer Society estimates that 45,000,000 Americans now living will eventually develop cancer. This means that malignant disease will strike two out of three families.

The situation with respect to children is even more deeply disturbing. A quarter century ago, cancer in children was considered a medical rarity. *Today, more American school children die of cancer than from any other disease.* So serious has this situation become that Boston has established the first hospital in the United States devoted exclusively to the treatment of children with cancer. Twelve per cent of all deaths in children between the ages of one and fourteen are caused by cancer. Large numbers of malignant tumors are discovered clinically in children under the age of five, but it is an even grimmer fact that significant numbers of such growths are present at or before birth. Dr. W. C. Hueper of the National Cancer Institute, a foremost authority on environmental cancer, has suggested that congenital cancers and cancers in infants may be related to the action of cancer-producing agents to which the mother has been exposed during pregnancy and which penetrate the placenta to act on the rapidly developing fetal tissues. Experiments show that the younger the animal is when it is subjected to a cancer-producing agent the more certain is the production of cancer. Dr. Francis Ray of the University of Florida has warned that "we may be initiating cancer in the children of today by the addition of chemicals [to food] . . . We will not know, perhaps for a generation or two, what the effects will be."

The problem that concerns us here is whether any of the chemicals we are using in our attempts to control nature play a direct or indirect role as causes of cancer. In terms of evidence gained from animal experiments we shall see that five or possibly six of the pesticides must definitely be rated as carcinogens.

The list is greatly lengthened if we add those considered by some physicians to cause leukemia in human patients. Here the evidence is circumstantial, as it must be since we do not experiment on human beings, but it is nonetheless impressive. Still other pesticides will be added as we include those whose action on living tissues or cells may be considered an indirect cause of malignancy.

One of the earliest pesticides associated with cancer is arsenic, occurring in sodium arsenite as a weed killer, and in calcium arsenate and various other compounds as insecticides. The association between arsenic and cancer in man and animals is historic. A fascinating example of the consequences of exposure to arsenic is related by Dr. Hueper in his *Occupational Tumors*, a classic monograph on the subject. The city of Reichenstein in Silesia had been for almost a thousand years the site of mining for gold and silver ores, and for several hundred years for arsenic ores. Over the centuries arsenic wastes accumulated in the vicinity of the mine shafts and were picked up by streams coming down from the mountains. The underground water also became contaminated, and arsenic entered the drinking water. For centuries many of the inhabitants of this region suffered from what came to be known as "the Reichenstein disease"—chronic arsenicism with accompanying disorders of the liver, skin, and gastrointestinal and nervous systems. Malignant tumors were a common accompaniment of the disease. Reichenstein's disease is now chiefly of historic interest, for new water supplies were provided a quarter of a century ago, from which arsenic was largely eliminated. In Córdoba Province in Argentina, however, chronic arsenic poisoning, accompanied by arsenical skin cancers, is endemic because of the contamination of drinking water derived from rock formations containing arsenic.

It would not be difficult to create conditions similar to those in Reichenstein and Córdoba by long continued use of arsenical insecticides. In the United States the arsenic-drenched soils of tobacco plantations, of many orchards in the Northwest, and of blueberry lands in the East may easily lead to pollution of water supplies.

An arsenic-contaminated environment affects not only man but animals as well. A report of great interest came from

Germany in 1936. In the area about Freiberg, Saxony, smelters for silver and lead poured arsenic fumes into the air, to drift out over the surrounding countryside and settle down upon the vegetation. According to Dr. Hueper, horses, cows, goats, and pigs, which of course fed on this vegetation, showed loss of hair and thickening of the skin. Deer inhabiting nearby forests sometimes had abnormal pigment spots and precancerous warts. One had a definitely cancerous lesion. Both domestic and wild animals were affected by "arsenical enteritis, gastric ulcers, and cirrhosis of the liver." Sheep kept near the smelters developed cancers of the nasal sinus; at their death arsenic was found in the brain, liver, and tumors. In the area there was also "an extraordinary mortality among insects, especially bees. After rainfalls which washed the arsenical dust from the leaves and carried it along into the water of brooks and pools, a great many fish died."

An example of a carcinogen belonging to the group of new, organic pesticides is a chemical widely used against mites and ticks. Its history provides abundant proof that, despite the supposed safeguards provided by legislation, the public can be exposed to a known carcinogen for several years before the slowly moving legal processes can bring the situation under control. The story is interesting from another standpoint, proving that what the public is asked to accept as "safe" today may turn out tomorrow to be extremely dangerous.

When this chemical was introduced in 1955, the manufacturer applied for a tolerance which would sanction the presence of small residues on any crops that might be sprayed. As required by law, he had tested the chemical on laboratory animals and submitted the results with his application. However, scientists of the Food and Drug Administration interpreted the tests as showing a possible cancer-producing tendency and the Commissioner accordingly recommended a "zero tolerance," which is a way of saying that no residues could legally occur on food shipped across state lines. But the manufacturer had the legal right to appeal and the case was accordingly reviewed by a committee. The committee's decision was a compromise: a tolerance of 1 part per million was to be established and the product marketed for two years, during which time further laboratory tests were to determine whether the chemical was actually a carcinogen.

Although the committee did not say so, its decision meant that the public was to act as guinea pigs, testing the suspected carcinogen along with the laboratory dogs and rats. But laboratory animals give more prompt results, and after the two years it was evident that this miticide was indeed a carcinogen. Even at that point, in 1957, the Food and Drug Administration could not instantly rescind the tolerance which allowed residues of a known carcinogen to contaminate food consumed by the public. Another year was required for various legal procedures. Finally, in December 1958 the zero tolerance which the Commissioner had recommended in 1955 became effective.

These are by no means the only known carcinogens among pesticides. In laboratory tests on animal subjects, DDT has produced suspicious liver tumors. Scientists of the Food and Drug Administration who reported the discovery of these tumors were uncertain how to classify them, but felt there was some "justification for considering them low grade hepatic cell carcinomas." Dr. Hueper now gives DDT the definite rating of a "chemical carcinogen."

Two herbicides belonging to the carbamate group, IPC and CIPC, have been found to play a role in producing skin tumors in mice. Some of the tumors were malignant. These chemicals seem to initiate the malignant change, which may then be completed by other chemicals of types prevalent in the environment.

The weed-killer aminotriazole has caused thyroid cancer in test animals. This chemical was misused by a number of cranberry growers in 1959, producing residues on some of the marketed berries. In the controversy that followed seizure of contaminated cranberries by the Food and Drug Administration, the fact that the chemical actually is cancer producing was widely challenged, even by many medical men. The scientific facts released by the Food and Drug Administration clearly indicate the carcinogenic nature of aminotriazole in laboratory rats. When these animals were fed this chemical at the rate of 100 parts per million in the drinking water (or one teaspoonful of chemical in ten thousand teaspoonfuls of water) they began to develop thyroid tumors at the 68th week. After two years, such tumors were present in more than half the rats examined. They were diagnosed as various types of benign and malignant growths. The tumors also appeared at lower levels of feeding—in fact, *a level that produced no effect was not found*. No one knows, of course, the level at which aminotriazole may be carcinogenic for man, but as a professor of medicine at Harvard University, Dr. David Rutstein, has pointed out, the level is just as likely to be to man's disfavor as to his advantage.

As yet insufficient time has elapsed to reveal the full effect of the new chlorinated hydrocarbon insecticides and of the modern herbicides. Most malignancies develop so slowly that they may require a considerable segment of the victim's life to reach the stage of showing clinical symptoms. In the early 1920's women who painted luminous figures on watch dials swallowed minute amounts of radium by touching the brushes to their lips; in some of these women bone cancers developed after a lapse of 15 or more years. A period of 15 to 30 years or even more has been demonstrated for some cancers caused by occupational exposures to chemical carcinogens.

In contrast to these industrial exposures to various carcinogens the first exposures to DDT date from about 1942 for military personnel and from about 1945 for civilians, and it was not until the early fifties that a wide variety of pesticidal chemicals came into use. The full maturing of whatever seeds of malignancy have been sown by these chemicals is yet to come.

There is, however, one presently known exception to the fact that a long period of latency is common to most malignancies. This exception is leukemia. Survivors of Hiroshima began to develop leukemia only three years after the atomic bombing, and there is now reason to believe the latent period may be considerably shorter. Other types of cancer may in time be found to have a relatively short latent period, also, but at present leukemia seems to be the exception to the general rule of extremely slow development.

Within the period covered by the rise of modern pesticides, the incidence of leukemia has been steadily rising. Figures available from the National Office of Vital Statistics clearly establish a disturbing rise in malignant diseases of the blood-forming tissues. In the year 1960, leukemia alone claimed 12,290 victims. Deaths from all types of malignancies of blood and lymph totaled 25,400, increasing sharply from the 16,690 figure of 1950. In terms of deaths per 100,000 of population, the increase is from 11.1 in 1950 to 14.1 in 1960. The increase is by no means confined to the United States; in all countries the recorded deaths from leukemia at all ages are rising at a rate of 4 to 5 per cent a year. What does it mean? To what lethal agent or agents, new to our environment, are people now exposed with increasing frequency?

Such world-famous institutions as the Mayo Clinic admit hundreds of victims of these diseases of the blood-forming organs. Dr. Malcolm Hargraves and his associates in the Hematology Department at the Mayo Clinic report that almost without exception these patients have had a history of exposure to various toxic chemicals, including sprays which contain DDT, chlordane, benzene, lindane, and petroleum distillates.

Environmental diseases related to the use of various toxic substances have been increasing, "particularly during the past ten years," Dr. Hargraves believes. From extensive clinical experience he believes that "the vast majority of patients suffering from the blood dyscrasias and lymphoid diseases have a significant history of exposure to the various hydrocarbons which in turn includes most of the pesticides of today. A careful medical history will almost invariably establish such a relationship." This specialist now has a large number of detailed case histories based on every patient he has seen with leukemias, aplastic anemias, Hodgkin's disease, and other disorders of the blood and blood-forming tissues. "They had all been exposed to these environmental agents, with a fair amount of exposure," he reports.

What do these case histories show? One concerned a housewife who abhorred spiders. In mid-August she had gone into her basement with an aerosol spray containing DDT and petroleum distillate. She sprayed the entire basement thoroughly, under the stairs, in the fruit cupboards and in all the protected areas around ceiling and rafters. As she finished the spraying she began to feel quite ill, with nausea and extreme anxiety and nervousness. Within the next few days she felt better, however, and apparently not suspecting the cause of her difficulty, she repeated the entire procedure in September, running through two more cycles of spraying, falling ill, recovering temporarily, spraying again. After the third use of the aerosol new symptoms developed: fever, pains in the joints and general malaise, acute phlebitis in one leg. When examined by Dr. Hargraves she was found to be suffering from acute leukemia. She died within the following month.

Another of Dr. Hargraves' patients was a professional man who had his office in an old building infested by roaches. Becoming embarrassed by the presence of these insects, he took control measures in his own hands. He spent most of one Sunday spraying the basement and all secluded areas. The spray was a 25 per cent DDT concentrate suspended in a solvent containing methylated naphthalenes. Within a short time he began to bruise and bleed. He entered the clinic bleeding from a number of hemorrhages. Studies of his blood revealed a severe depression of the bone marrow called aplastic anemia. During the next five and one half months he received 59 transfusions in addition to other therapy. There was partial recovery but about nine years later a fatal leukemia developed.

Where pesticides are involved, the chemicals that figure most prominently in the case histories are DDT, lindane, benzene hexachloride, the nitrophenols, the common moth crystal paradichlorobenzene, chlordane, and, of course, the solvents in which they are carried. As this physician emphasizes, pure exposure to a single chemical is the exception, rather than the rule. The commercial product usually contains combinations of several chemicals, suspended in a petroleum distillate plus some dispersing agent. The aromatic cyclic and unsaturated hydrocarbons of the vehicle may themselves be a major factor in the damage done the blood-forming organs. From the practical rather than the medical standpoint this distinction is of little importance, however, because these petroleum solvents are an inseparable part of most common spraying practices.

The medical literature of this and other countries contains many significant cases that support Dr. Hargraves' belief in a cause-and-effect relation between these chemicals and leukemia and other blood disorders. They concern such everyday people as farmers caught in the "fallout" of their own spray rigs or of planes, a college student who sprayed his study for ants and remained in the room to study, a woman who had installed a portable lindane vaporizer in her home, a worker in a cotton field that had been sprayed with chlordane and toxaphene. They carry, half concealed within their medical terminology, stories of such human tragedies as that of two young cousins in Czechoslovakia, boys who lived in the same town and had always worked and played together. Their last and most fateful employment was at a farm cooperative where it was their job to unload sacks of an insecticide (benzene hexachloride). Eight months later one of the boys was stricken with acute leukemia. In nine days he was dead. At about this time his cousin began to tire easily and to run a temperature. Within about three months his symptoms became more severe and he, too, was hospitalized. Again the diagnosis was acute leukemia, and again the disease ran its inevitably fatal course.

And then there is the case of a Swedish farmer, strangely reminiscent of that of the Japanese fisherman Kuboyama of the tuna vessel the *Lucky Dragon*. Like Kuboyama, the farmer had been a healthy man, gleaning his living from the land as Kuboyama had taken his from the sea. For each man a poison drifting out of the sky carried a death sentence. For one, it was radiation-poisoned ash; for the other, chemical dust. The farmer had treated about 60 acres of land with a dust containing DDT and benzene hexachloride. As he worked puffs of wind brought little clouds of dust swirling about him. "In the evening he felt unusually tired, and during the subsequent days he had a general feeling of weakness, with backache and aching legs as well as chills, and was obliged to take to his bed," says a report from the Medical Clinic at Lund. "His condition became worse, however, and on May 19 [a week after the spraying] he applied for admission to the local hospital." He had a high fever and his blood count was abnormal. He was transferred to the Medical Clinic, where, after an illness of two and one half months, he died. A post-mortem examination revealed a complete wasting away of the bone marrow.

How a normal and necessary process such as cell division can become altered so that it is alien and destructive is a problem that has engaged the attention of countless scientists and untold sums of money. What happens in a cell to change its orderly multiplication into the wild and uncontrolled proliferation of cancer?

When answers are found they will almost certainly be multiple. Just as cancer itself is a disease that wears many guises, appearing in various forms that differ in their origin, in the course of their development, and in the factors that influence their growth or regression, so there must be a corresponding variety of causes. Yet underlying them all, perhaps, only a few basic kinds of injuries to the cell are responsible. Here and there, in research widely scattered and sometimes not undertaken as a cancer study at all, we see glimmerings of the first light that may one day illuminate this problem.

Again we find that only by looking at some of the smallest units of life, the cell and its chromosomes, can we find that wider vision needed to penetrate such mysteries. Here, in this microcosm, we must look for those factors that somehow shift the marvelously functioning mechanisms of the cell out of their normal patterns.

One of the most impressive theories of the origin of cancer cells was developed by a German biochemist, Professor Otto Warburg of the Max Planck Institute of Cell Physiology. Warburg has devoted a lifetime of study to the complex processes of oxidation within the cell. Out of this broad background of understanding came a fascinating and lucid explanation of the way a normal cell can become malignant.

Warburg believes that either radiation or a chemical carcinogen acts by destroying the respiration of normal cells, thus depriving them of energy. This action may result from minute doses often repeated. The effect, once achieved, is irreversible. The cells not killed outright by the impact of such a respiratory poison struggle to compensate for the loss of energy. They can no longer carry on that extraordinary and efficient cycle by which vast amounts of ATP are produced, but are thrown back on a primitive and far less efficient method, that of fermentation. The struggle to survive by fermentation continues for a long period of time. It continues through ensuing cell divisions, so that all the descendant cells have this abnormal method of respiration. Once a cell has lost its normal respiration it cannot regain it—not in a year, not in a decade or in many decades. But little by little, in this grueling struggle to restore lost energy, those cells that survive begin to compensate by increased fermentation. It is a Darwinian struggle, in which only the most fit or adaptable survive. At last they reach the point where fermentation is able to produce as much energy as respiration. At this point, cancer cells may be said to have been created from normal body cells.

Warburg's theory explains many otherwise puzzling things. The long latent period of most cancers is the time required for the infinite number of cell divisions during which fermentation is gradually increasing after the initial damage to respiration. The time required for fermentation to become dominant varies in different species because of different fermentation rates: a short time in the rat, in which cancers appear quickly, a long time (decades even) in man, in whom the development of malignancy is a deliberate process.

The Warburg theory also explains why repeated small doses of a carcinogen are more dangerous under some circumstances than a single large dose. The latter may kill the cells outright, whereas the small doses allow some to survive, though in a damaged condition. These survivors may then develop into cancer cells. This is why there is no "safe" dose of a carcinogen.

In Warburg's theory we also find explanation of an otherwise incomprehensible fact—that one and the same agent can be useful in treating cancer and can also cause it. This, as everyone knows, is true of radiation, which kills cancer cells but may also cause cancer. It is also true of many of the chemicals now used against cancer. Why? Both types of agents damage respiration. Cancer cells already have a defective respiration, so with additional damage they die. The normal cells, suffering respiratory damage for the first time, are not killed but are set on the path that may eventually lead to malignancy.

Warburg's ideas received confirmation in 1953 when other workers were able to turn normal cells into cancer cells merely by depriving them of oxygen intermittently over long periods. Then in 1961 other confirmation came, this time from living animals rather than tissue cultures. Radioactive tracer substances were injected into cancerous mice. Then by careful measurements of their respiration, it was found that the fermentation rate was markedly above normal, just as Warburg had foreseen.

Measured by the standards established by Warburg, most pesticides meet the criterion of the perfect carcinogen too well for comfort. As we have seen in the preceding chapter, many of the chlorinated hydrocarbons, the phenols, and some herbicides interfere with oxidation and energy production within the cell. By these means they may be creating sleeping cancer cells, cells in which an irreversible malignancy will slumber long and undetected until finally—its cause long forgotten and even unsuspected—it flares into the open as recognizable cancer.

Another path to cancer may be by way of the chromosomes. Many of the most distinguished research men in this field look with suspicion on any agent that damages the chromosomes, interferes with cell division, or causes mutations. In the view of these men any mutation is a potential cause of cancer. Although discussions of mutations usually refer to those in the germ cells, which may then make their effect felt in future generations, there may also be mutations in the body cells. According to the mutation theory of the origin of cancer, a cell, perhaps under the influence of radiation or of a chemical, develops a mutation that allows it to escape the controls the body normally asserts over cell division. It is therefore able to multiply in a wild and unregulated manner. The new cells resulting from these divisions have the same ability to escape control, and in time enough such cells have accumulated to constitute a cancer.

Other investigators point to the fact that the chromosomes in cancer tissue are unstable; they tend to be broken or damaged, the number may be erratic, there may even be double sets.

The first investigators to trace chromosome abnormalities all the way to actual malignancy were Albert Levan and John J. Biesele, working at the Sloan-Kettering Institute in New York. As to which came first, the malignancy or the disturbance of the chromosomes, these workers say without hesitation that "the chromosomal irregularities precede the malignancy." Perhaps, they speculate, after the initial chromosome damage and the resulting instability there is a long period of trial and error through many cell generations (the long latent period of malignancy) during which a collection of mutations is finally accumulated which allow the cells to escape from control and embark on the unregulated multiplication that is cancer.

Ojvind Winge, one of the early proponents of the theory of chromosome instability, felt that chromosome doublings were especially significant. Is it coincidence, then, that benzene hexachloride and its relative, lindane, are known through repeated observations to double the chromosomes in experimental plants—and that these same chemicals have been implicated in many well-documented cases of fatal anemias? And what of the many other pesticides that interfere with cell division, break chromosomes, cause mutations?

It is easy to see why leukemia should be one of the most common diseases to result from exposure to radiation or to chemicals that imitate radiation. The principal targets of physical or chemical mutagenic agents are cells that are undergoing especially active division. This includes various tissues but most importantly those engaged in the production of blood. The bone marrow is the chief producer of red blood cells throughout life, sending some 10 million new cells per second into the bloodstream of man. White corpuscles are formed in the lymph glands and in some of the marrow cells at a variable, but still prodigious, rate.

Certain chemicals, again reminding us of radiation products like Strontium 90, have a peculiar affinity for the bone marrow. Benzene, a frequent constituent of insecticidal solvents, lodges in the marrow and remains deposited there for periods known to be as long as 20 months. Benzene itself has been recognized in medical literature for many years as a cause of leukemia.

The rapidly growing tissues of a child would also afford conditions most suitable for the development of malignant cells. Sir Macfarlane Burnet has pointed out that not only is leukemia increasing throughout the world but it has become most common in the three- to four-year age bracket, an age incidence shown by no other disease. According to this authority, "The peak between three and four years of age can hardly have any other interpretation than exposure of the young organism to a mutagenic stimulus around the time of birth."

Another mutagen known to produce cancer is urethane. When pregnant mice are treated with this chemical not only do they develop cancer of the lung but their young do, also. The only exposure of the infant mice to urethane was prenatal in these experiments, proving that the chemical must have passed through the placenta. In human populations exposed to urethane or related chemicals there is a possibility that tumors will develop in infants through prenatal exposure, as Dr. Hueper has warned.

Urethane as a carbamate is chemically related to the herbicides IPC and CIPC. Despite the warnings of cancer experts, carbamates are now widely used, not only as insecticides, weed killers, and fungicides, but also in a variety of products including plasticizers, medicines, clothing, and insulating materials.

The road to cancer may also be an indirect one. A substance that is not a carcinogen in the ordinary sense may disturb the normal functioning of some part of the body in such a way that malignancy results. Important examples are the cancers, especially of the reproductive system, that appear to be linked with disturbances of the balance of sex hormones; these disturbances, in turn, may in some cases be the result of something that affects the ability of the liver to preserve a proper level of these hormones. The chlorinated hydrocarbons are precisely the kind of agent that can bring about this kind of indirect carcinogenesis, because all of them are toxic in some degree to the liver.

The sex hormones are, of course, normally present in the body and perform a necessary growth-stimulating function in relation to the various organs of reproduction. But the body has a built-in protection against excessive accumulations, for the liver acts to keep a proper balance between male and female hormones (both are produced in the bodies of both sexes, although in different amounts) and to prevent an excess accumulation of either. It cannot do so, however, if it has been damaged by disease or chemicals, or if the supply of the B-complex vitamins has been reduced. Under these conditions the estrogens build up to abnormally high levels.

What are the effects? In animals, at least, there is abundant evidence from experiments. In one such, an investigator at the Rockefeller Institute for Medical Research found that rabbits with livers damaged by disease show a very high incidence of uterine tumors, thought to have developed because the liver was no longer able to inactivate the estrogens in the blood, so that they "subsequently rose to a carcinogenic level." Extensive experiments on mice, rats, guinea pigs, and monkeys show that prolonged administration of estrogens (not necessarily at high levels) has caused changes in the tissues of the reproductive organs, "varying from benign overgrowths to definite malignancy." Tumors of the kidneys have been induced in hamsters by administering estrogens.

Although medical opinion is divided on the question, much evidence exists to support the view that similar effects may occur in human tissues. Investigators at the Royal Victoria Hospital at McGill University found two thirds of 150 cases of uterine cancer studied by them gave evidence of abnormally high estrogen levels. In 90 per cent of a later series of 20 cases there was similar high estrogen activity.

It is possible to have liver damage sufficient to interfere with estrogen elimination without detection of the damage by any tests now available to the medical profession. This can easily be caused by the chlorinated hydrocarbons, which, as we have seen, set up changes in liver cells at very low levels of intake. They also cause loss of the B vitamins. This, too, is extremely important, for other chains of evidence show the protective role of these vitamins against cancer. The late C. P. Rhoads, onetime director of the Sloan-Kettering Institute for Cancer Research, found that test animals exposed to a very potent chemical carcinogen developed no cancer if they had been fed yeast, a rich source of the natural B vitamins. A deficiency of these vitamins has been found to accompany mouth cancer and perhaps cancer of other sites in the digestive tract. This has been observed not only in the United States but in the far northern parts of Sweden and Finland, where the diet is ordinarily deficient in vitamins. Groups prone to primary liver cancer, as for example the Bantu tribes of Africa, are typically subject to malnutrition. Cancer of the male breast is also prevalent in parts of Africa, associated with liver disease and malnutrition. In postwar Greece enlargement of the male breast was a common accompaniment of periods of starvation.

In brief, the argument for the indirect role of pesticides in cancer is based on their proven ability to damage the liver and to reduce the supply of B vitamins, thus leading to an increase in the "endogenous" estrogens, or those produced by the body itself. Added to these are the wide variety of synthetic estrogens to which we are increasingly exposed—those in cosmetics, drugs, foods, and occupational exposures. The combined effect is a matter that warrants the most serious concern.

Human exposures to cancer-producing chemicals (including pesticides) are uncontrolled and they are multiple. An individual may have many different exposures to the same chemical. Arsenic is an example. It exists in the environment of every individual in many different guises: as an air pollutant, a contaminant of water, a pesticide residue on food, in medicines, cosmetics, wood preservatives, or as a coloring agent in paints and inks. It is quite possible that no one of these exposures alone would be sufficient to precipitate malignancy—yet any single supposedly "safe dose" may be enough to tip the scales that are already loaded with other "safe doses."

Or again the harm may be done by two or more different carcinogens acting together, so that there is a summation of their effects. The individual exposed to DDT, for example, is almost certain to be exposed to other liver-damaging hydrocarbons, which are so widely used as solvents, paint removers, degreasing agents, dry-cleaning fluids, and anesthetics. What then can be a "safe dose" of DDT?

The situation is made even more complicated by the fact that one chemical may act on another to alter its effect. Cancer may sometimes require the complementary action of two chemicals, one of which sensitizes the cell or tissue so that it may later, under the action of another or promoting agent, develop true malignancy. Thus, the herbicides IPC and CIPC may act as initiators in the production of skin tumors, sowing the seeds of malignancy that may be brought into actual being by something else—perhaps a common detergent.

There may be interaction, too, between a physical and a chemical agent. Leukemia may occur as a two-step process, the malignant change being initiated by X-radiation, the promoting action being supplied by a chemical, as, for example, urethane. The growing exposure of the population to radiation from various sources, plus the many contacts with a host of chemicals suggest a grave new problem for the modern world.

The pollution of water supplies with radioactive materials poses another problem. Such materials, present as contaminants in water that also contains chemicals, may actually change the nature of the chemicals by the impact of ionizing radiation, rearranging their atoms in unpredictable ways to create new chemicals.

Water pollution experts throughout the United States are concerned by the fact that detergents are now a troublesome and practically universal contaminant of public water supplies. There is no practical way to remove them by treatment. Few detergents are known to be carcinogenic, but in an indirect way they may promote cancer by acting on the lining of the digestive tract, changing the tissues so that they more easily absorb dangerous chemicals, thereby aggravating their effect. But who can foresee and control this action? In the kaleidoscope of shifting conditions, what dose of a carcinogen can be "safe" except a zero dose?

We tolerate cancer-causing agents in our environment at our peril, as was clearly illustrated by a recent happening. In the spring of 1961 an epidemic of liver cancer appeared among rainbow trout in many federal, state, and private hatcheries. Trout in both eastern and western parts of the United States were affected; in some areas practically 100 per cent of the trout over three years of age developed cancer. This discovery was made because of a pre-existing arrangement between the Environmental Cancer Section of the National Cancer Institute and the Fish and Wildlife Service for the reporting of all fish with tumors, so that early warning might be had of a cancer hazard to man from water contaminants.

Although studies are still under way to determine the exact cause of this epidemic over so wide an area, the best evidence is said to point to some agent present in the prepared hatchery feeds. These contain an incredible variety of chemical additives and medicinal agents in addition to the basic foodstuffs.

The story of the trout is important for many reasons, but chiefly as an example of what can happen when a potent carcinogen is introduced into the environment of any species. Dr. Hueper has described this epidemic as a serious warning that greatly increased attention must be given to controlling the number and variety of environmental carcinogens. "If such preventive measures are not taken," says Dr. Hueper, "the stage will be set at a progressive rate for the future occurrence of a similar disaster to the human population."

The discovery that we are, as one investigator phrased it, living in a "sea of carcinogens" is of course dismaying and may easily lead to reactions of despair and defeatism. "Isn't it a hopeless situation?" is the common reaction. "Isn't it impossible even to attempt to eliminate these cancer-producing agents from our world? Wouldn't it be better not to waste time trying, but instead to put all our efforts into research to find a cure for cancer?"

When this question is put to Dr. Hueper, whose years of distinguished work in cancer make his opinion one to respect, his reply is given with the thoughtfulness of one who has pondered it long, and has a lifetime of research and experience behind his judgment. Dr. Hueper believes that our situation with regard to cancer today is very similar to that which faced mankind with regard to infectious diseases in the closing years of the 19th century. The causative relation between pathogenic organisms and many diseases had been established through the brilliant work of Pasteur and Koch. Medical men and even the general public were becoming aware that the human environment was inhabited by an enormous number of microorganisms capable of causing disease, just as today carcinogens pervade our surroundings. Most infectious diseases have now been brought under a reasonable degree of control and some have been practically eliminated. This brilliant medical achievement came about by an attack that was twofold—that stressed prevention as well as cure. Despite the prominence that "magic bullets" and "wonder drugs" hold in the layman's mind, most of the really decisive battles in the war against infectious disease consisted of measures to eliminate disease organisms from the environment. An example from history concerns the great outbreak of cholera in London more than one hundred years ago. A London physician, John Snow, mapped the occurrence of cases and found they originated in one area, all of whose inhabitants drew their water from one pump located on Broad Street. In a swift and decisive practice of preventive medicine, Dr. Snow removed the handle from the pump. The epidemic was thereby brought under control—not by a magic pill that killed the (then unknown) organism of cholera, but by eliminating the organism from the environment. Even therapeutic measures have the important result not only of curing the patient but of reducing the foci of infection. The present comparative rarity of tuberculosis results in large measure from the fact that the average person now seldom comes into contact with the tubercle bacillus.

Today we find our world filled with cancer-producing agents. An attack on cancer that is concentrated wholly or even largely on therapeutic measures (even assuming a "cure" could be found) in Dr. Hueper's opinion will fail because it leaves untouched the great reservoirs of carcinogenic agents which

would continue to claim new victims faster than the as yet elu-
sive "cure" could allay the disease.

Why have we been slow to adopt this common-sense approach to the cancer problem? Probably "the goal of curing
the victims of cancer is more exciting, more tangible, more
glamorous and rewarding than prevention," says Dr. Hueper.
Yet to prevent cancer from ever being formed is "definitely
more humane" and can be "much more effective than cancer
cures." Dr. Hueper has little patience with the wishful thinking
that promises "a magic pill that we shall take each morning before breakfast" as protection against cancer. Part of the public
trust in such an eventual outcome results from the misconception that cancer is a single, though mysterious disease, with a
single cause and, hopefully, a single cure. This of course is far
from the known truth. Just as environmental cancers are induced by a wide variety of chemical and physical agents, so the
malignant condition itself is manifested in many different and
biologically distinct ways.

The long promised "breakthrough," when or if it comes,
cannot be expected to be a panacea for all types of malignancy.
Although the search must be continued for therapeutic measures to relieve and to cure those who have already become
victims of cancer, it is a disservice to humanity to hold out the
hope that the solution will come suddenly, in a single master
stroke. It will come slowly, one step at a time. Meanwhile as
we pour our millions into research and invest all our hopes in
vast programs to find cures for established cases of cancer, we
are neglecting the golden opportunity to prevent, even while
we seek to cure.

The task is by no means a hopeless one. In one important
respect the outlook is more encouraging than the situation regarding infectious disease at the turn of the century. The world
was then full of disease germs, as today it is full of carcinogens.
But man did not put the germs into the environment and his
role in spreading them was involuntary. In contrast, man *has*
put the vast majority of carcinogens into the environment, and
he can, if he wishes, eliminate many of them. The chemical
agents of cancer have become entrenched in our world in two
ways: first, and ironically, through man's search for a better and
easier way of life; second, because the manufacture and sale of

such chemicals has become an accepted part of our economy and our way of life.

It would be unrealistic to suppose that all chemical carcinogens can or will be eliminated from the modern world. But a very large proportion are by no means necessities of life. By their elimination the total load of carcinogens would be enormously lightened, and the threat that one in every four will develop cancer would at least be greatly mitigated. The most determined effort should be made to eliminate those carcinogens that now contaminate our food, our water supplies, and our atmosphere, because these provide the most dangerous type of contact—minute exposures, repeated over and over throughout the years.

Among the most eminent men in cancer research are many others who share Dr. Hueper's belief that malignant diseases can be reduced significantly by determined efforts to identify the environmental causes and to eliminate them or reduce their impact. For those in whom cancer is already a hidden or a visible presence, efforts to find cures must of course continue. But for those not yet touched by the disease and certainly for the generations as yet unborn, prevention is the imperative need.

FIFTEEN

Nature Fights Back

TO HAVE RISKED so much in our efforts to mold nature to our satisfaction and yet to have failed in achieving our goal would indeed be the final irony. Yet this, it seems, is our situation. The truth, seldom mentioned but there for anyone to see, is that nature is not so easily molded and that the insects are finding ways to circumvent our chemical attacks on them.

"The insect world is nature's most astonishing phenomenon," said the Dutch biologist C. J. Briejèr. "Nothing is impossible to it; the most improbable things commonly occur there. One who penetrates deeply into its mysteries is continually breathless with wonder. He knows that anything can happen, and that the completely impossible often does."

The "impossible" is now happening on two broad fronts. By a process of genetic selection, the insects are developing strains resistant to chemicals. This will be discussed in the following chapter. But the broader problem, which we shall look at now, is the fact that our chemical attack is weakening the defenses inherent in the environment itself, defenses designed to keep

213

the various species in check. Each time we breach these defenses a horde of insects pours through.

From all over the world come reports that make it clear we
are in a serious predicament. At the end of a decade or more
of intensive chemical control, entomologists were finding that
problems they had considered solved a few years earlier had returned to plague them. And new problems had arisen as insects
once present only in insignificant numbers had increased to the
status of serious pests. By their very nature chemical controls
are self-defeating, for they have been devised and applied without taking into account the complex biological systems against
which they have been blindly hurled. The chemicals may have
been pretested against a few individual species, but not against
living communities.

In some quarters nowadays it is fashionable to dismiss the
balance of nature as a state of affairs that prevailed in an earlier, simpler world—a state that has now been so thoroughly
upset that we might as well forget it. Some find this a convenient assumption, but as a chart for a course of action it is
highly dangerous. The balance of nature is not the same today
as in Pleistocene times, but it is still there: a complex, precise,
and highly integrated system of relationships between living
things which cannot safely be ignored any more than the law
of gravity can be defied with impunity by a man perched on
the edge of a cliff. The balance of nature is not a *status quo*; it is
fluid, ever shifting, in a constant state of adjustment. Man, too,
is part of this balance. Sometimes the balance is in his favor;
sometimes—and all too often through his own activities—it is
shifted to his disadvantage.

Two critically important facts have been overlooked in
designing the modern insect control programs. The first is
that the really effective control of insects is that applied by
nature, not by man. Populations are kept in check by something the ecologists call the resistance of the environment, and
this has been so since the first life was created. The amount
of food available, conditions of weather and climate, the
presence of competing or predatory species, all are critically
important. "The greatest single factor in preventing insects
from overwhelming the rest of the world is the internecine
warfare which they carry out among themselves," said the

entomologist Robert Metcalf. Yet most of the chemicals now used kill all insects, our friends and enemies alike.

The second neglected fact is the truly explosive power of a species to reproduce once the resistance of the environment has been weakened. The fecundity of many forms of life is almost beyond our power to imagine, though now and then we have suggestive glimpses. I remember from student days the miracle that could be wrought in a jar containing a simple mixture of hay and water merely by adding to it a few drops of material from a mature culture of protozoa. Within a few days the jar would contain a whole galaxy of whirling, darting life— uncountable trillions of the slipper animalcule, *Paramecium*, each small as a dust grain, all multiplying without restraint in their temporary Eden of favorable temperatures, abundant food, absence of enemies. Or I think of shore rocks white with barnacles as far as the eye can see, or of the spectacle of passing through an immense school of jellyfish, mile after mile, with seemingly no end to the pulsing, ghostly forms scarcely more substantial than the water itself.

We see the miracle of nature's control at work when the cod move through winter seas to their spawning grounds, where each female deposits several millions of eggs. The sea does not become a solid mass of cod as it would surely do if all the progeny of all the cod were to survive. The checks that exist in nature are such that out of the millions of young produced by each pair only enough, on the average, survive to adulthood to replace the parent fish.

Biologists used to entertain themselves by speculating as to what would happen if, through some unthinkable catastrophe, the natural restraints were thrown off and all the progeny of a single individual survived. Thus Thomas Huxley a century ago calculated that a single female aphis (which has the curious power of reproducing without mating) could produce progeny in a single year's time whose total weight would equal that of the inhabitants of the Chinese empire of his day.

Fortunately for us such an extreme situation is only theoretical, but the dire results of upsetting nature's own arrangements are well known to students of animal populations. The stockman's zeal for eliminating the coyote has resulted in plagues of field mice, which the coyote formerly controlled. The oft repeated story of the Kaibab deer in Arizona is another case in point. At one time the deer population was in equilibrium with its environment. A number of predators—wolves, pumas, and coyotes—prevented the deer from outrunning their food supply. Then a campaign was begun to "conserve" the deer by killing off their enemies. Once the predators were gone, the deer increased prodigiously and soon there was not enough food for them. The browse line on the trees went higher and higher as they sought food, and in time many more deer were dying of starvation than had formerly been killed by predators. The whole environment, moreover, was damaged by their desperate efforts to find food.

The predatory insects of field and forests play the same role as the wolves and coyotes of the Kaibab. Kill them off and the population of the prey insect surges upward.

No one knows how many species of insects inhabit the earth because so many are yet to be identified. But more than 700,000 have already been described. This means that in terms of the number of species, 70 to 80 per cent of the earth's creatures are insects. The vast majority of these insects are held in check by natural forces, without any intervention by man. If this were not so, it is doubtful that any conceivable volume of chemicals—or any other methods—could possibly keep down their populations.

The trouble is that we are seldom aware of the protection afforded by natural enemies until it fails. Most of us walk unseeing through the world, unaware alike of its beauties, its wonders, and the strange and sometimes terrible intensity of the lives that are being lived about us. So it is that the activities of the insect predators and parasites are known to few. Perhaps we may have noticed an oddly shaped insect of ferocious mien on a bush in the garden and been dimly aware that the praying mantis lives at the expense of other insects. But we see with understanding eye only if we have walked in the garden at night and here and there with a flashlight have glimpsed the mantis stealthily creeping upon her prey. Then we sense something of the drama of the hunter and the hunted. Then we begin to feel something of that relentlessly pressing force by which nature controls her own.

The predators—insects that kill and consume other insects—are of many kinds. Some are quick and with the speed of swallows snatch their prey from the air. Others plod method-ically along a stem, plucking off and devouring sedentary in-sects like the aphids. The yellowjackets capture soft-bodied insects and feed the juices to their young. Muddauber wasps build columned nests of mud under the eaves of houses and stock them with insects on which their young will feed. The horseguard wasp hovers above herds of grazing cattle, destroy-ing the blood-sucking flies that torment them. The loudly buzzing syrphid fly, often mistaken for a bee, lays its eggs on leaves of aphis-infested plants; the hatching larvae then con-sume immense numbers of aphids. Ladybugs or lady beetles are among the most effective destroyers of aphids, scale insects, and other plant-eating insects. Literally hundreds of aphids are consumed by a single ladybug to stoke the little fires of energy which she requires to produce even a single batch of eggs.

Even more extraordinary in their habits are the parasitic in-sects. These do not kill their hosts outright. Instead, by a vari-ety of adaptations they utilize their victims for the nurture of their own young. They may deposit their eggs within the larvae or eggs of their prey, so that their own developing young may find food by consuming the host. Some attach their eggs to a caterpillar by means of a sticky solution; on hatching, the larval parasite bores through the skin of the host. Others, led by an instinct that simulates foresight, merely lay their eggs on a leaf so that a browsing caterpillar will eat them inadvertently.

Everywhere, in field and hedgerow and garden and forest, the insect predators and parasites are at work. Here, above a pond, the dragonflies dart and the sun strikes fire from their wings. So their ancestors sped through swamps where huge reptiles lived. Now, as in those ancient times, the sharp-eyed dragonflies capture mosquitoes in the air, scooping them in with basket-shaped legs. In the waters below, their young, the dragonfly nymphs, or naiads, prey on the aquatic stages of mosquitoes and other insects.

Or there, almost invisible against a leaf, is the lacewing, with green gauze wings and golden eyes, shy and secretive, descen-dant of an ancient race that lived in Permian times. The adult lacewing feeds mostly on plant nectars and the honeydew of aphids, and in time she lays her eggs, each on the end of a long stalk which she fastens to a leaf. From these emerge her children—strange, bristled larvae called aphis lions, which live by preying on aphids, scales, or mites, which they capture and suck dry of fluid. Each may consume several hundred aphids before the ceaseless turning of the cycle of its life brings the time when it will spin a white silken cocoon in which to pass the pupal stage.

And there are many wasps, and flies as well, whose very existence depends on the destruction of the eggs or larvae of other insects through parasitism. Some of the egg parasites are exceedingly minute wasps, yet by their numbers and their great activity they hold down the abundance of many crop-destroying species.

All these small creatures are working—working in sun and rain, during the hours of darkness, even when winter's grip has damped down the fires of life to mere embers. Then this vital force is merely smoldering, awaiting the time to flare again into activity when spring awakens the insect world. Meanwhile, under the white blanket of snow, below the frost-hardened soil, in crevices in the bark of trees, and in sheltered caves, the parasites and the predators have found ways to tide themselves over the season of cold.

The eggs of the mantis are secure in little cases of thin parchment attached to the branch of a shrub by the mother who lived her life span with the summer that is gone.

The female *Polistes* wasp, taking shelter in a forgotten corner of some attic, carries in her body the fertilized eggs, the heritage on which the whole future of her colony depends. She, the lone survivor, will start a small paper nest in the spring, lay a few eggs in its cells, and carefully rear a small force of workers. With their help she will then enlarge the nest and develop the colony. Then the workers, foraging ceaselessly through the hot days of summer, will destroy countless caterpillars.

Thus, through the circumstances of their lives, and the nature of our own wants, all these have been our allies in keeping the balance of nature tilted in our favor. Yet we have turned our artillery against our friends. The terrible danger is that we have grossly underestimated their value in keeping at bay a dark tide of enemies that, without their help, can overrun us.

The prospect of a general and permanent lowering of environmental resistance becomes grimly and increasingly real with each passing year as the number, variety, and destructiveness of insecticides grows. With the passage of time we may expect progressively more serious outbreaks of insects, both disease-carrying and crop-destroying species, in excess of anything we have ever known.

"Yes, but isn't this all theoretical?" you may ask. "Surely it won't really happen—not in my lifetime, anyway."

But it is happening, here and now. Scientific journals had already recorded some 50 species involved in violent dislocations of nature's balance by 1958. More examples are being found every year. A recent review of the subject contained references to 215 papers reporting or discussing unfavorable upsets in the balance of insect populations caused by pesticides.

Sometimes the result of chemical spraying has been a tremendous upsurge of the very insect the spraying was intended to control, as when blackflies in Ontario became 17 times more abundant after spraying than they had been before. Or when in England an enormous outbreak of the cabbage aphid—an outbreak that had no parallel on record—followed spraying with one of the organic phosphorus chemicals.

At other times spraying, while reasonably effective against the target insect, has let loose a whole Pandora's box of destructive pests that had never previously been abundant enough to cause trouble. The spider mite, for example, has become practically a worldwide pest as DDT and other insecticides have killed off its enemies. The spider mite is not an insect. It is a barely visible eight-legged creature belonging to the group that includes spiders, scorpions, and ticks. It has mouth parts adapted for piercing and sucking, and a prodigious appetite for the chlorophyll that makes the world green. It inserts these minute and stiletto-sharp mouth parts into the outer cells of leaves and evergreen needles and extracts the chlorophyll. A mild infestation gives trees and shrubbery a mottled or salt-and-pepper appearance; with a heavy mite population, foliage turns yellow and falls.

This is what happened in some of the western national forests a few years ago, when in 1956 the United States Forest Service sprayed some 885,000 acres of forested lands with DDT. The intention was to control the spruce budworm, but the following summer it was discovered that a problem worse than the budworm damage had been created. In surveying the forests from the air, vast blighted areas could be seen where the magnificent Douglas firs were turning brown and dropping their needles. In the Helena National Forest and on the western slopes of the Big Belt Mountains, then in other areas of Montana and down into Idaho the forests looked as though they had been scorched. It was evident that this summer of 1957 had brought the most extensive and spectacular infestation of spider mites in history. Almost all of the sprayed area was affected. Nowhere else was the damage evident. Searching for precedents, the foresters could remember other scourges of spider mites, though less dramatic than this one. There had been similar trouble along the Madison River in Yellowstone Park in 1929, in Colorado 20 years later, and then in New Mexico in 1956. *Each of these outbreaks had followed forest spraying with insecticides.* (The 1929 spraying, occurring before the DDT era, employed lead arsenate.)

Why does the spider mite appear to thrive on insecticides? Besides the obvious fact that it is relatively insensitive to them, there seem to be two other reasons. In nature it is kept in check by various predators such as ladybugs, a gall midge, predaceous mites and several pirate bugs, all of them extremely sensitive to insecticides. The third reason has to do with population pressure within the spider mite colonies. An undisturbed colony of mites is a densely settled community, huddled under a protective webbing for concealment from its enemies. When sprayed, the colonies disperse as the mites, irritated though not killed by the chemicals, scatter out in search of places where they will not be disturbed. In so doing they find a far greater abundance of space and food than was available in the former colonies. Their enemies are now dead so there is no need for the mites to spend their energy in secreting protective webbing. Instead, they pour all their energies into producing more mites. It is not uncommon for their egg production to be increased threefold—all through the beneficent effect of insecticides.

In the Shenandoah Valley of Virginia, a famous apple-growing region, hordes of a small insect called the red-banded leaf roller arose to plague the growers as soon as DDT began to replace arsenate of lead. Its depredations had never before been important; soon its toll rose to 50 per cent of the crop and it achieved the status of the most destructive pest of apples, not only in this region but throughout much of the East and Midwest, as the use of DDT increased.

The situation abounds in ironies. In the apple orchards of Nova Scotia in the late 1940's the worst infestations of the codling moth (cause of "wormy apples") were in the orchards regularly sprayed. In unsprayed orchards the moths were not abundant enough to cause real trouble.

Diligence in spraying had a similarly unsatisfactory reward in the eastern Sudan, where cotton growers had a bitter experience with DDT. Some 60,000 acres of cotton were being grown under irrigation in the Gash Delta. Early trials of DDT having given apparently good results, spraying was intensified. It was then that trouble began. One of the most destructive enemies of cotton is the bollworm. But the more cotton was sprayed, the more bollworms appeared. The unsprayed cotton suffered less damage to fruits and later to mature bolls than the sprayed, and in twice-sprayed fields the yield of seed cotton dropped significantly. Although some of the leaf-feeding insects were eliminated, any benefit that might thus have been gained was more than offset by bollworm damage. In the end the growers were faced with the unpleasant truth that their cotton yield would have been greater had they saved themselves the trouble and expense of spraying.

In the Belgian Congo and Uganda the results of heavy applications of DDT against an insect pest of the coffee bush were almost "catastrophic." The pest itself was found to be almost completely unaffected by the DDT, while its predator was extremely sensitive.

In America, farmers have repeatedly traded one insect enemy for a worse one as spraying upsets the population dynamics of the insect world. Two of the mass-spraying programs recently carried out have had precisely this effect. One was the fire ant eradication program in the South; the other was the spraying for the Japanese beetle in the Midwest. (See Chapters 10 and 7.)

When a wholesale application of heptachlor was made to the farmlands in Louisiana in 1957, the result was the unleashing of one of the worst enemies of the sugarcane crop—the sugarcane borer. Soon after the heptachlor treatment, damage by borers increased sharply. The chemical aimed at the fire ant had killed off the enemies of the borer. The crop was so severely damaged that farmers sought to bring suit against the state for negligence in not warning them that this might happen.

The same bitter lesson was learned by Illinois farmers. After the devastating bath of dieldrin recently administered to the farmlands in eastern Illinois for the control of the Japanese beetle, farmers discovered that corn borers had increased enormously in the treated area. In fact, corn grown in fields within this area contained almost twice as many of the destructive larvae of this insect as did the corn grown outside. The farmers may not yet be aware of the biological basis of what has happened, but they need no scientists to tell them they have made a poor bargain. In trying to get rid of one insect, they have brought on a scourge of a much more destructive one. According to Department of Agriculture estimates, total damage by the Japanese beetle in the United States adds up to about 10 million dollars a year, while damage by the corn borer runs to about 85 million.

It is worth noting that natural forces had been heavily relied on for control of the corn borer. Within two years after this insect was accidentally introduced from Europe in 1917, the United States Government had mounted one of its most intensive programs for locating and importing parasites of an insect pest. Since that time 24 species of parasites of the corn borer have been brought in from Europe and the Orient at considerable expense. Of these, 5 are recognized as being of distinct value in control. Needless to say, the results of all this work are now jeopardized as the enemies of the corn borer are killed off by the sprays.

If this seems absurd, consider the situation in the citrus groves of California, where the world's most famous and successful experiment in biological control was carried out in the 1880's. In 1872 a scale insect that feeds on the sap of citrus trees appeared in California and within the next 15 years developed into a pest so destructive that the fruit crop in many orchards was a complete loss. The young citrus industry was threatened with destruction. Many farmers gave up and pulled out their trees. Then a parasite of the scale insect was imported from Australia, a small lady beetle called the vedalia. Within only two years after the first shipment of the beetles, the scale was under complete control throughout the citrus-growing sections of California. From that time on one could search for days among the orange groves without finding a single scale insect.

Then in the 1940's the citrus growers began to experiment with glamorous new chemicals against other insects. With the advent of DDT and the even more toxic chemicals to follow, the populations of the vedalia in many sections of California were wiped out. Its importation had cost the government a mere $5000. Its activities had saved the fruit growers several millions of dollars a year, but in a moment of heedlessness the benefit was canceled out. Infestations of the scale insect quickly reappeared and damage exceeded anything that had been seen for fifty years.

"This possibly marked the end of an era," said Dr. Paul DeBach of the Citrus Experiment Station in Riverside. Now control of the scale has become enormously complicated. The vedalia can be maintained only by repeated releases and by the most careful attention to spray schedules, to minimize their contact with insecticides. And regardless of what the citrus growers do, they are more or less at the mercy of the owners of adjacent acreages, for severe damage has been done by insecticidal drift.

All these examples concern insects that attack agricultural crops. What of those that carry disease? There have already been warnings. On Nissan Island in the South Pacific, for example, spraying had been carried on intensively during the Second World War, but was stopped when hostilities came to an end. Soon swarms of a malaria-carrying mosquito reinvaded the island. All of its predators had been killed off and there had not been time for new populations to become established. The way was therefore clear for a tremendous population explosion. Marshall Laird, who has described this incident, compares chemical control to a treadmill; once we have set foot on it we are unable to stop for fear of the consequences.

In some parts of the world disease can be linked with spraying in quite a different way. For some reason, snail-like mollusks seem to be almost immune to the effects of insecticides. This has been observed many times. In the general holocaust that followed the spraying of salt marshes in eastern Florida (pages 130–131), aquatic snails alone survived. The scene as described was a macabre picture—something that might have been created by a surrealist brush. The snails moved among the bodies of the dead fishes and the moribund crabs, devouring the victims of the death rain of poison.

But why is this important? It is important because many aquatic snails serve as hosts of dangerous parasitic worms that spend part of their life cycle in a mollusk, part in a human being. Examples are the blood flukes, or schistosoma, that cause serious disease in man when they enter the body by way of drinking water or through the skin when people are bathing in infested waters. The flukes are released into the water by the host snails. Such diseases are especially prevalent in parts of Asia and Africa. Where they occur, insect control measures that favor a vast increase of snails are likely to be followed by grave consequences.

And of course man is not alone in being subject to snail-borne disease. Liver disease in cattle, sheep, goats, deer, elk, rabbits, and various other warm-blooded animals may be caused by liver flukes that spend part of their life cycles in fresh-water snails. Livers infested with these worms are unfit for use as human food and are routinely condemned. Such rejections cost American cattlemen about 3½ million dollars annually. Anything that acts to increase the number of snails can obviously make this problem an even more serious one.

Over the past decade these problems have cast long shadows, but we have been slow to recognize them. Most of those best fitted to develop natural controls and assist in putting them into effect have been too busy laboring in the more exciting vineyards of chemical control. It was reported in 1960 that only 2 per cent of all the economic entomologists in the country were then working in the field of biological controls. A substantial number of the remaining 98 per cent were engaged in research on chemical insecticides.

Why should this be? The major chemical companies are pouring money into the universities to support research on insecticides. This creates attractive fellowships for graduate students and attractive staff positions. Biological-control studies, on the other hand, are never so endowed—for the simple reason that they do not promise anyone the fortunes that are to be made in the chemical industry. These are left to state and federal agencies, where the salaries paid are far less.

This situation also explains the otherwise mystifying fact that certain outstanding entomologists are among the leading advocates of chemical control. Inquiry into the background of some of these men reveals that their entire research program is supported by the chemical industry. Their professional prestige, sometimes their very jobs depend on the perpetuation of chemical methods. Can we then expect them to bite the hand that literally feeds them? But knowing their bias, how much credence can we give to their protests that insecticides are harmless?

Amid the general acclaim for chemicals as the principal method of insect control, minority reports have occasionally been filed by those few entomologists who have not lost sight of the fact that they are neither chemists nor engineers, but biologists.

F. H. Jacob in England has declared that "the activities of many so-called economic entomologists would make it appear that they operate in the belief that salvation lies at the end of a spray nozzle . . . that when they have created problems of resurgence or resistance or mammalian toxicity, the chemist will be ready with another pill. That view is not held here . . . Ultimately only the biologist will provide the answers to the basic problems of pest control."

"Economic entomologists must realize," wrote A. D. Pickett of Nova Scotia, "that they are dealing with living things . . . their work must be more than simply insecticide testing or a quest for highly destructive chemicals." Dr. Pickett himself was a pioneer in the field of working out sane methods of insect control that take full advantage of the predatory and parasitic species. The method which he and his associates evolved is today a shining model but one too little emulated. Only in the integrated control programs developed by some California entomologists do we find anything comparable in this country.

Dr. Pickett began his work some thirty-five years ago in the apple orchards of the Annapolis Valley in Nova Scotia, once one of the most concentrated fruit-growing areas in Canada.

At that time it was believed that insecticides—then inorganic chemicals—would solve the problems of insect control, that the only task was to induce fruit growers to follow the recommended methods. But the rosy picture failed to materialize. Somehow the insects persisted. New chemicals were added, better spraying equipment was devised, and the zeal for spraying increased, but the insect problem did not get any better. Then DDT promised to "obliterate the nightmare" of codling moth outbreaks. What actually resulted from its use was an unprecedented scourge of mites. "We move from crisis to crisis, merely trading one problem for another," said Dr. Pickett.

At this point, however, Dr. Pickett and his associates struck out on a new road instead of going along with other entomologists who continued to pursue the will-o'-the-wisp of the ever more toxic chemical. Recognizing that they had a strong ally in nature, they devised a program that makes maximum use of natural controls and minimum use of insecticides. Whenever insecticides are applied only minimum dosages are used— barely enough to control the pest without avoidable harm to beneficial species. Proper timing also enters in. Thus, if nicotine sulphate is applied before rather than after the apple blossoms turn pink one of the important predators is spared, probably because it is still in the egg stage.

Dr. Pickett uses special care to select chemicals that will do as little harm as possible to insect parasites and predators. "When we reach the point of using DDT, parathion, chlordane, and other new insecticides as routine control measures in the same way we have used the inorganic chemicals in the past, entomologists interested in biological control may as well throw in the sponge," he says. Instead of these highly toxic, broad-spectrum insecticides, he places chief reliance on ryania (derived from ground stems of a tropical plant), nicotine sulphate, and lead arsenate. In certain situations very weak concentrations of DDT or malathion are used (1 or 2 ounces per 100 gallons—in contrast to the usual 1 or 2 pounds per 100 gallons). Although these two are the least toxic of the modern insecticides, Dr. Pickett hopes by further research to replace them with safer and more selective materials.

How well has this program worked? Nova Scotia orchardists who are following Dr. Pickett's modified spray program are producing as high a proportion of first-grade fruit as are those who are using intensive chemical applications. They are also getting as good production. They are getting these results, moreover, at a substantially lower cost. The outlay for insecticides in Nova Scotia apple orchards is only from 10 to 20 per cent of the amount spent in most other apple-growing areas.

More important than even these excellent results is the fact that the modified program worked out by these Nova Scotian entomologists is not doing violence to nature's balance. It is well on the way to realizing the philosophy stated by the Canadian entomologist G. C. Ullyett a decade ago: "We must change our philosophy, abandon our attitude of human superiority and admit that in many cases in natural environments we find ways and means of limiting populations of organisms in a more economical way than we can do it ourselves."

SIXTEEN

The Rumblings of an Avalanche

IF DARWIN were alive today the insect world would delight and astound him with its impressive verification of his theories of the survival of the fittest. Under the stress of intensive chemical spraying the weaker members of the insect populations are being weeded out. Now, in many areas and among many species only the strong and fit remain to defy our efforts to control them.

Nearly half a century ago, a professor of entomology at Washington State College, A. L. Melander, asked the now purely rhetorical question, "Can insects become resistant to sprays?" If the answer seemed to Melander unclear, or slow in coming, that was only because he asked his question too soon—in 1914 instead of 40 years later. In the pre-DDT era, inorganic chemicals, applied on a scale that today would seem extraordinarily modest, produced here and there strains of insects that could survive chemical spraying or dusting. Melander himself had run into

difficulty with the San José scale, for some years satisfactorily controlled by spraying with lime sulfur. Then in the Clarkston area of Washington the insects became refractory—they were harder to kill than in the orchards of the Wenatchee and Yakima valleys and elsewhere.

Suddenly the scale insects in other parts of the country seemed to have got the same idea: it was not necessary for them to die under the sprayings of lime sulfur, diligently and liberally applied by orchardists. Throughout much of the Midwest thousands of acres of fine orchards were destroyed by insects now impervious to spraying.

Then in California the time-honored method of placing canvas tents over trees and fumigating them with hydrocyanic acid began to yield disappointing results in certain areas, a problem that led to research at the California Citrus Experiment Station, beginning about 1915 and continuing for a quarter of a century. Another insect to learn the profitable way of resistance was the codling moth, or appleworm, in the 1920's, although lead arsenate had been used successfully against it for some 40 years.

But it was the advent of DDT and all its many relatives that ushered in the true Age of Resistance. It need have surprised no one with even the simplest knowledge of insects or of the dynamics of animal populations that within a matter of a very few years an ugly and dangerous problem had clearly defined itself. Yet awareness of the fact that insects possess an effective counterweapon to aggressive chemical attack seems to have dawned slowly. Only those concerned with disease-carrying insects seem by now to have been thoroughly aroused to the alarming nature of the situation; the agriculturists still for the most part blithely put their faith in the development of new and ever more toxic chemicals, although the present difficulties have been born of just such specious reasoning.

If understanding of the phenomenon of insect resistance developed slowly, it was far otherwise with resistance itself. Before 1945 only about a dozen species were known to have developed resistance to any of the pre-DDT insecticides. With the new organic chemicals and new methods for their intensive application, resistance began a meteoric rise that reached the alarming level of 137 species in 1960. No one believes the end is in sight. More than 1000 technical papers have now been published on the subject. The World Health Organization has enlisted the aid of some 300 scientists in all parts of the world, declaring that "resistance is at present the most important single problem facing vector-control programmes." A distinguished British student of animal populations, Dr. Charles Elton, has said, "We are hearing the early rumblings of what may become an avalanche in strength."

Sometimes resistance develops so rapidly that the ink is scarcely dry on a report hailing successful control of a species with some specified chemical when an amended report has to be issued. In South Africa, for example, cattlemen had long been plagued by the blue tick, from which, on one ranch alone, 600 head of cattle had died in one year. The tick had for some years been resistant to arsenical dips. Then benzene hexachloride was tried, and for a very short time all seemed to be well. Reports issued early in the year 1949 declared that the arsenic-resistant ticks could be controlled readily with the new chemical; later in the same year, a bleak notice of developing resistance had to be published. The situation prompted a writer in the *Leather Trades Review* to comment in 1950: "News such as this quietly trickling through scientific circles and appearing in small sections of the overseas press is enough to make headlines as big as those concerning the new atomic bomb if only the significance of the matter were properly understood."

Although insect resistance is a matter of concern in agriculture and forestry, it is in the field of public health that the most serious apprehensions have been felt. The relation between various insects and many diseases of man is an ancient one. Mosquitoes of the genus *Anopheles* may inject into the human bloodstream the single-celled organism of malaria. Other mosquitoes transmit yellow fever. Still others carry encephalitis. The housefly, which does not bite, nevertheless by contact may contaminate human food with the bacillus of dysentery, and in many parts of the world may play an important part in the transmission of eye diseases. The list of diseases and their insect carriers, or vectors, includes typhus and body lice, plague and rat fleas, African sleeping sickness and tsetse flies, various fevers and ticks, and innumerable others.

These are important problems and must be met. No responsible person contends that insect-borne disease should be ignored. The question that has now urgently presented itself is whether it is either wise or responsible to attack the problem by methods that are rapidly making it worse. The world has heard much of the triumphant war against disease through the control of insect vectors of infection, but it has heard little of the other side of the story—the defeats, the short-lived triumphs that now strongly support the alarming view that the insect enemy has been made actually stronger by our efforts. Even worse, we may have destroyed our very means of fighting.

A distinguished Canadian entomologist, Dr. A. W. A. Brown, was engaged by the World Health Organization to make a comprehensive survey of the resistance problem. In the resulting monograph, published in 1958, Dr. Brown has this to say: "Barely a decade after the introduction of the potent synthetic insecticides in public health programmes, the main technical problem is the development of resistance to them by the insects they formerly controlled." In publishing his monograph, the World Health Organization warned that "the vigorous offensive now being pursued against arthropod-borne diseases such as malaria, typhus fever, and plague risks a serious setback unless this new problem can be rapidly mastered."

What is the measure of this setback? The list of resistant species now includes practically all of the insect groups of medical importance. Apparently the blackflies, sand flies, and tsetse flies have not yet become resistant to chemicals. On the other hand, resistance among houseflies and body lice has now developed on a global scale. Malaria programs are threatened by resistance among mosquitoes. The oriental rat flea, the principal vector of plague, has recently demonstrated resistance to DDT, a most serious development. Countries reporting resistance among a large number of other species represent every continent and most of the island groups.

Probably the first medical use of modern insecticides occurred in Italy in 1943 when the Allied Military Government launched a successful attack on typhus by dusting enormous numbers of people with DDT. This was followed two years later by extensive application of residual sprays for the control of malaria mosquitoes. Only a year later the first signs of trouble appeared. Both houseflies and mosquitoes of the genus *Culex* began to show resistance to the sprays. In 1948 a new chemical, chlordane, was tried as a supplement to DDT. This time good control was obtained for two years, but by August of 1950 chlordane-resistant flies appeared, and by the end of that year all of the houseflies as well as the *Culex* mosquitoes seemed to be resistant to chlordane. As rapidly as new chemicals were brought into use, resistance developed. By the end of 1951, DDT, methoxychlor, chlordane, heptachlor, and benzene hexachloride had joined the list of chemicals no longer effective. The flies, meanwhile, had become "fantastically abundant."

The same cycle of events was being repeated in Sardinia during the late 1940's. In Denmark, products containing DDT were first used in 1944; by 1947 fly control had failed in many places. In some areas of Egypt, flies had already become resistant to DDT by 1948; BHC was substituted but was effective for less than a year. One Egyptian village in particular symbolizes the problem. Insecticides gave good control of flies in 1950 and during this same year the infant mortality rate was reduced by nearly 50 per cent. The next year, nevertheless, flies were resistant to DDT and chlordane. The fly population returned to its former level; so did infant mortality.

In the United States, DDT resistance among flies had become widespread in the Tennessee Valley by 1948. Other areas followed. Attempts to restore control with dieldrin met with little success, for in some places the flies developed strong resistance to this chemical *within only two months*. After running through all the available chlorinated hydrocarbons, control agencies turned to the organic phosphates, but here again the story of resistance was repeated. The present conclusion of experts is that "housefly control has escaped insecticidal techniques and once more must be based on general sanitation."

The control of body lice in Naples was one of the earliest and most publicized achievements of DDT. During the next few years its success in Italy was matched by the successful control of lice affecting some two million people in Japan and Korea in the winter of 1945–46. Some premonition of trouble ahead might have been gained by the failure to control a typhus epidemic in Spain in 1948. Despite this failure in actual practice, encouraging laboratory experiments led entomologists to believe lice were unlikely to develop resistance. Events in Korea in the winter of 1950–51 were therefore startling. When DDT powder was applied to a group of Korean soldiers the extraordinary result was an actual increase in the infestation of lice. When lice were collected and tested, it was found that 5 per cent DDT powder caused no increase in their natural mortality rate. Similar results among lice collected from vagrants in Tokyo, from an asylum in Itabashi, and from refugee camps in Syria, Jordan, and eastern Egypt, confirmed the ineffectiveness of DDT for the control of lice and typhus. When by 1957 the list of countries in which lice had become resistant to DDT was extended to include Iran, Turkey, Ethiopia, West Africa, South Africa, Peru, Chile, France, Yugoslavia, Afghanistan, Uganda, Mexico, and Tanganyika, the initial triumph in Italy seemed dim indeed.

The first malaria mosquito to develop resistance to DDT was *Anopheles sacharovi* in Greece. Extensive spraying was begun in 1946 with early success; by 1949, however, observers noticed that adult mosquitoes were resting in large numbers under road bridges, although they were absent from houses and stables that had been treated. Soon this habit of outside resting was extended to caves, outbuildings, and culverts and to the foliage and trunks of orange trees. Apparently the adult mosquitoes had become sufficiently tolerant of DDT to escape from sprayed buildings and rest and recover in the open. A few months later they were able to remain in houses, where they were found resting on treated walls.

This was a portent of the extremely serious situation that has now developed. Resistance to insecticides by mosquitoes of the anophelene group has surged upward at an astounding rate, being created by the thoroughness of the very house-spraying programs designed to eliminate malaria. In 1956, only 5 species of these mosquitoes displayed resistance; by early 1960 the number had risen from 5 to 28! The number includes very dangerous malaria vectors in West Africa, the Middle East, Central America, Indonesia, and the eastern European region.

Among other mosquitoes, including carriers of other diseases, the pattern is being repeated. A tropical mosquito that carries parasites responsible for such diseases as elephantiasis has become strongly resistant in many parts of the world. In some areas of the United States the mosquito vector of western equine encephalitis has developed resistance. An even more serious problem concerns the vector of yellow fever, for centuries one of the great plagues of the world. Insecticide-resistant strains of this mosquito have occurred in Southeast Asia and are now common in the Caribbean region.

The consequences of resistance in terms of malaria and other diseases are indicated by reports from many parts of the world. An outbreak of yellow fever in Trinidad in 1954 followed failure to control the vector mosquito because of resistance. There has been a flare-up of malaria in Indonesia and Iran. In Greece, Nigeria, and Liberia the mosquitoes continue to harbor and transmit the malaria parasite. A reduction of diarrheal disease achieved in Georgia through fly control was wiped out within about a year. The reduction in acute conjunctivitis in Egypt, also attained through temporary fly control, did not last beyond 1950.

Less serious in terms of human health, but vexatious as man measures economic values, is the fact that salt-marsh mosquitoes in Florida also are showing resistance. Although these are not vectors of disease, their presence in bloodthirsty swarms had rendered large areas of coastal Florida uninhabitable until control—of an uneasy and temporary nature—was established. But this was quickly lost.

The ordinary house mosquito is here and there developing resistance, a fact that should give pause to many communities that now regularly arrange for wholesale spraying. This species is now resistant to several insecticides, among which is the almost universally used DDT, in Italy, Israel, Japan, France, and parts of the United States, including California, Ohio, New Jersey, and Massachusetts.

Ticks are another problem. The woodtick, vector of spotted fever, has recently developed resistance; in the brown dog tick the ability to escape a chemical death has long been thoroughly and widely established. This poses problems for human beings as well as for dogs. The brown dog tick is a semitropical species and when it occurs as far north as New Jersey it must live over winter in heated buildings rather than out of doors. John C. Pallister of the American Museum of Natural History reported in the summer of 1959 that his department had been getting a number of calls from neighboring apartments on Central Park West. "Every now and then," Mr. Pallister said, "a whole apartment house gets infested with young ticks, and they're hard to get rid of. A dog will pick up ticks in Central Park, and then the ticks lay eggs and they hatch in the apartment. They seem immune to DDT or chlordane or most of our modern sprays. It used to be very unusual to have ticks in New York City, but now they're all over here and on Long Island, in Westchester and on up into Connecticut. We've noticed this particularly in the past five or six years."

The German cockroach throughout much of North America has become resistant to chlordane, once the favorite weapon of exterminators who have now turned to the organic phosphates. However, the recent development of resistance to these insecticides confronts the exterminators with the problem of where to go next.

Agencies concerned with vector-borne disease are at present coping with their problems by switching from one insecticide to another as resistance develops. But this cannot go on indefinitely, despite the ingenuity of the chemists in supplying new materials. Dr. Brown has pointed out that we are traveling "a one-way street." No one knows how long the street is. If the dead end is reached before control of disease-carrying insects is achieved, our situation will indeed be critical.

With insects that infest crops the story is the same.

To the list of about a dozen agricultural insects showing resistance to the inorganic chemicals of an earlier era there is now added a host of others resistant to DDT, BHC, lindane, toxaphene, dieldrin, aldrin, and even to the phosphates from which so much was hoped. The total number of resistant species among crop-destroying insects had reached 65 in 1960.

The first cases of DDT resistance among agricultural insects appeared in the United States in 1951, about six years after its first use. Perhaps the most troublesome situation concerns the codling moth, which is now resistant to DDT in practically all of the world's apple-growing regions. Resistance in cabbage insects is creating another serious problem. Potato insects are escaping chemical control in many sections of the United States. Six species of cotton insects, along with an assortment of thrips, fruit moths, leaf hoppers, caterpillars, mites, aphids, wireworms, and many others now are able to ignore the farmer's assault with chemical sprays.

The chemical industry is perhaps understandably loath to face up to the unpleasant fact of resistance. Even in 1959, with more than 100 major insect species showing definite resistance to chemicals, one of the leading journals in the field of agricultural chemistry spoke of "real or imagined" insect resistance.

Yet hopefully as the industry may turn its face the other way, the problem simply does not go away, and it presents some unpleasant economic facts. One is that the cost of insect control by chemicals is increasing steadily. It is no longer possible to stockpile materials well in advance; what today may be the most promising of insecticidal chemicals may be the dismal failure of tomorrow. The very substantial financial investment involved in backing and launching an insecticide may be swept away as the insects prove once more that the effective approach to nature is not through brute force. And however rapidly technology may invent new uses for insecticides and new ways of applying them, it is likely to find the insects keeping a lap ahead.

Darwin himself could scarcely have found a better example of the operation of natural selection than is provided by the way the mechanism of resistance operates. Out of an original population, the members of which vary greatly in qualities of structure, behavior, or physiology, it is the "tough" insects that survive chemical attack. Spraying kills off the weaklings. The only survivors are insects that have some inherent quality that allows them to escape harm. These are the parents of the new generation, which, by simple inheritance, possesses all the qualities of "toughness" inherent in its forebears. Inevitably it follows that intensive spraying with powerful chemicals only makes worse the problem it is designed to solve. After a few generations, instead of a mixed population of strong and weak insects, there results a population consisting entirely of tough, resistant strains.

The means by which insects resist chemicals probably vary and as yet are not thoroughly understood. Some of the insects that defy chemical control are thought to be aided by a structural advantage, but there seems to be little actual proof of this. That immunity exists in some strains is clear, however, from observations like those of Dr. Briejèr, who reports watching flies at the Pest Control Institute at Springforbi, Denmark, "disporting themselves in DDT as much at home as primitive sorcerers cavorting over red-hot coals."

Similar reports come from other parts of the world. In Malaya, at Kuala Lumpur, mosquitoes at first reacted to DDT by leaving the treated interiors. As resistance developed, however, they could be found at rest on surfaces where the deposit of

DDT beneath them was clearly visible by torchlight. And in an army camp in southern Taiwan samples of resistant bed-bugs were found actually carrying a deposit of DDT powder on their bodies. When these bedbugs were experimentally placed in cloth impregnated with DDT, they lived for as long as a month; they proceeded to lay their eggs; and the resulting young grew and thrived.

Nevertheless, the quality of resistance does not necessar-ily depend on physical structure. DDT-resistant flies possess an enzyme that allows them to detoxify the insecticide to the less toxic chemical DDE. This enzyme occurs only in flies that possess a genetic factor for DDT resistance. This factor is, of course, hereditary. How flies and other insects detoxify the or-ganic phosphorus chemicals is less clearly understood.

Some behavioral habit may also place the insect out of reach of chemicals. Many workers have noticed the tendency of resis-tant flies to rest more on untreated horizontal surfaces than on treated walls. Resistant houseflies may have the stable-fly habit of sitting still in one place, thus greatly reducing the frequency of their contact with residues of poison. Some malaria mosqui-toes have a habit that so reduces their exposure to DDT as to make them virtually immune. Irritated by the spray, they leave the huts and survive outside.

Ordinarily resistance takes two or three years to develop, al-though occasionally it will do so in only one season, or even less. At the other extreme it may take as long as six years. The number of generations produced by an insect population in a year is important, and this varies with species and climate. Flies in Canada, for example, have been slower to develop resistance than those in southern United States, where long hot summers favor a rapid rate of reproduction.

The hopeful question is sometimes asked, "If insects can become resistant to chemicals, could human beings do the same thing?" Theoretically they could; but since this would take hundreds or even thousands of years, the comfort to those living now is slight. Resistance is not something that develops in an individual. If he possesses at birth some qualities that make him less susceptible than others to poisons he is more likely to survive and produce children. Resistance, therefore, is something that develops in a population after time measured in several or many generations. Human populations reproduce at the rate of roughly three generations per century, but new insect generations arise in a matter of days or weeks.

"It is more sensible in some cases to take a small amount of damage in preference to having none for a time but paying for it in the long run by losing the very means of fighting," is the advice given in Holland by Dr. Briejèr in his capacity as director of the Plant Protection Service. "Practical advice should be 'Spray as little as you possibly can' rather than 'Spray to the limit of your capacity.' . . . Pressure on the pest population should always be as slight as possible."

Unfortunately, such vision has not prevailed in the corresponding agricultural services of the United States. The Department of Agriculture's *Yearbook* for 1952, devoted entirely to insects, recognizes the fact that insects become resistant but says, "More applications or greater quantities of the insecticides are needed then for adequate control." The Department does not say what will happen when the only chemicals left untried are those that render the earth not only insectless but lifeless. But in 1959, only seven years after this advice was given, a Connecticut entomologist was quoted in the *Journal of Agricultural and Food Chemistry* to the effect that on at least one or two insect pests *the last available* new material was then being used.

Dr. Briejèr says:

> It is more than clear that we are traveling a dangerous road. . . . *We are going to have to do some very energetic research on other control measures, measures that will have to be biological, not chemical. Our aim should be to guide natural processes as cautiously as possible in the desired direction rather than to use brute force. . . .*
>
> We need a more high-minded orientation and a deeper insight, which I miss in many researchers. Life is a miracle beyond our comprehension, and we should reverence it even where we have to struggle against it. . . . The resort to weapons such as insecticides to control it is a proof of insufficient knowledge and of an incapacity so to guide the processes of nature that brute force becomes unnecessary. Humbleness is in order; there is no excuse for scientific conceit here.

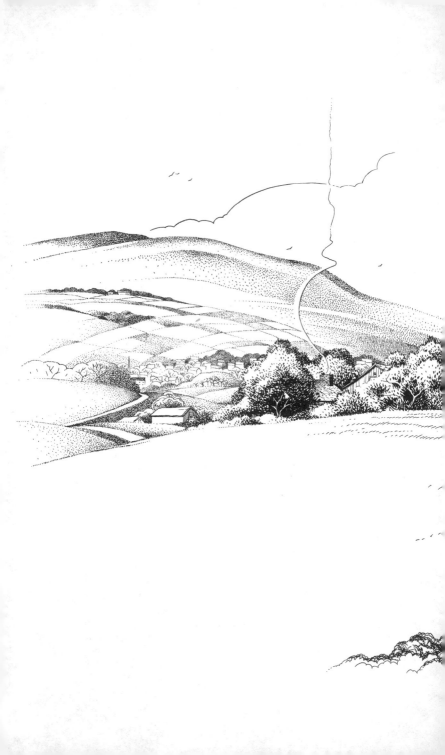

SEVENTEEN

The Other Road

W̶E STAND NOW where two roads diverge. But unlike the roads in Robert Frost's familiar poem, they are not equally fair. The road we have long been traveling is deceptively easy, a smooth superhighway on which we progress with great speed, but at its end lies disaster. The other fork of the road—the one "less traveled by"—offers our last, our only chance to reach a destination that assures the preservation of our earth.

The choice, after all, is ours to make. If, having endured much, we have at last asserted our "right to know," and if, knowing, we have concluded that we are being asked to take senseless and frightening risks, then we should no longer accept the counsel of those who tell us that we must fill our world with poisonous chemicals; we should look about and see what other course is open to us.

A truly extraordinary variety of alternatives to the chemical control of insects is available. Some are already in use and have achieved brilliant success. Others are in the stage of laboratory testing. Still others are little more than ideas in the minds of imaginative scientists, waiting for the opportunity to put them to the test. All have this in common: they are *biological* solutions, based on understanding of the living organisms they seek to control, and of the whole fabric of life to which these organisms belong. Specialists representing various areas of the vast field of biology are contributing—entomologists, pathologists, geneticists, physiologists, biochemists, ecologists—all pouring their knowledge and their creative inspirations into the formation of a new science of biotic controls.

"Any science may be likened to a river," says a Johns Hopkins biologist, Professor Carl P. Swanson. "It has its obscure and unpretentious beginning; its quiet stretches as well as its rapids; its periods of drought as well as of fullness. It gathers momentum with the work of many investigators and as it is fed by other streams of thought; it is deepened and broadened by the concepts and generalizations that are gradually evolved."

So it is with the science of biological control in its modern sense. In America it had its obscure beginnings a century ago with the first attempts to introduce natural enemies of insects that were proving troublesome to farmers, an effort that sometimes moved slowly or not at all, but now and again gathered speed and momentum under the impetus of an outstanding success. It had its period of drought when workers in applied entomology, dazzled by the spectacular new insecticides of the 1940's, turned their backs on all biological methods and set foot on "the treadmill of chemical control." But the goal of an insect-free world continued to recede. Now at last, as it has become apparent that the heedless and unrestrained use of chemicals is a greater menace to ourselves than to the targets, the river which is the science of biotic control flows again, fed by new streams of thought.

Some of the most fascinating of the new methods are those that seek to turn the strength of a species against itself—to use the drive of an insect's life forces to destroy it. The most spectacular of these approaches is the "male sterilization" technique developed by the chief of the United States Department of Agriculture's Entomology Research Branch, Dr. Edward Knipling, and his associates.

About a quarter of a century ago Dr. Knipling startled his colleagues by proposing a unique method of insect control. If it were possible to sterilize and release large numbers of insects, he theorized, the sterilized males would, under certain conditions, compete with the normal wild males so successfully that, after repeated releases, only infertile eggs would be produced and the population would die out.

The proposal was met with bureaucratic inertia and with skepticism from scientists, but the idea persisted in Dr. Knipling's mind. One major problem remained to be solved before it could be put to the test—a practical method of insect sterilization had to be found. Academically, the fact that insects could be sterilized by exposure to X-ray had been known since 1916, when an entomologist by the name of G. A. Runner reported such sterilization of cigarette beetles. Hermann Muller's pioneering work on the production of mutations by X-ray opened up vast new areas of thought in the late 1920's, and by the middle of the century various workers had reported the sterilization by X-rays or gamma rays of at least a dozen species of insects.

But these were laboratory experiments, still a long way from practical application. About 1950, Dr. Knipling launched a serious effort to turn insect sterilization into a weapon that would wipe out a major insect enemy of livestock in the South, the screw-worm fly. The females of this species lay their eggs in any open wound of a warm-blooded animal. The hatching larvae are parasitic, feeding on the flesh of the host. A full-grown steer may succumb to a heavy infestation in 10 days, and livestock losses in the United States have been estimated at $40,000,000 a year. The toll of wildlife is harder to measure, but it must be great. Scarcity of deer in some areas of Texas is attributed to the screw-worm. This is a tropical or subtropical insect, inhabiting South and Central America and Mexico, and in the United States normally restricted to the Southwest. About 1933, however, it was accidentally introduced into Florida, where the climate allowed it to survive over winter and to establish populations. It even pushed into southern Alabama and Georgia, and soon the livestock industry of the southeastern states was faced with annual losses running to $20,000,000.

A vast amount of information on the biology of the screwworm had been accumulated over the years by Agriculture Department scientists in Texas. By 1954, after some preliminary field trials on Florida islands, Dr. Knipling was ready for a full-scale test of his theory. For this, by arrangement with the Dutch Government, he went to the island of Curaçao in the Caribbean, cut off from the mainland by at least 50 miles of sea.

Beginning in August 1954, screw-worms reared and sterilized in an Agriculture Department laboratory in Florida were flown to Curaçao and released from airplanes at the rate of about 400 per square mile per week. Almost at once the number of egg masses deposited on experimental goats began to decrease, as did their fertility. Only seven weeks after the releases were started, all eggs were infertile. Soon it was impossible to find a single egg mass, sterile or otherwise. The screw-worm had indeed been eradicated on Curaçao.

The resounding success of the Curaçao experiment whetted the appetites of Florida livestock raisers for a similar feat that would relieve them of the scourge of screw-worms. Although the difficulties here were relatively enormous—an area 300 times as large as the small Caribbean island—in 1957 the United States Department of Agriculture and the State of Florida joined in providing funds for an eradication effort. The project involved the weekly production of about 50 million screw-worms at a specially constructed "fly factory," the use of 20 light airplanes to fly pre-arranged flight patterns, five to six hours daily, each plane carrying a thousand paper cartons, each carton containing 200 to 400 irradiated flies.

The cold winter of 1957–58, when freezing temperatures gripped northern Florida, gave an unexpected opportunity to start the program while the screw-worm populations were reduced and confined to a small area. By the time the program was considered complete at the end of 17 months, 3½ billion artificially reared, sterilized flies had been released over Florida and sections of Georgia and Alabama. The last-known animal wound infestation that could be attributed to screw-worms occurred in February 1959. In the next few weeks several adults were taken in traps. Thereafter no trace of the screw-worm could be discovered. Its extinction in the Southeast had been accomplished—a triumphant demonstration of the worth of scientific creativity, aided by thorough basic research, persistence, and determination.

Now a quarantine barrier in Mississippi seeks to prevent the re-entrance of the screw-worm from the Southwest, where it is firmly entrenched. Eradication there would be a formidable undertaking, considering the vast areas involved and the probability of re-invasion from Mexico. Nevertheless, the stakes are high and the thinking in the Department seems to be that some sort of program, designed at least to hold the screw-worm populations at very low levels, may soon be attempted in Texas and other infested areas of the Southwest.

The brilliant success of the screw-worm campaign has stimulated tremendous interest in applying the same methods to other insects. Not all, of course, are suitable subjects for this technique, much depending on details of the life history, population density, and reactions to radiation.

Experiments have been undertaken by the British in the hope that the method could be used against the tsetse fly in Rhodesia. This insect infests about a third of Africa, posing a menace to human health and preventing the keeping of livestock in an area of some 4½ million square miles of wooded grasslands. The habits of the tsetse differ considerably from those of the screw-worm fly, and although it can be sterilized by radiation some technical difficulties remain to be worked out before the method can be applied.

The British have already tested a large number of other species for susceptibility to radiation. United States scientists have had some encouraging early results with the melon fly and the oriental and Mediterranean fruit flies in laboratory tests in Hawaii and field tests on the remote island of Rota. The corn borer and the sugarcane borer are also being tested. There are possibilities, too, that insects of medical importance might be controlled by sterilization. A Chilean scientist has pointed out that malaria-carrying mosquitoes persist in his country in spite of insecticide treatment; the release of sterile males might then provide the final blow needed to eliminate this population.

The obvious difficulties of sterilizing by radiation have led to search for an easier method of accomplishing similar results, and there is now a strongly running tide of interest in chemical sterilants.

Scientists at the Department of Agriculture laboratory in Orlando, Florida, are now sterilizing the housefly in laboratory experiments and even in some field trials, using chemicals incorporated in suitable foods. In a test on an island in the Florida Keys in 1961, a population of flies was nearly wiped out within a period of only five weeks. Repopulation of course followed from nearby islands, but as a pilot project the test was successful. The Department's excitement about the promise of this method is easily understood. In the first place, as we have seen, the housefly has now become virtually uncontrollable by insecticides. A completely new method of control is undoubtedly needed. One of the problems of sterilization by radiation is that this requires not only artificial rearing but the release of sterile males in larger number than are present in the wild population. This could be done with the screw-worm, which is actually not an abundant insect. With the housefly, however, more than doubling the population through releases could be highly objectionable, even though the increase would be only temporary. A chemical sterilant, on the other hand, could be combined with a bait substance and introduced into the natural environment of the fly; insects feeding on it would become sterile and in the course of time the sterile flies would predominate and the insects would breed themselves out of existence.

The testing of chemicals for a sterilizing effect is much more difficult than the testing of chemical poisons. It takes 30 days to evaluate one chemical—although, of course, a number of tests can be run concurrently. Yet between April 1958 and December 1961 several hundred chemicals were screened at the Orlando laboratory for a possible sterilizing effect. The Department of Agriculture seems happy to have found among these even a handful of chemicals that show promise.

Now other laboratories of the Department are taking up the problem, testing chemicals against stable flies, mosquitoes, boll weevils, and an assortment of fruit flies. All this is presently experimental but in the few years since work began on chemosterilants the project has grown enormously. In theory it has many attractive features. Dr. Knipling has pointed out that effective chemical insect sterilization "might easily outdo some of the best of known insecticides." Take an imaginary situation in which a population of a million insects is multiplying five times in each generation. An insecticide might kill 90 per cent of each generation, leaving 125,000 insects alive after the third generation. In contrast, a chemical that would produce 90 per cent sterility would leave only 125 insects alive.

On the other side of the coin is the fact that some extremely potent chemicals are involved. It is fortunate that at least during these early stages most of the men working with chemosterilants seem mindful of the need to find safe chemicals and safe methods of application. Nonetheless, suggestions are heard here and there that these sterilizing chemicals might be applied as aerial sprays—for example, to coat the foliage chewed by gypsy moth larvae. To attempt any such procedure without thorough advance research on the hazards involved would be the height of irresponsibility. If the potential hazards of the chemosterilants are not constantly borne in mind we could easily find ourselves in even worse trouble than that now created by the insecticides.

The sterilants currently being tested fall generally into two groups, both of which are extremely interesting in their mode of action. The first are intimately related to the life processes, or metabolism, of the cell; i.e., they so closely resemble a substance the cell or tissue needs that the organism "mistakes" them for the true metabolite and tries to incorporate them in its normal building processes. But the fit is wrong in some detail and the process comes to a halt. Such chemicals are called antimetabolites.

The second group consists of chemicals that act on the chromosomes, probably affecting the gene chemicals and causing the chromosomes to break up. The chemosterilants of this group are alkylating agents, which are extremely reactive chemicals, capable of intense cell destruction, damage to chromosomes, and production of mutations. It is the view of Dr. Peter Alexander of the Chester Beatty Research Institute in London that "any alkylating agent which is effective in steriliz- ing insects would also be a powerful mutagen and carcinogen." Dr. Alexander feels that any conceivable use of such chemicals in insect control would be "open to the most severe objec- tions." It is to be hoped, therefore, that the present experi- ments will lead not to actual use of these particular chemicals but to the discovery of others that will be safe and also highly specific in their action on the target insect.

Some of the most interesting of the recent work is concerned with still other ways of forging weapons from the insect's own life processes. Insects produce a variety of venoms, attractants, repellants. What is the chemical nature of these secretions? Could we make use of them as, perhaps, very selective insecti- cides? Scientists at Cornell University and elsewhere are trying to find answers to some of these questions, studying the de- fense mechanisms by which many insects protect themselves from attack by predators, working out the chemical struc- ture of insect secretions. Other scientists are working on the so-called "juvenile hormone," a powerful substance which pre- vents metamorphosis of the larval insect until the proper stage of growth has been reached.

Perhaps the most immediately useful result of this explora- tion of insect secretion is the development of lures, or attrac- tants. Here again, nature has pointed the way. The gypsy moth is an especially intriguing example. The female moth is too heavy-bodied to fly. She lives on or near the ground, fluttering about among low vegetation or creeping up tree trunks. The male, on the contrary, is a strong flier and is attracted even from considerable distances by a scent released by the female from special glands. Entomologists have taken advantage of this fact for a good many years, laboriously preparing this sex attractant from the bodies of the female moths. It was then used in traps set for the males in census operations along the fringe of the insect's range. But this was an extremely expen- sive procedure. Despite the much publicized infestations in the northeastern states, there were not enough gypsy moths to provide the material, and hand-collected female pupae had to be imported from Europe, sometimes at a cost of half a dollar per tip. It was a tremendous breakthrough, therefore, when, after years of effort, chemists of the Agriculture Department recently succeeded in isolating the attractant. Following upon this discovery was the successful preparation of a closely related synthetic material from a constituent of castor oil; this not only deceives the male moths but is apparently fully as attractive as the natural substance. As little as one microgram (1/1000 gram) in a trap is an effective lure.

All this is of much more than academic interest, for the new and economical "gyplure" might be used not merely in census operations but in control work. Several of the more attractive possibilities are now being tested. In what might be termed an experiment in psychological warfare, the attractant is combined with a granular material and distributed by planes. The aim is to confuse the male moth and alter the normal behavior so that, in the welter of attractive scents, he cannot find the true scent trail leading to the female. This line of attack is being carried even further in experiments aimed at deceiving the male into attempting to mate with a spurious female. In the laboratory, male gypsy moths have attempted copulation with chips of wood, vermiculite, and other small, inanimate objects, so long as they were suitably impregnated with gyplure. Whether such diversion of the mating instinct into nonproductive channels would actually serve to reduce the population remains to be tested, but it is an interesting possibility.

The gypsy moth lure was the first insect sex attractant to be synthesized, but probably there will soon be others. A number of agricultural insects are being studied for possible attractants that man could imitate. Encouraging results have been obtained with the Hessian fly and the tobacco hornworm.

Combinations of attractants and poisons are being tried against several insect species. Government scientists have developed an attractant called methyl-eugenol, which males of the oriental fruit fly and the melon fly find irresistible. This has been combined with a poison in tests in the Bonin Islands 450 miles south of Japan. Small pieces of fiberboard were impregnated with the two chemicals and were distributed by air over the entire island chain to attract and kill the male flies. This program of "male annihilation" was begun in 1960: a year later the Agriculture Department estimated that more than 99 per cent of the population had been eliminated. The method as here applied seems to have marked advantages over the conventional broadcasting of insecticides. The poison, an organic phosphorus chemical, is confined to squares of fiberboard which are unlikely to be eaten by wildlife; its residues, moreover, are quickly dissipated and so are not potential contaminants of soil or water.

But not all communication in the insect world is by scents that lure or repel. Sound also may be a warning or an attraction. The constant stream of ultrasonic sound that issues from a bat in flight (serving as a radar system to guide it through darkness) is heard by certain moths, enabling them to avoid capture. The wing sounds of approaching parasitic flies warn the larvae of some sawflies to herd together for protection. On the other hand, the sounds made by certain wood-boring insects enable their parasites to find them, and to the male mosquito the wingbeat of the female is a siren song.

What use, if any, can be made of this ability of the insect to detect and react to sound? As yet in the experimental stage, but nonetheless interesting, is the initial success in attracting male mosquitoes to playback recordings of the flight sound of the female. The males were lured to a charged grid and so killed. The repellant effect of bursts of ultrasonic sound is being tested in Canada against corn borer and cutworm moths. Two authorities on animal sound, Professors Hubert and Mable Frings of the University of Hawaii, believe that a field method of influencing the behavior of insects with sound only awaits discovery of the proper key to unlock and apply the vast existing knowledge of insect sound production and reception. Repellant sounds may offer greater possibilities than attractants. The Fringses are known for their discovery that starlings scatter in alarm before a recording of the distress cry of one of their fellows; perhaps somewhere in this fact is a central truth that may be applied to insects. To practical men of industry the possibilities seem real enough so that at least one major electronic corporation is preparing to set up a laboratory to test them.

Sound is also being tested as an agent of direct destruction. Ultrasonic sound will kill all mosquito larvae in a laboratory tank; however, it kills other aquatic organisms as well. In other experiments, blowflies, mealworms, and yellow fever mosquitoes have been killed by airborne ultrasonic sound in a matter of seconds. All such experiments are first steps toward wholly new concepts of insect control which the miracles of electronics may some day make a reality.

The new biotic control of insects is not wholly a matter of electronics and gamma radiation and other products of man's inventive mind. Some of its methods have ancient roots, based on the knowledge that, like ourselves, insects are subject to disease. Bacterial infections sweep through their populations like the plagues of old; under the onset of a virus their hordes sicken and die. The occurrence of disease in insects was known before the time of Aristotle; the maladies of the silkworm were celebrated in medieval poetry; and through study of the diseases of this same insect the first understanding of the principles of infectious disease came to Pasteur.

Insects are beset not only by viruses and bacteria but also by fungi, protozoa, microscopic worms, and other beings from all that unseen world of minute life that, by and large, befriends mankind. For the microbes include not only disease organisms but those that destroy waste matter, make soils fertile, and enter into countless biological processes like fermentation and nitrification. Why should they not also aid us in the control of insects?

One of the first to envision such use of microorganisms was the 19th-century zoologist Elie Metchnikoff. During the concluding decades of the 19th and the first half of the 20th centuries the idea of microbial control was slowly taking form. The first conclusive proof that an insect could be brought under control by introducing a disease into its environment came in the late 1930's with the discovery and use of milky disease for the Japanese beetle, which is caused by the spores of a bacterium belonging to the genus *Bacillus*. This classic example of bacterial control has a long history of use in the eastern part of the United States, as I have pointed out in Chapter 7.

High hopes now attend tests of another bacterium of this genus—*Bacillus thuringiensis*—originally discovered in Germany in 1911 in the province of Thuringia, where it was found to cause a fatal septicemia in the larvae of the flour moth. This bacterium actually kills by poisoning rather than by disease. Within its vegetative rods there are formed, along with spores, peculiar crystals composed of a protein substance highly toxic to certain insects, especially to the larvae of the mothlike lepidopteras. Shortly after eating foliage coated with this toxin the larva suffers paralysis, stops feeding, and soon dies. For practical purposes, the fact that feeding is interrupted promptly is of course an enormous advantage, for crop damage stops almost as soon as the pathogen is applied. Compounds containing spores of *Bacillus thuringiensis* are now being manufactured by several firms in the United States under various trade names. Field tests are being made in several countries: in France and Germany against larvae of the cabbage butterfly, in Yugoslavia against the fall webworm, in the Soviet Union against a tent caterpillar. In Panama, where tests were begun in 1961, this bacterial insecticide may be the answer to one or more of the serious problems confronting banana growers. There the root borer is a serious pest of the banana, so weakening its roots that the trees are easily toppled by wind. Dieldrin has been the only chemical effective against the borer, but it has now set in motion a chain of disaster. The borers are becoming resistant. The chemical has also destroyed some important insect predators and so has caused an increase in the tortricids—small, stout-bodied moths whose larvae scar the surface of the bananas. There is reason to hope the new microbial insecticide will eliminate both the tortricids and the borers and that it will do so without upsetting natural controls.

In eastern forests of Canada and the United States bacterial insecticides may be one important answer to the problems of such forest insects as the budworms and the gypsy moth. In 1960 both countries began field tests with a commercial preparation of *Bacillus thuringiensis.* Some of the early results have been encouraging. In Vermont, for example, the end results of bacterial control were as good as those obtained with DDT. The main technical problem now is to find a carrying solution that will stick the bacterial spores to the needles of the evergreens. On crops this is not a problem—even a dust can be used. Bacterial insecticides have already been tried on a wide variety of vegetables, especially in California.

Meanwhile, other perhaps less spectacular work is concerned with viruses. Here and there in California fields of young alfalfa are being sprayed with a substance as deadly as any insecticide for the destructive alfalfa caterpillar—a solution containing a virus obtained from the bodies of caterpillars that have died because of infection with this exceedingly virulent disease. The bodies of only five diseased caterpillars provide enough virus to treat an acre of alfalfa. In some Canadian forests a virus that affects pine sawflies has proved so effective in control that it has replaced insecticides.

Scientists in Czechoslovakia are experimenting with protozoa against webworms and other insect pests, and in the United States a protozoan parasite has been found to reduce the egg-laying potential of the corn borer.

To some the term microbial insecticide may conjure up pictures of bacterial warfare that would endanger other forms of life. This is not true. In contrast to chemicals, insect pathogens are harmless to all but their intended targets. Dr. Edward Steinhaus, an outstanding authority on insect pathology, has stated emphatically that there is "no authenticated recorded instance of a true insect pathogen having caused an infectious disease in a vertebrate animal either experimentally or in nature." The insect pathogens are so specific that they infect only a small group of insects—sometimes a single species. Biologically they do not belong to the type of organisms that cause disease in higher animals or in plants. Also, as Dr. Steinhaus points out, outbreaks of insect disease in nature always remain confined to insects, affecting neither the host plants nor animals feeding on them.

Insects have many natural enemies—not only microbes of many kinds but other insects. The first suggestion that an insect might be controlled by encouraging its enemies is generally credited to Erasmus Darwin about 1800. Probably because it was the first generally practiced method of biological control, this setting of one insect against another is widely but erroneously thought to be the only alternative to chemicals.

In the United States the true beginnings of conventional biological control date from 1888 when Albert Koebele, the first of a growing army of entomologist explorers, went to Australia to search for natural enemies of the cottony cushion scale that threatened the California citrus industry with destruction. As we have seen in Chapter 15, the mission was crowned with spectacular success, and in the century that followed the world has been combed for natural enemies to control the insects that have come uninvited to our shores. In all, about 100 species of imported predators and parasites have become established. Besides the vedalia beetles brought in by Koebele, other importations have been highly successful. A wasp imported from Japan established complete control of an insect attacking eastern apple orchards. Several natural enemies of the spotted alfalfa aphid, an accidental import from the Middle East, are credited with saving the California alfalfa industry. Parasites and predators of the gypsy moth achieved good control, as did the *Tiphia* wasp against the Japanese beetle. Biological control of scales and mealy bugs is estimated to save California several millions of dollars a year—indeed, one of the leading entomologists of that state, Dr. Paul DeBach, has estimated that for an investment of $4,000,000 in biological control work California has received a return of $100,000,000.

Examples of successful biological control of serious pests by importing their natural enemies are to be found in some 40 countries distributed over much of the world. The advantages of such control over chemicals are obvious: it is relatively inexpensive, it is permanent, it leaves no poisonous residues. Yet biological control has suffered from lack of support. California is virtually alone among the states in having a formal program in biological control, and many states have not even one entomologist who devotes full time to it. Perhaps for want of support biological control through insect enemies has not always been carried out with the scientific thoroughness it requires—exacting studies of its impact on the populations of insect prey have seldom been made, and releases have not always been made with the precision that might spell the difference between success and failure.

The predator and the preyed upon exist not alone, but as part of a vast web of life, all of which needs to be taken into account. Perhaps the opportunities for the more conventional types of biological control are greatest in the forests. The farmlands of modern agriculture are highly artificial, unlike anything nature ever conceived. But the forests are a different world, much closer to natural environments. Here, with a minimum of help and a maximum of noninterference from man, Nature can have her way, setting up all that wonderful and intricate system of checks and balances that protects the forest from undue damage by insects.

In the United States our foresters seem to have thought of biological control chiefly in terms of introducing insect parasites and predators. The Canadians take a broader view, and some of the Europeans have gone farthest of all to develop the science of "forest hygiene" to an amazing extent. Birds, ants, forest spiders, and soil bacteria are as much a part of a forest as the trees, in the view of European foresters, who take care to inoculate a new forest with these protective factors. The encouragement of birds is one of the first steps. In the modern era of intensive forestry the old hollow trees are gone and with them homes for woodpeckers and other tree-nesting birds. This lack is met by nesting boxes, which draw the birds back into the forest. Other boxes are specially designed for owls and for bats, so that these creatures may take over in the dark hours the work of insect hunting performed in daylight by the small birds.

But this is only the beginning. Some of the most fascinating control work in European forests employs the forest red ant as an aggressive insect predator—a species which, unfortunately, does not occur in North America. About 25 years ago Professor Karl Gösswald of the University of Würzburg developed a method of cultivating this ant and establishing colonies. Under his direction more than 10,000 colonies of the red ant have been established in about 90 test areas in the German Federal Republic. Dr. Gösswald's method has been adopted in Italy and other countries, where ant farms have been established to supply colonies for distribution in the forests. In the Apennines, for example, several hundred nests have been set out to protect reforested areas.

"Where you can obtain in your forest a combination of birds' and ants' protection together with some bats and owls, the biological equilibrium has already been essentially improved," says Dr. Heinz Ruppertshofen, a forestry officer in Mölln, Germany, who believes that a single introduced predator or parasite is less effective than an array of the "natural companions" of the trees.

New ant colonies in the forests at Mölln are protected from woodpeckers by wire netting to reduce the toll. In this way the woodpeckers, which have increased by 400 per cent in 10 years in some of the test areas, do not seriously reduce the ant colonies, and pay handsomely for what they take by picking harmful caterpillars off the trees. Much of the work of caring for the ant colonies (and the birds' nesting boxes as well) is assumed by a youth corps from the local school, children 10 to 14 years old. The costs are exceedingly low; the benefits amount to permanent protection of the forests.

Another extremely interesting feature of Dr. Ruppertshofen's work is his use of spiders, in which he appears to be a pioneer. Although there is a large literature on the classification and natural history of spiders, it is scattered and fragmentary and deals not at all with their value as an agent of biological control. Of the 22,000 known kinds of spiders, 760 are native to Germany (and about 2000 to the United States). Twenty-nine families of spiders inhabit German forests.

To a forester the most important fact about a spider is the kind of net it builds. The wheel-net spiders are most important, for the webs of some of them are so narrow-meshed that they can catch all flying insects. A large web (up to 16 inches in diameter) of the cross spider bears some 120,000 adhesive nodules on its strands. A single spider may destroy in her life of 18 months an average of 2000 insects. A biologically sound forest has 50 to 150 spiders to the square meter (a little more than a square yard). Where there are fewer, the deficiency may be remedied by collecting and distributing the baglike cocoons containing the eggs. "Three cocoons of the wasp spider [which occurs also in America] yield a thousand spiders, which can catch 200,000 flying insects," says Dr. Ruppertshofen. The tiny and delicate young of the wheel-net spiders that emerge in the spring are especially important, he says, "as they spin in a teamwork a net umbrella above the top shoots of the trees and thus protect the young shoots against the flying insects." As the spiders molt and grow, the net is enlarged.

Canadian biologists have pursued rather similar lines of investigation, although with differences dictated by the fact that North American forests are largely natural rather than planted, and that the species available as aids in maintaining a healthy forest are somewhat different. The emphasis in Canada is on small mammals, which are amazingly effective in the control of certain insects, especially those that live within the spongy soil of the forest floor. Among such insects are the sawflies, so-called because the female has a saw-shaped ovipositor with which she slits open the needles of evergreen trees in order to deposit her eggs. The larvae eventually drop to the ground and form cocoons in the peat of tamarack bogs or the duff under spruce or pines. But beneath the forest floor is a world honey-combed with the tunnels and runways of small mammals—whitefooted mice, voles, and shrews of various species. Of all these small burrowers, the voracious shrews find and consume the largest number of sawfly cocoons. They feed by placing a forefoot on the cocoon and biting off the end, showing an extraordinary ability to discriminate between sound and empty cocoons. And for their insatiable appetite the shrews have no rivals. Whereas a vole can consume about 200 cocoons a day, a shrew, depending on the species, may devour up to 800! This may result, according to laboratory tests, in destruction of 75 to 98 per cent of the cocoons present.

It is not surprising that the island of Newfoundland, which has no native shrews but is beset with sawflies, so eagerly de-sired some of these small, efficient mammals that in 1958 the introduction of the masked shrew—the most efficient sawfly predator—was attempted. Canadian officials report in 1962 that the attempt has been successful. The shrews are multiply-ing and are spreading out over the island, some marked indi-viduals having been recovered as much as ten miles from the point of release.

There is, then, a whole battery of armaments available to the forester who is willing to look for permanent solutions that preserve and strengthen the natural relations in the forest. Chemical pest control in the forest is at best a stopgap measure bringing no real solution, at worst killing the fishes in the for-est streams, bringing on plagues of insects, and destroying the natural controls and those we may be trying to introduce. By such violent measures, says Dr. Ruppertshofen, "the partner-ship for life of the forest is entirely being unbalanced, and the catastrophes caused by parasites repeat in shorter and shorter periods . . . We, therefore, have to put an end to these unnatural manipulations brought into the most important and almost last natural living space which has been left for us."

Through all these new, imaginative, and creative approaches to the problem of sharing our earth with other creatures there runs a constant theme, the awareness that we are dealing with life—with living populations and all their pressures and counterpressures, their surges and recessions. Only by taking account of such life forces and by cautiously seeking to guide them into channels favorable to ourselves can we hope to achieve a reasonable accommodation between the insect hordes and ourselves.

The current vogue for poisons has failed utterly to take into account these most fundamental considerations. As crude a weapon as the cave man's club, the chemical barrage has been hurled against the fabric of life—a fabric on the one hand delicate and destructible, on the other miraculously tough and resilient, and capable of striking back in unexpected ways. These extraordinary capacities of life have been ignored by the practitioners of chemical control who have brought to their task no "high-minded orientation," no humility before the vast forces with which they tamper.

The "control of nature" is a phrase conceived in arrogance, born of the Neanderthal age of biology and philosophy, when it was supposed that nature exists for the convenience of man. The concepts and practices of applied entomology for the most part date from that Stone Age of science. It is our alarming misfortune that so primitive a science has armed itself with the most modern and terrible weapons, and that in turning them against the insects it has also turned them against the earth.

List of Principal Sources

CHAPTER 2: THE OBLIGATION TO ENDURE

Page 15

"Report on Environmental Health Problems," *Hearings*, 86th Congress, Subcom. of Com. on Appropriations, March 1960, p. 170.

Page 16

The Pesticide Situation for 1957–58, U.S. Dept of Agric., Commodity Stabilization Service, April 1958, p. 10.

Page 17

Elton, Charles S., *The Ecology of Invasions by Animals and Plants*. New York: Wiley, 1958.

Page 19

Shepard, Paul, "The Place of Nature in Man's World," *Atlantic Naturalist*, Vol. 13 (April–June 1958), pp. 85–89.

CHAPTER 3: ELIXIRS OF DEATH

Pages 21–39

Gleason, Marion, et al., *Clinical Toxicology of Commercial Products*. Baltimore: Williams and Wilkins, 1957.

Pages 21–39

Gleason, Marion, et al., *Bulletin of Supplementary Material: Clinical Toxicology of Commercial Products*, Vol. IV, No. 9. Univ. of Rochester.

Page 22

The Pesticide Situation for 1958–59, U.S. Dept. of Agric., Commodity Stabilization Service, April 1959, pp. 1–24.

Page 22

The Pesticide Situation for 1960–61, U.S. Dept. of Agric., Commodity Stabilization Service, July 1961, pp. 1–23.

Page 23

Hueper, W. C., *Occupational Tumors and Allied Diseases*. Springfield, Ill.: Thomas, 1942.

Page 23
  Todd, Frank E., and S. E. McGregor, "Insecticides and
  Bees," *Yearbook of Agric.*, U.S. Dept. of Agric., 1952, pp.
  131–35.
Page 23
  Hueper, *Occupational Tumors.*
Page 25
  Bowen, C. V., and S. A. Hall, "The Organic Insecticides,"
  Yearbook of Agric., U.S. Dept. of Agric., 1952, pp. 209–18.
Page 26
  Van Oettingen, W. F., *The Halogenated Aliphatic, Olefinic,
  Cyclic, Aromatic, and Aliphatic-Aromatic Hydrocarbons:
  Including the Halogenated Insecticides, Their Toxicity and
  Potential Dangers.* U.S. Dept. of Health, Education, and
  Welfare. Public Health Service Publ. No. 414 (1955), pp.
  341–42.
Page 26
  Laug, Edwin P., et al., "Occurrence of DDT in Human Fat
  and Milk," *A.M.A. Archives Indus. Hygiene and Occupat.
  Med.*, Vol. 3 (1951), pp. 245–46.
Page 26
  Biskind, Morton S., "Public Health Aspects of the New In-
  secticides," *Am. Jour. Diges. Diseases*, Vol. 20 (1953), No. 11,
  pp. 331–41.
Page 26
  Laug, Edwin P., et al., "Liver Cell Alteration and DDT Stor-
  age in the Fat of the Rat Induced by Dietary Levels of 1 to 50
  p.p.m. DDT," *Jour. Pharmacol. and Exper. Therapeut.*, Vol.
  98 (1950), p. 268.
Page 26
  Ortega, Paul, et al., "Pathologic Changes in the Liver of Rats
  after Feeding Low Levels of Various Insecticides," *A.M.A.
  Archives Path.*, Vol. 64 (Dec. 1957), pp. 614–22.
Page 27
  Fitzhugh, O. Garth, and A. A. Nelson, "The Chronic Oral
  Toxicity of DDT (2,2-BIS p-CHLOROPHENYL-1,1,1-TRI-
  CHLOROETHANE)," *Jour. Pharmacol. and Exper. Thera-
  peut.*, Vol. 89 (1947), No. 1, pp. 18–30.
Page 27
  Laug et al., "Occurrence of DDT in Human Fat and Milk."

Page 27

Hayes, Wayland J., Jr., et al., "Storage of DDT and DDE in People with Different Degrees of Exposure to DDT," *A.M.A. Archives Indus. Health*, Vol. 18 (Nov. 1958), pp. 398–406.

Page 27

Durham, William F., et al., "Insecticide Content of Diet and Body Fat of Alaskan Natives," *Science*, Vol. 134 (1961), No. 3493, pp. 1880–81.

Page 27

Van Oettingen, *Halogenated . . . Hydrocarbons*, p. 363.

Page 27

Smith, Ray F., et al., "Secretion of DDT in Milk of Dairy Cows Fed Low Residue Alfalfa," *Jour. Econ. Entomol.*, Vol. 41 (1948), pp. 759–63.

Pages 27–28

Laug et al., "Occurrence of DDT in Human Fat and Milk."

Page 28

Finnegan, J. K., et al., "Tissue Distribution and Elimination of DDD and DDT Following Oral Administration to Dogs and Rats," *Proc. Soc. Exper. Biol. and Med.*, Vol. 72 (1949), pp. 356–57.

Page 28

Laug et al., "Liver Cell Alteration."

Page 28

"Chemicals in Food Products," *Hearings*, H.R. 74, House Select Com. to Investigate Use of Chemicals in Food Products, Pt. 1 (1951), p. 275.

Pages 28–29

Van Oettingen, *Halogenated . . . Hydrocarbons*, p. 322.

Page 29

"Chemicals in Food Products," *Hearings*, 81st Congress, H.R. 323, Com. to Investigate Use of Chemicals in Food Products, Pt. 1 (1950), pp. 388–90.

Page 29

Clinical Memoranda on Economic Poisons. U.S. Public Health Service Publ. No. 476 (1956), p. 28.

Page 29

Gannon, Norman, and J. H. Bigger, "The Conversion of Aldrin and Heptachlor to Their Epoxides in Soil," *Jour. Econ. Entomol.*, Vol. 51 (Feb. 1958), pp. 1–2.

Page 29
   Davidow, B., and J. L. Radomski, "Isolation of an Epoxide
   Metabolite from Fat Tissues of Dogs Fed Heptachlor," *Jour.
   Pharmacol. and Exper. Therapeut.*, Vol. 107 (March 1953),
   pp. 259–65.
Page 29
   Van Oettingen, *Halogenated . . . Hydrocarbons*, p. 310.
Page 29
   Drinker, Cecil K., et al., "The Problem of Possible Systemic
   Effects from Certain Chlorinated Hydrocarbons," *Jour. In-
   dus. Hygiene and Toxicol.*, Vol. 19 (Sept. 1937), p. 283.
Page 29
   "Occupational Dieldrin Poisoning," Com. on Toxicology,
   Jour. Am. Med. Assn., Vol. 172 (April 1960), pp. 2077–80.
Page 29
   Scott, Thomas G., et al., "Some Effects of a Field Appli-
   cation of Dieldrin on Wildlife," *Jour. Wildlife Management*,
   Vol. 23 (Oct. 1959), pp. 409–27.
Page 30
   Paul, A. H., "Dieldrin Poisoning—a Case Report," *New
   Zealand Med. Jour.*, Vol. 58 (1959), p. 393.
Page 30
   Hayes, Wayland J., Jr., "The Toxicity of Dieldrin to Man,"
   Bull. World Health Organ., Vol. 20 (1959), pp. 891–912.
Page 30
   Gannon, Norman, and G. C. Decker, "The Conversion of
   Aldrin to Dieldrin on Plants," *Jour. Econ. Entomol.*, Vol. 51
   (Feb. 1958), pp. 8–11.
Page 30
   Kitselman, C. H., et al., "Toxicological Studies of Aldrin
   (Compound 118) on Large Animals," *Am. Jour. Vet. Re-
   search*, Vol. 11 (1950), p. 378.
Pages 30–31
   Dahlen, James H., and A. O. Haugen, "Effect of Insecti-
   cides on Quail and Doves," *Alabama Conservation*, Vol. 26
   (1954), No. 1, pp. 21–23.
Pages 30–31
   DeWitt, James B., "Chronic Toxicity to Quail and Pheasants
   of Some Chlorinated Insecticides," *Jour. Agric. and Food
   Chem.*, Vol. 4 (1956), No. 10, pp. 863–66.

Pages 30–31
Kitselman, C. H., "Long Term Studies on Dogs Fed Aldrin and Dieldrin in Sublethal Doses, with Reference to the Histopathological Findings and Reproduction," *Jour. Am. Vet. Med. Assn.*, Vol. 123 (1953), p. 28.

Pages 30–31
Treon, J. F., and A. R. Borgmann, "The Effects of the Complete Withdrawal of Food from Rats Previously Fed Diets Containing Aldrin or Dieldrin." Kettering Lab., Univ. of Cincinnati; mimeo. Quoted from Robert L. Rudd and Richard E. Genelly, *Pesticides: Their Use and Toxicity in Relation to Wildlife*. Calif. Dept. of Fish and Game, Game Bulletin No. 7 (1956), p. 52.

Page 31
Myers, C. S., "Endrin and Related Pesticides: A Review." Penna. Dept. of Health Research Report No. 45 (1958). Mimeo.

Page 31
Jacobziner, Harold, and H. W. Raybin, "Poisoning by Insecticide (Endrin)," *New York State Jour. Med.*, Vol. 59 (May 15, 1959), pp. 2017–22.

Page 31
"Care in Using Pesticide Urged," *Clean Streams*, No. 46 (June 1959). Penna. Dept. of Health.

Page 32
Metcalf, Robert L., "The Impact of the Development of Organophosphorus Insecticides upon Basic and Applied Science," *Bull. Entomol. Soc. Am.*, Vol. 5 (March 1959), pp. 3–15.

Pages 32–33
Mitchell, Philip H., *General Physiology*. New York: McGraw-Hill, 1958. Pp. 14–15.

Page 33
Brown, A. W. A., *Insect Control by Chemicals*. New York: Wiley, 1951.

Page 33
Toivonen, T., et al., "Parathion Poisoning Increasing Frequency in Finland," *Lancet*, Vol. 2 (1959), No. 7095, pp. 175–76.

Pages 33–34
Hayes, Wayland J., Jr., "Pesticides in Relation to Public Health," *Annual Rev. Entomol.*, Vol. 5 (1960), pp. 379–404.

Page 34

Occupational Disease in California Attributed to Pesticides and Other Agricultural Chemicals. Calif. Dept. of Public Health, 1957, 1958, 1959, and 1960.

Page 34

Quinby, Griffith E., and A. B. Lemmon, "Parathion Residues As a Cause of Poisoning in Crop Workers," Jour. Am. Med. Assn., Vol. 166 (Feb. 15, 1958), pp. 740–46.

Page 34

Carman, G. C., et al., "Absorption of DDT and Parathion by Fruits," Abstracts, 115th Meeting Am. Chem. Soc. (1949), p. 30A.

Page 34

Clinical Memoranda on Economic Poisons, p. 11.

Page 35

Frawley, John P., et al., "Marked Potentiation in Mammalian Toxicity from Simultaneous Administration of Two Anticholinesterase Compounds," Jour. Pharmacol. and Exper. Therapeut., Vol. 121 (1957), No. 1, pp. 96–106.

Page 35

Rosenberg, Philip, and J. M. Coon, "Potentiation between Cholinesterase Inhibitors," Proc. Soc. Exper. Biol. and Med., Vol. 97 (1958), pp. 836–39.

Page 35

Dubois, Kenneth P., "Potentiation of the Toxicity of Insecticidal Organic Phosphates," A.M.A. Archives Indus. Health, Vol. 18 (Dec. 1958), pp. 488–96.

Page 35

Murphy, S. D., et al., "Potentiation of Toxicity of Malathion by Triorthotolyl Phosphate," Proc. Soc. Exper. Biol. and Med., Vol. 100 (March 1959), pp. 483–87.

Page 35

Graham, R. C. B., et al., "The Effect of Some Organophosphorus and Chlorinated Hydrocarbon Insecticides on the Toxicity of Several Muscle Relaxants," Jour. Pharm. and Pharmacol., Vol. 9 (1957), pp. 312–19.

Page 35

Rosenberg, Philip, and J. M. Coon, "Increase of Hexobarbital Sleeping Time by Certain Anticholinesterases," Proc. Soc. Exper. Biol. and Med., Vol. 98 (1958), pp. 650–52.

Page 35
Dubois, "Potentiation of Toxicity."

Page 36
Hurd-Karrer, A. M., and F. W. Poos, "Toxicity of Selenium-Containing Plants to Aphids," *Science*, Vol. 84 (1936), p. 252.

Page 36
Ripper, W. E., "The Status of Systemic Insecticides in Pest Control Practices," *Advances in Pest Control Research.* New York: Interscience, 1957. Vol. 1, pp. 305–52.

Page 36
Occupational Disease in California, 1959.

Page 37
Glynne-Jones, G. D., and W. D. E. Thomas, "Experiments on the Possible Contamination of Honey with Schradan," *Annals Appl. Biol.*, Vol. 40 (1953), p. 546.

Page 37
Radeleff, R. D., et al., *The Acute Toxicity of Chlorinated Hydrocarbon and Organic Phosphorus Insecticides to Livestock.* U.S. Dept. of Agric. Technical Bulletin 1122 (1955).

Page 38
Brooks, F. A., "The Drifting of Poisonous Dusts Applied by Airplanes and Land Rigs," *Agric. Engin.*, Vol. 28 (1947), No. 6, pp. 233–39.

Page 38
Stevens, Donald B., "Recent Developments in New York State's Program Regarding Use of Chemicals to Control Aquatic Vegetation," paper presented at 13th Annual Meeting Northeastern Weed Control Conf. (Jan. 8, 1959).

Page 38
Anon., "No More Arsenic," *Economist*, Oct. 10, 1959.

Page 38
"Arsenites in Agriculture," *Lancet*, Vol. 1 (1960), p. 178.

Page 38
Horner, Warren D., "Dinitrophenol and Its Relation to Formation of Cataract," (A.M.A.) *Archives Ophthalmol.*, Vol. 27 (1942), pp. 1097–1121.

Page 39
Weinbach, Eugene C., "Biochemical Basis for the Toxicity of

Pentachlorophenol," *Proc. Natl. Acad. Sci.*, Vol. 43 (1957), No. 5, pp. 393–97.

CHAPTER 4: SURFACE WATERS
AND UNDERGROUND SEAS

Page 42
Biological Problems in Water Pollution. Transactions, 1959 seminar. U.S. Public Health Service Technical Report W60-3 (1960).

Page 42
"Report on Environmental Health Problems," *Hearings*, 86th Congress, Subcom. of Com. on Appropriations, March 1960, p. 78.

Page 42
Tarzwell, Clarence M., "Pollutional Effects of Organic In-secticides to Fishes," *Transactions*, 24th North Am. Wildlife Conf. (1959), Washington, D.C., pp. 132–42. Pub. by Wild-life Management Inst.

Pages 42–43
Nicholson, H. Page, "Insecticide Pollution of Water Re-sources," *Jour. Am. Waterworks Assn.*, Vol. 51 (1959), pp. 981–86.

Pages 42–43
Woodward, Richard L., "Effects of Pesticides in Water Sup-plies," *Jour. Am. Waterworks Assn.*, Vol. 52 (1960), No. 11, pp. 1367–72.

Page 43
Cope, Oliver B., "The Retention of DDT by Trout and White-fish," in *Biological Problems in Water Pollution*, pp. 72–75.

Pages 43–44
Kuenen, P. H., *Realms of Water.* New York: Wiley, 1955.

Pages 43–44
Gilluly, James, et al., *Principles of Geology.* San Francisco: Freeman, 1951.

Pages 44–45
Walton, Graham, "Public Health Aspects of the Contamina-tion of Ground Water in South Platte River Basin in Vicinity of Henderson, Colorado, August, 1959." U.S. Public Health Service, Nov. 2, 1959. Mimeo.

Pages 44–45
"Report on Environmental Health Problems."

Page 45
Hueper, W. C., "Cancer Hazards from Natural and Artificial Water Pollutants," *Proc.*, Conf. on Physiol. Aspects of Water Quality, Washington, D.C., Sept. 8–9, 1960. U.S. Public Health Service.

Pages 47–50
Hunt, E. G., and A. I. Bischoff, "Inimical Effects on Wildlife of Periodic DDD Applications to Clear Lake," *Calif. Fish and Game*, Vol. 46 (1960), No. 1, pp. 91–106.

Pages 49–50
Woodard, G., et al., "Effects Observed in Dogs Following the Prolonged Feeding of DDT and Its Analogues," *Federation Proc.*, Vol. 7 (1948), No. 1, p. 266.

Pages 49–50
Nelson, A. A., and G. Woodard, "Severe Adrenal Cortical Atrophy (Cytotoxic) and Hepatic Damage Produced in Dogs by Feeding DDD or TDE," (A.M.A.) *Archives Path.*, Vol. 48 (1949), p. 387.

Pages 49–50
Zimmermann, B., et al., "The Effects of DDD on the Human Adrenal; Attempts to Use an Adrenal-Destructive Agent in the Treatment of Disseminated Mammary and Prostatic Cancer," *Cancer*, Vol. 9 (1956), pp. 940–48.

Page 50
Cohen, Jesse M., et al., "Effect of Fish Poisons on Water Supplies. I. Removal of Toxic Materials," *Jour. Am. Waterworks Assn.*, Vol. 52 (1960), No. 12, pp. 1551–65. "II. Odor Problems," Vol. 53 (1960), No. 1, pp. 49–61. "III. Field Study, Dickinson, North Dakota," Vol. 53 (1961), No. 2, pp. 233–46.

Pages 50–51
Hueper, "Cancer Hazards from Water Pollutants."

CHAPTER 5: REALMS OF THE SOIL

Page 54
Simonson, Roy W., "What Soils Are," *Yearbook of Agric.*, U.S. Dept. of Agric., 1957, pp. 17–31.

Page 54
   Clark, Francis E., "Living Organisms in the Soil," *Yearbook of Agric.*, U.S. Dept. of Agric., 1957, pp. 157–65.
Pages 54–55
   Farb, Peter, *Living Earth.* New York: Harper, 1959.
Page 56
   Lichtenstein, E. P., and K. R. Schulz, "Persistence of Some Chlorinated Hydrocarbon Insecticides As Influenced by Soil Types, Rate of Application and Temperature," *Jour. Econ. Entomol.*, Vol. 52 (1959), No. 1, pp. 124–31.
Page 57
   Thomas, F. J. D., "The Residual Effects of Crop-Protection Chemicals in the Soil," in *Proc.*, 2nd Internatl. Plant Protection Conf. (1956), Fernhurst Research Station, England.
Page 57
   Eno, Charles F., "Chlorinated Hydrocarbon Insecticides: What Have They Done to Our Soil?" *Sunshine State Agric. Research Report* for July 1959.
Page 57
   Mader, Donald L., "Effect of Humus of Different Origin in Moderating the Toxicity of Biocides." Doctorate thesis, Univ. of Wisc., 1960.
Page 57
   Sheals, J. G., "Soil Population Studies. I. The Effects of Cultivation and Treatment with Insecticides," *Bull. Entomol. Research*, Vol. 47 (Dec. 1956), pp. 803–22.
Pages 57–58
   Hetrick, L. A., "Ten Years of Testing Organic Insecticides As Soil Poisons against the Eastern Subterranean Termite," *Jour. Econ. Entomol.*, Vol. 50 (1957), p. 316.
Page 58
   Lichtenstein, E. P., and J. B. Polivka, "Persistence of Insecticides in Turf Soils," *Jour. Econ. Entomol.*, Vol. 52 (1959), No. 2, pp. 289–93.
Page 58
   Ginsburg, J. M., and J. P. Reed, "A Survey on DDT-Accumulation in Soils in Relation to Different Crops," *Jour. Econ. Entomol.*, Vol. 47 (1954), No. 3, pp. 467–73.
Page 58
   Cullinan, F. P., "Some New Insecticides—Their Effect on

Plants and Soils," *Jour. Econ. Entomol.*, Vol. 42 (1949), pp. 387–91.

Page 58

Satterlee, Henry S., "The Problem of Arsenic in American Cigarette Tobacco," *New Eng. Jour. Med.*, Vol. 254 (June 21, 1956), pp. 1149–54

Page 58

Lichtenstein, E. P., "Absorption of Some Chlorinated Hydrocarbon Insecticides from Soils into Various Crops," *Jour. Agric. and Food Chem.*, Vol. 7 (1959), No. 6, pp. 430–33.

Pages 58–59

"Chemicals in Foods and Cosmetics," *Hearings*, 81st Congress, H.R. 74 and 447, House Select Com. to Investigate Use of Chemicals in Foods and Cosmetics, Pt. 3 (1952), pp. 1385–1416. Testimony of L. G. Cox.

Page 59

Klostermeyer, E. C., and C. B. Skotland, *Pesticide Chemicals As a Factor in Hop Die-out*. Washington Agric. Exper. Stations Circular 362 (1959).

Pages 59–60

Stegeman, LeRoy C., "The Ecology of the Soil." Transcription of a seminar, New York State Univ. College of Forestry, 1960.

CHAPTER 6: EARTH'S GREEN MANTLE

Pages 62–63

Patterson, Robert L., *The Sage Grouse in Wyoming*. Denver: Sage Books, Inc., for Wyoming Fish and Game Commission, 1952.

Pages 63–64

Murie, Olaus J., "The Scientist and Sagebrush," *Pacific Discovery*, Vol. 13 (1960), No. 4, p. 1.

Page 64

Pechanec, Joseph, et al., *Controlling Sagebrush on Rangelands*. U.S. Dept. of Agric. Farmers' Bulletin No. 2072 (1960).

Pages 64–65

Douglas, William O., *My Wilderness: East to Katahdin*. New York: Doubleday, 1961.

Page 65
   Egler, Frank E., *Herbicides: 60 Questions and Answers Concerning Roadside and Rightofway Vegetation Management.* Litchfield, Conn.: Litchfield Hills Audubon Soc., 1961.
Page 65
   Fisher, C. E., et al., *Control of Mesquite on Grazing Lands.* Texas Agric. Exper. Station Bulletin 935 (Aug. 1959).
Page 65
   Goodrum, Phil D., and V. H. Reid, "Wildlife Implications of Hardwood and Brush Controls," *Transactions,* 21st North Am. Wildlife Conf. (1956).
Page 65
   A Survey of Extent and Cost of Weed Control and Specific Weed Problems. U.S. Dept. of Agric. ARS 34-23 (March 1962).
Page 67
   Barnes, Irston R., "Sprays Mar Beauty of Nature," *Washington Post,* Sept. 25, 1960.
Page 67
   Goodwin, Richard H., and William A. Niering, *A Roadside Crisis: The Use and Abuse of Herbicides.* Connecticut Arboretum Bulletin No. 11 (March 1959), pp. 1–13.
Page 67
   Boardman, William, "The Dangers of Weed Spraying," *Veterinarian,* Vol. 6 (Jan. 1961), pp. 9–19.
Page 68
   Willard, C. J., "Indirect Effects of Herbicides," *Proc.,* 7th Annual Meeting North Central Weed Control Conf. (1950), pp. 110–12.
Pages 68–69
   Douglas, William O., *My Wilderness: The Pacific West.* New York: Doubleday, 1960.
Page 69
   Egler, Frank E., *Vegetation Management for Rights-of-Way and Roadsides.* Smithsonian Report for 1953 (Smithsonian Inst., Washington, D.C.), pp. 299–322.
Pages 69–70
   Bohart, George E., "Pollination by Native Insects," *Yearbook of Agric.,* U.S. Dept. of Agric., 1952, pp. 107–21.
Page 70
   Egler, *Vegetation Management.*

Pages 70–71

Niering, William A., and Frank E. Egler, "A Shrub Community of *Viburnum lentago*, Stable for Twenty-five Years," *Ecology*, Vol. 36 (April 1955), pp. 356–60.

Pages 70–71

Pound, Charles E., and Frank E. Egler, "Brush Control in Southeastern New York: Fifteen Years of Stable Tree-Less Communities," *Ecology*, Vol. 34 (Jan. 1953), pp. 63–73.

Pages 70–71

Egler, Frank E., "Science, Industry, and the Abuse of Rights of Way," *Science*, Vol. 127 (1958), No. 3298, pp. 573–80.

Page 71

Niering, William A., "Principles of Sound Right-of-Way Vegetation Management," *Econ. Botany*, Vol. 12 (April–June 1958), pp. 140–44.

Page 71

Hall, William C., and William A. Niering, "The Theory and Practice of Successful Selective Control of 'Brush' by Chemicals," *Proc.*, 13th Annual Meeting Northeastern Weed Control Conf. (Jan. 8, 1959).

Page 71

Egler, Frank E., "Fifty Million More Acres for Hunting?" *Sports Afield*, Dec. 1954.

Pages 71–72

McQuilkin, W. E., and L. R. Strickenberg, *Roadside Brush Control with 2,4,5-T on Eastern National Forests*. Northeastern Forest Exper. Station Paper No. 148. Upper Darby, Penna., 1961.

Page 72

Goldstein, N. P., et al., "Peripheral Neuropathy after Exposure to an Ester of Dichlorophenoxyacetic Acid," *Jour. Am. Med. Assn.*, Vol. 171 (1959), pp. 1306–9.

Page 72

Brody, T. M., "Effect of Certain Plant Growth Substances on Oxidative Phosphorylation in Rat Liver Mitochondria," *Proc. Soc. Exper. Biol. and Med.*, Vol. 80 (1952), pp. 533–36.

Page 72

Croker, Barbara H., "Effects of 2,4-D and 2,4,5-T on Mitosis in *Allium cepa*," *Bot. Gazette*, Vol. 114 (1953), pp. 274–83.

Page 72
Willard, "Indirect Effects of Herbicides."

Pages 72–73
Stahler, L. M., and E. J. Whitehead, "The Effect of 2,4-D on Potassium Nitrate Levels in Leaves of Sugar Beets," *Science*, Vol. 112 (1950), No. 2921, pp. 749–51.

Page 73
Olson, O., and E. Whitehead, "Nitrate Content of Some South Dakota Plants," *Proc.*, South Dakota Acad. of Sci., Vol. 20 (1940), p. 95.

Page 73
What's New in Farm Science. Univ. of Wisc. Agric. Exper. Station Annual Report, Pt. II, Bulletin 527 (July 1957), p. 18.

Page 73
Stahler and Whitehead, "The Effect of 2,4-D on Potassium Nitrate Levels."

Page 73
Grayson, R. R., "Silage Gas Poisoning: Nitrogen Dioxide Pneumonia, a New Disease in Agricultural Workers," *Annals Internal Med.*, Vol. 45 (1956), pp. 393–408.

Page 73
Crawford, R. F., and W. K. Kennedy, *Nitrates in Forage Crops and Silage: Benefits, Hazards, Precautions.* New York State College of Agric., Cornell Misc. Bulletin 37 (June 1960).

Page 74
Briejèr, C. J., To author.

Page 75
Knake, Ellery L., and F. W. Slife, "Competition of *Setaria faterii* with Corn and Soybeans," *Weeds*, Vol. 10 (1962), No. 1, pp. 26–29.

Page 75
Goodwin and Niering, *A Roadside Crisis.*

Page 75
Egler, Frank E., To author.

Page 75
DeWitt, James B., To author.

Page 77
Holloway, James K., "Weed Control by Insect," *Sci. American*, Vol. 197 (1957), No. 1, pp. 56–62.

Page 77

Holloway, James K., and C. B. Huffaker, "Insects to Control a Weed," *Yearbook of Agric.*, U.S. Dept. of Agric., 1952, pp. 135–40.

Page 77

Huffaker, C. B., and C. E. Kennett, "A Ten-Year Study of Vegetational Changes Associated with Biological Control of Klamath Weed," *Jour. Range Management*, Vol. 12 (1959), No. 2, pp. 69–82.

Pages 77–78

Bishopp, F. C., "Insect Friends of Man," *Yearbook of Agric.*, U.S. Dept. of Agric., 1952, pp. 79–87.

CHAPTER 7: NEEDLESS HAVOC

Page 81

Nickell, Walter, To author.

Page 81

Here Is Your 1959 Japanese Beetle Control Program. Release, Michigan State Dept. of Agric., Oct. 19, 1959.

Page 81

Hadley, Charles H., and Walter E. Fleming, "The Japanese Beetle," *Yearbook of Agric.*, U.S. Dept. of Agric., 1952, pp. 567–73.

Page 82

Here Is Your 1959 Japanese Beetle Control Program.

Page 82

"No Bugs in Plane Dusting," *Detroit News*, Nov. 10, 1959.

Page 83

Michigan Audubon Newsletter, Vol. 9 (Jan. 1960).

Page 83

"No Bugs in Plane Dusting."

Page 83

Hickey, Joseph J., "Some Effects of Insecticides on Terrestrial Birdlife," *Report* of Subcom. on Relation of Chemicals to Forestry and Wildlife, Madison, Wisc., Jan. 1961. Special Report No. 6.

Pages 84–85

Scott, Thomas G., To author, Dec. 14, 1961.

Page 85

"Coordination of Pesticides Programs," *Hearings*, 86th Congress, H.R. 11502, Com. on Merchant Marine and Fisheries, May 1960, p. 66.

Pages 85–87

Scott, Thomas G., et al., "Some Effects of a Field Application of Dieldrin on Wildlife," *Jour. Wildlife Management*, Vol. 23 (1959), No. 4, pp. 409–27.

Page 87

Hayes, Wayland J., Jr., "The Toxicity of Dieldrin to Man," *Bull. World Health Organ.*, Vol. 20 (1959), pp. 891–912.

Page 87

Scott, Thomas G., To author, Dec. 14, 1961, Jan. 8, Feb. 15, 1962.

Pages 89–90

Hawley, Ira M., "Milky Diseases of Beetles," *Yearbook of Agric.*, U.S. Dept. of Agric., 1952, pp. 394–401.

Pages 89–90

Fleming, Walter E., "Biological Control of the Japanese Beetle Especially with Entomogenous Diseases," *Proc.*, 10th Internatl. Congress of Entomologists (1956), Vol. 3 (1958), pp. 115–25.

Page 90

Chittick, Howard A. (Fairfax Biological Lab.), To author, Nov. 30, 1960.

Page 91

Scott et al., "Some Effects of a Field Application of Dieldrin on Wildlife."

CHAPTER 8: AND NO BIRDS SING

Page 94

Audubon Field Notes. "Fall Migration—Aug. 16 to Nov. 30, 1958." Vol. 13 (1959), No. 1, pp. 1–68.

Page 95

Swingle, R. U., et al., "Dutch Elm Disease," *Yearbook of Agric.*, U.S. Dept. of Agric., 1949, pp. 451–52.

Pages 95–96

Mehner, John F., and George J. Wallace, "Robin Populations

and Insecticides," *Atlantic Naturalist*, Vol. 14 (1959), No. 1, pp. 4–10.

Pages 95–96

Wallace, George J., "Insecticides and Birds," *Audubon Mag.*, Jan.–Feb. 1959.

Pages 96–97

Barker, Roy J., "Notes on Some Ecological Effects of DDT Sprayed on Elms," *Jour. Wildlife Management*, Vol. 22 (1958), No. 3, pp. 269–74.

Page 97

Hickey, Joseph J., and L. Barrie Hunt, "Songbird Mortality Following Annual Programs to Control Dutch Elm Disease," *Atlantic Naturalist*, Vol. 15 (1960), No. 2, pp. 87–92.

Page 97

Wallace, "Insecticides and Birds."

Page 97

Wallace, George J., "Another Year of Robin Losses on a University Campus," *Audubon Mag.*, March–April 1960.

Pages 97–98

"Coordination of Pesticides Programs," *Hearings*, H.R. 11502, 86th Congress, Com. on Merchant Marine and Fisheries, May 1960, pp. 10, 12.

Page 98

Hickey, Joseph J., and L. Barrie Hunt, "Initial Songbird Mortality Following a Dutch Elm Disease Control Program," *Jour. Wildlife Management*, Vol. 24 (1960), No. 3, pp. 259–65.

Page 98

Wallace, George J., et al., *Bird Mortality in the Dutch Elm Disease Program in Michigan*. Cranbrook Inst. of Science Bulletin 41 (1961).

Page 98

Hickey, Joseph J., "Some Effects of Insecticides on Terrestrial Birdlife," *Report* of Subcom. on Relation of Chemicals to Forestry and Wildlife, State of Wisconsin, Jan. 1961, pp. 2–43.

Pages 98–99

Walton, W. R., *Earthworms As Pests and Otherwise*. U.S. Dept. of Agric. Farmers' Bulletin No. 1569 (1928).

Pages 98–99
 Wright, Bruce S., "Woodcock Reproduction in DDT-Sprayed Areas of New Brunswick," *Jour. Wildlife Management*, Vol. 24 (1960), No. 4, pp. 419–20.
Page 99
 Dexter, R. W., "Earthworms in the Winter Diet of the Opossum and the Raccoon," *Jour. Mammal.*, Vol. 32 (1951), p. 464.
Page 100
 Wallace et al., *Bird Mortality in the Dutch Elm Disease Program.*
Page 100
 "Coordination of Pesticides Programs." Testimony of George J. Wallace, p. 10.
Page 100
 Wallace, "Insecticides and Birds."
Page 101
 Bent, Arthur C., *Life Histories of North American Jays, Crows, and Titmice*. Smithsonian Inst., U.S. Natl. Museum Bulletin 191 (1946).
Page 101
 MacLellan, C. R., "Woodpecker Control of the Codling Moth in Nova Scotia Orchards," *Atlantic Naturalist*, Vol. 16 (1961), No. 1, pp. 17–25.
Page 101
 Knight, F. B., "The Effects of Woodpeckers on Populations of the Engelmann Spruce Beetle," *Jour. Econ. Entomol.*, Vol. 51 (1958), pp. 603–7.
Page 102
 Carter, J. C., To author, June 16, 1960.
Page 103
 Sweeney, Joseph A., To author, March 7, 1960.
Page 103
 Welch, D. S., and J. G. Matthysse, *Control of the Dutch Elm Disease in New York State*. New York State College of Agric., Cornell Ext. Bulletin No. 932 (June 1960), pp. 3–16.
Page 103
 Matthysse, J. G., *An Evaluation of Mist Blowing and Sanitation in Dutch Elm Disease Control Programs*. New York

State College of Agric., Cornell Ext. Bulletin No. 30 (July 1959), pp. 2–16.
Page 104
Miller, Howard, To author, Jan. 17, 1962.
Page 104
Matthysse, *An Evaluation of Mist Blowing and Sanitation.*
Page 105
Elton, Charles S., *The Ecology of Invasions by Animals and Plants.* New York: Wiley, 1958.
Pages 105–106
Broley, Charles E., "The Bald Eagle in Florida," *Atlantic Naturalist*, July 1957, pp. 230–31.
Pages 105–106
——, "The Plight of the American Bald Eagle," *Audubon Mag.*, July–Aug. 1958, pp. 162–63.
Pages 105–106
Cunningham, Richard L., "The Status of the Bald Eagle in Florida," *Audubon Mag.*, Jan.–Feb. 1960, pp. 24–43.
Page 106
"Vanishing Bald Eagle Gets Champion," *Florida Naturalist*, April 1959, p. 64.
Page 106
McLaughlin, Frank, "Bald Eagle Survey in New Jersey," *New Jersey Nature News*, Vol. 16 (1959), No. 2, p. 25. Interim Report, Vol. 16 (1959), No. 3, p. 51.
Page 107
Broun, Maurice, To author, May 22, 30, 1960.
Page 107
Beck, Herbert H., To author, July 30, 1959.
Page 108
Rudd, Robert L., and Richard E. Genelly, *Pesticides: Their Use and Toxicity in Relation to Wildlife.* Calif. Dept. of Fish and Game, Game Bulletin No. 7 (1956), p. 57.
Page 108
DeWitt, James B., "Effects of Chlorinated Hydrocarbon Insecticides upon Quail and Pheasants," *Jour. Agric. and Food Chem.*, Vol. 3 (1955), No. 8, p. 672.
Page 108
——, "Chronic Toxicity to Quail and Pheasants of Some

Chlorinated Insecticides," *Jour. Agric. and Food Chem.*, Vol. 4 (1956), No. 10, p. 863.

Page 109

Imler, Ralph H., and E. R. Kalmbach, *The Bald Eagle and Its Economic Status.* U.S. Fish and Wildlife Service Circular 30 (1955).

Page 109

Mills, Herbert R., "Death in the Florida Marshes," *Audubon Mag.*, Sept.–Oct. 1952.

Page 109

Bulletin, Internatl. Union for the Conservation of Nature, May and Oct. 1957.

Pages 109–110

The Deaths of Birds and Mammals Connected with Toxic Chemicals in the First Half of 1960. Report No. 1 of the British Trust for Ornithology and Royal Soc. for the Protection of Birds. Com. on Toxic Chemicals, Royal Soc. Protect. Birds.

Pages 110–112

Sixth Report from the Estimates Com., Ministry of Agric., Fisheries and Food, Sess. 1960–61, House of Commons.

Page 111

Christian, Garth, "Do Seed Dressings Kill Foxes?" *Country Life*, Jan. 12, 1961.

Page 111

Rudd, Robert L., and Richard E. Genelly, "Avian Mortality from DDT in Californian Rice Fields," *Condor*, Vol. 57 (March–April 1955), pp. 117–18.

Pages 111–112

Rudd and Genelly, *Pesticides.*

Pages 112–113

Dykstra, Walter W., "Nuisance Bird Control," *Audubon Mag.*, May–June 1960, pp. 118–19.

Pages 112–113

Buchheister, Carl W., "What About Problem Birds?" *Audubon Mag.*, May–June 1960, pp. 116–18.

Page 113

Quinby, Griffith E., and A. B. Lemmon, "Parathion Residues As a Cause of Poisoning in Crop Workers," *Jour. Am. Med. Assn.*, Vol. 166 (Feb. 15, 1958), pp. 740–46.

CHAPTER 9: RIVERS OF DEATH

Pages 115–120
Kerswill, C. J., "Effects of DDT Spraying in New Brunswick on Future Runs of Adult Salmon," *Atlantic Advocate*, Vol. 48 (1958), pp. 65–68.

Pages 115–120
Keenleyside, M. H. A., "Insecticides and Wildlife," *Canadian Audubon*, Vol. 21 (1959), No. 1, pp. 1–7.

Pages 115–120
——, "Effects of Spruce Budworm Control on Salmon and Other Fishes in New Brunswick," *Canadian Fish Culturist*, Issue 24 (1959), pp. 17–22.

Pages 115–120
Kerswill, C. J., *Investigation and Management of Atlantic Salmon in 1956* (also for 1957, 1958, 1959–60; in 4 parts). Federal-Provincial Co-ordinating Com. on Atlantic Salmon (Canada).

Page 117
Ide, F. P., "Effect of Forest Spraying with DDT on Aquatic Insects of Salmon Streams," *Transactions*, Am. Fisheries Soc., Vol. 86 (1957), pp. 208–19.

Pages 118–119
Kerswill, C. J., To author, May 9, 1961.

Page 119
——, To author, June 1, 1961.

Page 120
Warner, Kendall, and O. C. Fenderson, "Effects of Forest Insect Spraying on Northern Maine Trout Streams." Maine Dept. of Inland Fisheries and Game. Mimeo., n.d.

Page 120
Alderdice, D. F., and M. E. Worthington, "Toxicity of a DDT Forest Spray to Young Salmon." *Canadian Fish Culturist*, Issue 24 (1959), pp. 41–48.

Pages 120–121
Hourston, W. R., To author, May 23, 1961.

Page 121
Graham, R. J., and D. O. Scott, *Effects of Forest Insect Spraying on Trout and Aquatic Insects in Some Montana Streams*. Final Report, Mont. State Fish and Game Dept., 1958.

Pages 121–122

Graham, R. J., "Effects of Forest Insect Spraying on Trout and Aquatic Insects in Some Montana Streams," in *Biological Problems in Water Pollution*. Transactions, 1959 seminar. U.S. Public Health Service Technical Report W60-3 (1960).

Page 122

Crouter, R. A., and E. H. Vernon, "Effects of Black-headed Budworm Control on Salmon and Trout in British Columbia," *Canadian Fish Culturist*, Issue 24 (1959), pp. 23–40.

Page 123

Whiteside, J. M., "Spruce Budworm Control in Oregon and Washington, 1949–1956," *Proc.*, 10th Internatl. Congress of Entomologists (1956), Vol. 4 (1958), pp. 291–302.

Page 124

Pollution-Caused Fish Kills in 1960. U.S. Public Health Service Publ. No. 847 (1961), pp. 1–20.

Page 124

"U.S. Anglers—Three Billion Dollars," *Sport Fishing Inst. Bull.*, No. 119 (Oct. 1961).

Pages 124–125

Powers, Edward (Bur. of Commercial Fisheries), To author.

Pages 124–125

Rudd, Robert L., and Richard E. Genelly, *Pesticides: Their Use and Toxicity in Relation to Wildlife*. Calif. Dept. of Fish and Game, Game Bulletin No. 7 (1956), p. 88.

Pages 124–125

Biglane, K. E., To author, May 8, 1961.

Pages 124–125

Release No. 58-38, Penna. Fish Commission, Dec. 8, 1958.

Pages 124–125

Rudd and Genelly, *Pesticides*, p. 60.

Pages 124–125

Henderson, C., et al., "The Relative Toxicity of Ten Chlorinated Hydrocarbon Insecticides to Four Species of Fish," paper presented at 88th Annual Meeting Am. Fisheries Soc. (1958).

Pages 124–125

"The Fire Ant Eradication Program and How It Affects Wildlife," subject of *Proc. Symposium*, 12th Annual Conf.

Southeastern Assn. Game and Fish Commissioners, Louisville, Ky. (1958). Pub. by the Assn., Columbia, S.C., 1958.

Pages 124–125

"Effects of the Fire Ant Eradication Program on Wildlife," report, U.S. Fish and Wildlife Service, May 25, 1958. Mimeo.

Pages 124–125

Pesticide-Wildlife Review, 1959. Bur. Sport Fisheries and Wildlife Circular 84 (1960), U.S. Fish and Wildlife Service, pp. 1–36.

Pages 124–125

Baker, Maurice F., "Observations of Effects of an Application of Heptachlor or Dieldrin on Wildlife," in *Proc. Symposium*, pp. 18–20.

Page 125

Glasgow, L. L., "Studies on the Effect of the Imported Fire Ant Control Program on Wildlife in Louisiana," in *Proc. Symposium*, pp. 24–29.

Page 125

Pesticide-Wildlife Review, 1959.

Page 125

Progress in Sport Fishery Research, 1960. Bur. Sport Fisheries and Wildlife Circular 101 (1960), U.S. Fish and Wildlife Service.

Page 125

"Resolution Opposing Fire-Ant Program Passed by American Society of Ichthyologists and Herpetologists," *Copeia* (1959), No. 1, p. 89.

Pages 125–127

Young, L. A., and H. P. Nicholson, "Stream Pollution Resulting from the Use of Organic Insecticides," *Progressive Fish Culturist*, Vol. 13 (1951), No. 4, pp. 193–98.

Page 127

Rudd and Genelly, *Pesticides.*

Page 127

Lawrence, J. M., "Toxicity of Some New Insecticides to Several Species of Pondfish," *Progressive Fish Culturist*, Vol. 12 (1950), No. 4, pp. 141–46.

Page 127

Pielow, D. P., "Lethal Effects of DDT on Young Fish," *Nature*, Vol. 158 (1946), No. 4011, p. 378.

Page 128

Herald, E. S., "Notes on the Effect of Aircraft-Distributed DDT-Oil Spray upon Certain Philippine Fishes," *Jour. Wildlife Management*, Vol. 13 (1949), No. 3, p. 316.

Pages 128–130

"Report of Investigation of the Colorado River Fish Kill, January, 1961." Texas Game and Fish Commission, 1961. Mimeo.

Pages 130–132

Harrington, R. W., Jr., and W. L. Bidlingmayer, "Effects of Dieldrin on Fishes and Invertebrates of a Salt Marsh," *Jour. Wildlife Management*, Vol. 22 (1958), No. 1, pp. 76–82.

Pages 131–132

Mills, Herbert R., "Death in the Florida Marshes," *Audubon Mag.*, Sept.–Oct. 1952.

Page 132

Springer, Paul F., and John R. Webster, *Effects of DDT on Salt-marsh Wildlife: 1949.* U.S. Fish and Wildlife Service, Special Scientific Report, Wildlife No. 10 (1949).

Page 132

John C. Pearson, To author.

Pages 133–134

Butler, Philip A., "Effects of Pesticides on Commercial Fisheries," *Proc.*, 13th Annual Session (Nov. 1960), Gulf and Caribbean Fisheries Inst., pp. 168–71.

CHAPTER 10: INDISCRIMINATELY FROM THE SKIES

Page 138

Perry, C. C., *Gypsy Moth Appraisal Program and Proposed Plan to Prevent Spread of the Moths.* U.S. Dept. of Agric. Technical Bulletin No. 1124 (Oct. 1955).

Pages 138–139

Corliss, John M., "The Gypsy Moth," *Yearbook of Agric.*, U.S. Dept. of Agric., 1952, pp. 694–98.

Page 139

Worrell, Albert C., "Pests, Pesticides, and People," offprint from *Am. Forests Mag.*, July 1960.

Page 139

Clausen, C. P., "Parasites and Predators," *Yearbook of Agric.*, U.S. Dept. of Agric., 1952, pp. 380–88.

Page 139
Perry, *Gypsy Moth Appraisal Program.*
Page 139
Worrell, "Pests, Pesticides, and People."
Page 140
"USDA Launches Large-Scale Effort to Wipe Out Gypsy Moth," press release, U.S. Dept. of Agric., March 20, 1957.
Page 140
Worrell, "Pests, Pesticides, and People."
Page 140
Robert Cushman Murphy et al. v. *Ezra Taft Benson et al.* U.S. District Court, Eastern District of New York, Oct. 1959, Civ. No. 17610.
Page 140
Murphy et al. v. *Benson et al.* Petition for a Writ of Certiorari to the U.S. Court of Appeals for the Second Circuit, Oct. 1959.
Page 140
Waller, W. K., "Poison on the Land," *Audubon Mag.*, March–April 1958, pp. 68–71.
Page 140
Murphy et al. v. *Benson et al.* U.S. Supreme Court Reports, Memorandum Cases, No. 662, March 28, 1960.
Page 140
Waller, "Poison on the Land."
Page 141
Am. Bee Jour., June 1958, p. 224.
Page 142
Murphy et al. v. *Benson et al.* U.S. Court of Appeals, Second Circuit. Brief for Defendant-Appellee Butler, No. 25,448, March 1959.
Page 142
Brown, William L., Jr., "Mass Insect Control Programs: Four Case Histories," *Psyche*, Vol. 68 (1961), Nos. 2–3, pp. 75–111.
Pages 142–143
Arant, F. S., et al., "Facts about the Imported Fire Ant," *Highlights of Agric. Research*, Vol. 5 (1958), No. 4.
Page 143
Brown, "Mass Insect Control Programs."

Page 143
"Pesticides: Hedgehopping into Trouble?" *Chemical Week*, Feb. 8, 1958, p. 97.

Pages 143–144
Arant et al., "Facts about the Imported Fire Ant."

Page 144
Byrd, I. B., "What Are the Side Effects of the Imported Fire Ant Control Program?" in *Biological Problems in Water Pollution*. Transactions, 1959 seminar. U.S. Public Health Service Technical Report W60-3 (1960), pp. 46–50.

Page 144
Hays, S. B., and K. L. Hays, "Food Habits of *Solenopsis saevissima richteri* Forel," *Jour. Econ. Entomol.*, Vol. 52 (1959), No. 3, pp. 455–57.

Page 144
Caro, M. R., et al., "Skin Responses to the Sting of the Imported Fire Ant," *A.M.A. Archives Dermat.*, Vol. 75 (1957), pp. 475–88.

Page 145
Byrd, "Side Effects of Fire Ant Program."

Page 145
Baker, Maurice F., in *Virginia Wildlife*, Nov. 1958.

Page 145
Brown, "Mass Insect Control Programs."

Page 146
Pesticide-Wildlife Review, 1959. Bur. Sport Fisheries and Wildlife Circular 84 (1960), U.S. Fish and Wildlife Service, pp. 1–36.

Page 146
"The Fire Ant Eradication Program and How It Affects Wildlife," subject of *Proc. Symposium*, 12th Annual Conf. Southeastern Assn. Game and Fish Commissioners, Louisville, Ky. (1958). Pub. by the Assn., Columbia, S.C., 1958.

Pages 146–147
Wright, Bruce S., "Woodcock Reproduction in DDT-Sprayed Areas of New Brunswick," *Jour. Wildlife Management*, Vol. 24 (1960), No. 4, pp. 419–20.

Page 147
Clawson, Sterling G., "Fire Ant Eradication—and Quail," *Alabama Conservation*, Vol. 30 (1959), No. 4, p. 14.

Page 147

Rosene, Walter, "Whistling-Cock Counts of Bobwhite Quail on Areas Treated with Insecticide and on Untreated Areas, Decatur County, Georgia," in *Proc. Symposium*, pp. 14–18.

Page 147

Pesticide-Wildlife Review, 1959.

Pages 147–148

Cottam, Clarence, "The Uncontrolled Use of Pesticides in the Southeast," address to Southeastern Assn. Fish, Game and Conservation Commissioners, Oct. 1959.

Pages 148–149

Poitevint, Otis L., Address to Georgia Sportsmen's Fed., Oct. 1959.

Page 149

Ely, R. E., et al., "Excretion of Heptachlor Epoxide in the Milk of Dairy Cows Fed Heptachlor-Sprayed Forage and Technical Heptachlor," *Jour. Dairy Sci.*, Vol. 38 (1955), No. 6, pp. 669–72.

Page 149

Gannon, N., et al., "Storage of Dieldrin in Tissues and Its Excretion in Milk of Dairy Cows Fed Dieldrin in Their Diets," *Jour. Agric. and Food Chem.*, Vol. 7 (1959), No. 12, pp. 824–32.

Page 150

Insecticide Recommendations of the Entomology Research Division for the Control of Insects Attacking Crops and Livestock for 1961. U.S. Dept. of Agric. Handbook No. 120 (1961).

Page 150

Peckinpaugh, H. S. (Ala. Dept. of Agric. and Indus.), To author, March 24, 1959.

Page 150

Hartman, H. L. (La. State Board of Health), To author, March 23, 1959.

Page 150

Lakey, J. F. (Texas Dept. of Health), To author, March 23, 1959.

Page 150

Davidow, B., and J. L. Radomski, "Metabolite of Heptachlor, Its Analysis, Storage, and Toxicity," *Federation Proc.*, Vol. 11 (1952), No. 1, p. 336.

Page 150

Food and Drug Administration, U.S. Dept. of Health, Education, and Welfare, in *Federal Register*, Oct. 27, 1959.

Page 151

Burgess, E. D. (U.S. Dept. of Agric.), To author, June 23, 1961.

Page 151

"Fire Ant Control is Parley Topic," *Beaumont [Texas] Journal*, Sept. 24, 1959.

Page 151

"Coordination of Pesticides Programs," *Hearings*, 86th Congress, H.R. 11502, Com. on Merchant Marine and Fisheries, May 1960, p. 45.

Page 151

Newsom, L. D. (Head, Entomol. Research, La. State Univ.), To author, March 23, 1962.

Page 151

Green, H. B., and R. E. Hutchins, *Economical Method for Control of Imported Fire Ant in Pastures and Meadows*. Miss. State Univ. Agric. Exper. Station Information Sheet 586 (May 1958).

CHAPTER 11: BEYOND THE DREAMS OF THE BORGIAS

Page 154

"Chemicals in Food Products," *Hearings*, 81st Congress, H.R. 323, Com. to Investigate Use of Chemicals in Food Products, Pt. I (1950), pp. 388–90.

Page 155

Clothes Moths and Carpet Beetles. U.S. Dept. of Agric., Home and Garden Bulletin No. 24 (1961).

Page 155

Mulrennan, J. A., To author, March 15, 1960.

Page 156

New York Times, May 22, 1960.

Page 156

Petty, Charles S., "Organic Phosphate Insecticide Poisoning. Residual Effects in Two Cases," *Am. Jour. Med.*, Vol. 24 (1958), pp. 467–70.

Pages 156–157

Miller, A. C., et al., "Do People Read Labels on Household

Insecticides?" *Soap and Chem. Specialties*, Vol. 34 (1958), No. 7, pp. 61–63.

Page 157

Hayes, Wayland J., Jr., et al., "Storage of DDT and DDE in People with Different Degrees of Exposure to DDT," *A.M.A. Archives Indus. Health*, Vol. 18 (Nov. 1958), pp. 398–406.

Page 157

Walker, Kenneth C., et al., "Pesticide Residues in Foods. Dichlorodiphenyltrichloroethane and Dichlorodiphenyldichloroethylene Content of Prepared Meals," *Jour. Agric. and Food Chem.*, Vol. 2 (1954), No. 20, pp. 1034–37.

Page 158

Hayes, Wayland J., Jr., et al., "The Effect of Known Repeated Oral Doses of Chlorophenothane (DDT) in Man," *Jour. Am. Med. Assn.*, Vol. 162 (1956), No. 9, pp. 890–97.

Page 158

Milstead, K. L., "Highlights in Various Areas of Enforcement," address to 64th Annual Conf. Assn. of Food and Drug Officials of U.S., Dallas (June 1960).

Pages 158–159

Durham, William, et al., "Insecticide Content of Diet and Body Fat of Alaskan Natives," *Science*, Vol. 134 (1961), No. 3493, pp. 1880–81.

Page 159

"Pesticides—1959," *Jour. Agric. and Food Chem.*, Vol. 7 (1959), No. 10, pp. 674–88.

Page 159

Annual Reports, Food and Drug Administration, U.S. Dept. of Health, Education, and Welfare. For 1957, pp. 196, 197; 1956, p. 203.

Pages 159–160

Markarian, Haig, et al., "Insecticide Residues in Foods Subjected to Fogging under Simulated Warehouse Conditions," *Abstracts*, 135th Meeting Am. Chem. Soc. (April 1959).

CHAPTER 12: THE HUMAN PRICE

Page 166

Price, David E., "Is Man Becoming Obsolete?" *Public Health Reports*, Vol. 74 (1959), No. 8, pp. 693–99.

Page 166

"Report on Environmental Health Problems," *Hearings*, 86th Congress, Subcom. of Com. on Appropriations, March 1960, p. 34.

Page 166

Dubos, René, *Mirage of Health*. New York: Harper, 1959. World Perspectives Series. P. 171.

Page 167

Medical Research: A Midcentury Survey. Vol. 2, *Unsolved Clinical Problems in Biological Perspective*. Boston: Little, Brown, 1955. P. 4.

Page 167

"Chemicals in Food Products," *Hearings*, 81st Congress, H.R. 323, Com. to Investigate Use of Chemicals in Food Products, 1950, p. 5. Testimony of A. J. Carlson.

Page 168

Paul, A. H., "Dieldrin Poisoning—a Case Report," *New Zealand Med. Jour.*, Vol. 58 (1959), p. 393.

Page 168

"Insecticide Storage in Adipose Tissue," editorial, *Jour. Am. Med. Assn.*, Vol. 145 (March 10, 1951), pp. 735–36.

Page 168

Mitchell, Philip H., *A Textbook of General Physiology*. New York: McGraw-Hill, 1956. 5th ed.

Pages 168–169

Miller, B. F., and R. Goode, *Man and His Body: The Wonders of the Human Mechanism*. New York: Simon and Schuster, 1960.

Pages 168–169

Dubois, Kenneth P., "Potentiation of the Toxicity of Insecticidal Organic Phosphates," *A.M.A. Archives Indus. Health*, Vol. 18 (Dec. 1958), pp. 488–96.

Page 169

Gleason, Marion, et al., *Clinical Toxicology of Commercial Products*. Baltimore: Williams and Wilkins, 1957.

Page 170

Case, R. A. M., "Toxic Effects of DDT in Man," *Brit. Med. Jour.*, Vol. 2 (Dec. 15, 1945), pp. 842–45.

Page 170

Wigglesworth, V. D., "A Case of DDT Poisoning in Man," *Brit. Med. Jour.*, Vol. 1 (April 14, 1945), p. 517.

Page 170
Hayes, Wayland J., Jr., et al., "The Effect of Known Repeated Oral Doses of Chlorophenothane (DDT) in Man," *Jour. Am. Med. Assn.*, Vol. 162 (Oct. 27, 1956), pp. 890–97.

Page 171
Hargraves, Malcolm M., "Chemical Pesticides and Conservation Problems," address to 23rd Annual Conv. Natl. Wildlife Fed. (Feb. 27, 1959). Mimeo.

Page 171
——, and D. G. Hanlon, "Leukemia and Lymphoma—Environmental Diseases?" paper presented at Internatl. Congress of Hematology, Japan, Sept. 1960. Mimeo.

Page 171
"Chemicals in Food Products," *Hearings*, 81st Congress, H.R. 323, Com. to Investigate Use of Chemicals in Food Products, 1950. Testimony of Dr. Morton S. Biskind.

Page 172
Thompson, R. H. S., "Cholinesterases and Anticholinesterases," *Lectures on the Scientific Basis of Medicine*, Vol. II (1952–53), Univ. of London. London: Athlone Press, 1954.

Page 172
Laug, E. P., and F. M. Keenz, "Effect of Carbon Tetrachloride on Toxicity and Storage of Methoxychlor in Rats," *Federation Proc.*, Vol. 10 (March 1951), p. 318.

Pages 172–173
Hayes, Wayland J., Jr., "The Toxicity of Dieldrin to Man," *Bull. World Health Organ.*, Vol. 20 (1959), pp. 891–912.

Page 173
"Abuse of Insecticide Fumigating Devices," *Jour. Am. Med. Assn.*, Vol. 156 (Oct. 9, 1954), pp. 607–8.

Page 173
"Chemicals in Food Products." Testimony of Dr. Paul B. Dunbar, pp. 28–29.

Pages 173–174
Smith, M. I., and E. Elrove, "Pharmacological and Chemical Studies of the Cause of So-Called Ginger Paralysis," *Public Health Reports*, Vol. 45 (1930), pp. 1703–16.

Page 174
Durham, W. F., et al., "Paralytic and Related Effects of Certain Organic Phosphorus Compounds," *A.M.A. Archives Indus. Health*, Vol. 13 (1956), pp. 326–30.

Page 174

Bidstrup, P. L., et al., "Anticholinesterases (Paralysis in Man Following Poisoning by Cholinesterase Inhibitors)," *Chem. and Indus.*, Vol. 24 (1954), pp. 674–76.

Page 174

Gershon, S., and F. H. Shaw, "Psychiatric Sequelae of Chronic Exposure to Organophosphorus Insecticides," *Lancet*, Vol. 7191 (June 24, 1961), pp. 1371–74.

CHAPTER 13:
THROUGH A NARROW WINDOW

Page 175

Wald, George, "Life and Light," *Sci. American*, Oct. 1959, pp. 40–42.

Page 176

Rabinowitch, E. I., Quoted in *Medical Research: A Midcentury Survey*. Vol. 2, *Unsolved Clinical Problems in Biological Perspective*. Boston: Little, Brown, 1955. P. 25.

Page 177

Ernster, L., and O. Lindberg, "Animal Mitochondria," *Annual Rev. Physiol.*, Vol. 20 (1958), pp. 13–42.

Page 177

Siekevitz, Philip, "Powerhouse of the Cell," *Sci. American*, Vol. 197 (1957), No. 1, pp. 131–40.

Page 177

Green, David E., "Biological Oxidation," *Sci. American*, Vol. 199 (1958), No. 1, pp. 56–62.

Page 177

Lehninger, Albert L., "Energy Transformation in the Cell," *Sci. American*, Vol. 202 (1960), No. 5, pp. 102–14.

Page 177

——, *Oxidative Phosphorylation*. Harvey Lectures (1953–54), Ser. XLIX, Harvard University. Cambridge: Harvard Univ. Press, 1955. Pp. 176–215.

Page 178

Siekevitz, "Powerhouse of the Cell."

Page 178

Simon, E. W., "Mechanisms of Dinitrophenol Toxicity," *Biol. Rev.*, Vol. 28 (1953), pp. 453–79.

Page 179

Yost, Henry T., and H. H. Robson, "Studies on the Effects of Irradiation of Cellular Particulates. III. The Effect of Combined Radiation Treatments on Phosphorylation," *Biol. Bull.*, Vol. 116 (1959), No. 3, pp. 498–506.

Page 179

Loomis, W. F., and Lipmann, F., "Reversible Inhibition of the Coupling between Phosphorylation and Oxidation," *Jour. Biol. Chem.*, Vol. 173 (1948), pp. 807–8.

Page 179

Brody, T. M., "Effect of Certain Plant Growth Substances on Oxidative Phosphorylation in Rat Liver Mitochondria," *Proc. Soc. Exper. Biol. and Med.*, Vol. 80 (1952), pp. 533–36.

Page 179

Sacklin, J. A., et al., "Effect of DDT on Enzymatic Oxidation and Phosphorylation," *Science*, Vol. 122 (1955), pp. 377–78.

Page 179

Danziger, L., "Anoxia and Compounds Causing Mental Disorders in Man," *Diseases Nervous System*, Vol. 6 (1945), No. 12, pp. 365–70.

Page 179

Goldblatt, Harry, and G. Cameron, "Induced Malignancy in Cells from Rat Myocardium Subjected to Intermittent Anaerobiosis During Long Propagation in Vitro," *Jour. Exper. Med.*, Vol. 97 (1953), No. 4, pp. 525–52.

Pages 179–180

Warburg, Otto, "On the Origin of Cancer Cells," *Science*, Vol. 123 (1956), No. 3191, pp. 309–14.

Page 180

"Congenital Malformations Subject of Study," *Registrar*, U.S. Public Health Service, Vol. 24, No. 12 (Dec. 1959), p. 1.

Page 180

Brachet, J., *Biochemical Cytology.* New York: Academic Press, 1957. P. 516.

Page 181

Genelly, Richard E., and Robert L. Rudd, "Effects of DDT, Toxaphene, and Dieldrin on Pheasant Reproduction," *Auk*, Vol. 73 (Oct. 1956), pp. 529–39.

Page 181
Wallace, George J., To author, June 2, 1960.
Page 181
Cottam, Clarence, "Some Effects of Sprays on Crops and Livestock," address to Soil Conservation Soc. of Am., Aug. 1961. Mimeo.
Page 181
Bryson, M. J., et al., "DDT in Eggs and Tissues of Chickens Fed Varying Levels of DDT," *Advances in Chem.*, Ser. No. 1, 1950.
Page 182
Genelly, Richard E., and Robert L. Rudd, "Chronic Toxicity of DDT, Toxaphene, and Dieldrin to Ring-necked Pheasants," *Calif. Fish and Game*, Vol. 42 (1956), No. 1, pp. 5–14.
Page 182
Emmel, L., and M. Krupe, "The Mode of Action of DDT in Warm-blooded Animals," *Zeits. für Naturforschung*, Vol. 1 (1946), pp. 691–95.
Page 182
Wallace, George J., To author.
Page 182
Pillmore, R. E., "Insecticide Residues in Big Game Animals," U.S. Fish and Wildlife Service, pp. 1–10. Denver, 1961. Mimeo.
Page 182
Hodge, C. H., et al., "Short-Term Oral Toxicity Tests of Methoxychlor in Rats and Dogs," *Jour. Pharmacol. and Exper. Therapeut.*, Vol. 99 (1950), p. 140.
Page 182
Burlington, H., and V. F. Lindeman, "Effect of DDT on Testes and Secondary Sex Characters of White Leghorn Cockerels," *Proc. Soc. Exper. Biol. and Med.*, Vol. 74 (1950), pp. 48–51.
Page 182
Lardy, H. A., and P. H. Phillips, "The Effect of Thyroxine and Dinitrophenol on Sperm Metabolism," *Jour. Biol. Chem.*, Vol. 149 (1943), p. 177.
Page 183
"Occupational Oligospermia," letter to Editor, *Jour. Am. Med. Assn.*, Vol. 140, No. 1249 (Aug. 13, 1949).

Page 183

Burnet, F. Macfarlane, "Leukemia As a Problem in Preventive Medicine," *New Eng. Jour. Med.*, Vol. 259 (1958), No. 9, pp. 423–31.

Page 183

Alexander, Peter, "Radiation-Imitating Chemicals," *Sci. American*, Vol. 202 (1960), No. 1, pp. 99–108.

Page 184

Simpson, George G., C. S. Pittendrigh, and L. H. Tiffany, *Life: An Introduction to Biology.* New York: Harcourt, Brace, 1957.

Page 185

Burnet, "Leukemia As a Problem in Preventive Medicine."

Page 185

Beam, A. G., and J. L. German III, "Chromosomes and Disease," *Sci. American*, Vol. 205 (1961), No. 5, pp. 66–76.

Page 185

"The Nature of Radioactive Fall-out and Its Effects on Man," *Hearings*, 85th Congress, Joint Com. on Atomic Energy, Pt. 2 (June 1957), p. 1062. Testimony of Dr. Hermann J. Muller.

Pages 185–186

Alexander, "Radiation-Imitating Chemicals."

Pages 185–186

Muller, Hermann J., "Radiation and Human Mutation," *Sci. American*, Vol. 193 (1955), No. 11, pp. 58–68.

Page 186

Conen, P. E., and G. S. Lansky, "Chromosome Damage during Nitrogen Mustard Therapy," *Brit. Med. Jour.*, Vol. 2 (Oct. 21, 1961), pp. 1055–57.

Page 186

Blasquez, J., and J. Maier, "Ginandromorfismo en *Culex fatigans* sometidos por generaciones sucesivas a exposiciones de DDT," *Revista de Sanidad y Assistencia Social* (Caracas), Vol. 16 (1951), pp. 607–12.

Page 186

Levan, A., and J. H. Tjio, "Induction of Chromosome Fragmentation by Phenols," *Hereditas*, Vol. 34 (1948), pp. 453–84.

Page 186

Loveless, A., and S. Revell, "New Evidence on the Mode of

Action of 'Mitotic Poisons,'" *Nature*, Vol. 164 (1949), pp. 938–44.

Page 186

Hadorn, E., et al., Quoted by Charlotte Auerbach in "Chemical Mutagenesis," *Biol. Rev.*, Vol. 24 (1949), pp. 355–91.

Page 186

Wilson, S. M., et al., "Cytological and Genetical Effects of the Defoliant Endothal," *Jour. of Heredity*, Vol. 47 (1956), No. 4, pp. 151–55.

Page 186

Vogt, quoted by W. J. Burdette in "The Significance of Mutation in Relation to the Origin of Tumors: A Review," *Cancer Research*, Vol. 15 (1955), No. 4, pp. 201–26.

Page 187

Swanson, Carl, *Cytology and Cytogenetics.* Englewood Cliffs, N.J.: Prentice-Hall, 1957.

Page 187

Kostoff, D., "Induction of Cytogenic Changes and Atypical Growth by Hexachlorcyclohexane," *Science*, Vol. 109 (May 6, 1949), pp. 467–68.

Page 187

Sass, John E., "Response of Meristems of Seedlings to Benzene Hexachloride Used As a Seed Protectant," *Science*, Vol. 114 (Nov. 2, 1951), p. 466.

Page 187

Shenefelt, R. D., "What's Behind Insect Control?" in *What's New in Farm Science.* Univ. of Wisc. Agric. Exper. Station Bulletin 512 (Jan. 1955).

Page 187

Croker, Barbara H., "Effects of 2,4-D and 2,4,5-T on Mitosis in *Allium cepa*," *Bot. Gazette*, Vol. 114 (1953), pp. 274–83.

Page 187

Mühling, G. N., et al., "Cytological Effects of Herbicidal Substituted Phenols," *Weeds*, Vol. 8 (1960), No. 2, pp. 173–81.

Page 187

Davis, David E., To author, Nov. 24, 1961.

Page 187

Jacobs, Patricia A., et al., "The Somatic Chromosomes in Mongolism," *Lancet*, No. 7075 (April 4, 1959), p. 710.

Page 188

Ford, C. E., and P. A. Jacobs, "Human Somatic Chromosomes," *Nature*, June 7, 1958, pp. 1565–68.

Page 188

"Chromosome Abnormality in Chronic Myeloid Leukaemia," editorial, *Brit. Med. Jour.*, Vol. 1 (Feb. 4, 1961), p. 347.

Page 188

Bearn and German, "Chromosomes and Disease."

Page 188

Patau, K., et al., "Partial-Trisomy Syndromes. I. Sturge-Weber's Disease," *Am. Jour. Human Genetics*, Vol. 13 (1961), No. 3, pp. 287–98.

Page 189

———, "Partial-Trisomy Syndromes. II. An Insertion As Cause of the OFD Syndrome in Mother and Daughter," *Chromosoma* (Berlin), Vol. 12 (1961), pp. 573–84.

Page 189

Therman, E., et al., "The D Trisomy Syndrome and XO Gonadal Dysgenesis in Two Sisters," *Am. Jour. Human Genetics*, Vol. 13 (1961), No. 2, pp. 193–204.

CHAPTER 14: ONE IN EVERY FOUR

Page 191

Hueper, W. C., "Newer Developments in Occupational and Environmental Cancer," *A.M.A. Archives Inter. Med.*, Vol. 100 (Sept. 1957), pp. 487–503.

Page 192

———, *Occupational Tumors and Allied Diseases.* Springfield, Ill.: Thomas, 1942.

Page 193

———, "Environmental Cancer Hazards: A Problem of Community Health," *Southern Med. Jour.*, Vol. 50 (1957), No. 7, pp. 923–33.

Page 193

"Estimated Numbers of Deaths and Death Rates for Selected Causes: United States," Annual Summary for 1959, Pt. 2, *Monthly Vital Statistics Report*, Vol. 7, No. 13 (July 22, 1959), p. 14. Natl. Office of Vital Statistics, Public Health Service.

Page 193
1962 Cancer Facts and Figures, American Cancer Society.

Page 193
Vital Statistics of the United States, 1959. Natl. Office of Vital Statistics, Public Health Service. Vol. I, Sec. 6, Mortality Statistics. Table 6-K.

Page 193
Hueper, W. C., *Environmental and Occupational Cancer.* Public Health Reports, Supplement 209 (1948).

Page 193
"Food Additives," *Hearings*, 85th Congress, Subcom. of Com. on Interstate and Foreign Commerce, July 19, 1957. Testimony of Dr. Francis E. Ray, p. 200.

Page 194
Hueper, *Occupational Tumors and Allied Diseases.*

Page 195
——, "Potential Role of Non-Nutritive Food Additives and Contaminants as Environmental Carcinogens," *A.M.A. Archives Path.*, Vol. 62 (Sept. 1956), pp. 218–49.

Page 195
"Tolerances for Residues of Aramite," *Federal Register*, Sept. 30, 1955. Food and Drug Administration, U.S. Dept. of Health, Education, and Welfare.

Page 196
"Notice of Proposal to Establish Zero Tolerances for Aramite," *Federal Register*, April 26, 1958. Food and Drug Administration.

Page 196
"Aramite—Revocation of Tolerances; Establishment of Zero Tolerances," *Federal Register*, Dec. 24, 1958. Food and Drug Administration.

Page 196
Van Oettingen, W. F., *The Halogenated Aliphatic, Olefinic, Cyclic, Aromatic, and Aliphatic-Aromatic Hydrocarbons: Including the Halogenated Insecticides, Their Toxicity and Potential Dangers.* U.S. Dept. of Health, Education, and Welfare. Public Health Service Publ. No. 414 (1955).

Page 196
Hueper, W. C., and W. W. Payne, "Observations on the Occurrence of Hepatomas in Rainbow Trout," *Jour. Natl. Cancer Inst.*, Vol. 27 (1961), pp. 1123–43.

Page 196

VanEsch, G. J., et al., "The Production of Skin Tumours in Mice by Oral Treatment with Urethane-Isopropyl-N-Phenyl Carbamate or Isopropyl-N-Chlorophenyl Carbamate in Combination with Skin Painting with Croton Oil and Tween 60," *Brit. Jour. Cancer*, Vol. 12 (1958), pp. 355–62.

Pages 196–197

"Scientific Background for Food and Drug Administration Action against Aminotriazole in Cranberries." Food and Drug Administration, U.S. Dept. of Health, Education, and Welfare, Nov. 17, 1959. Mimeo.

Page 197

Rutstein, David, Letter to *New York Times*, Nov. 16, 1959.

Page 197

Hueper, W. C., "Causal and Preventive Aspects of Environmental Cancer," *Minnesota Med.*, Vol. 39 (Jan. 1956), pp. 5–11, 22.

Pages 197–198

"Estimated Numbers of Deaths and Death Rates for Selected Causes: United States," Annual Summary for 1960, Pt. 2, *Monthly Vital Statistics Report*, Vol. 9, No. 13 (July 28, 1961), Table 3.

Page 198

Robert Cushman Murphy et al. v. *Ezra Taft Benson et al.* U.S. District Court, Eastern District of New York, Oct. 1959, Civ. No. 17610. Testimony of Dr. Malcolm M. Hargraves.

Page 199

Hargraves, Malcolm M., "Chemical Pesticides and Conservation Problems," address to 23rd Annual Conv. Natl. Wildlife Fed. (Feb. 27, 1959). Mimeo.

Page 199

——, and D. G. Hanlon, "Leukemia and Lymphoma— Environmental Diseases?" paper presented at Internatl. Congress of Hematology, Japan, Sept. 1960. Mimeo.

Page 199

Wright, C., et al., "Agranulocytosis Occurring after Exposure to a DDT Pyrethrum Aerosol Bomb," *Am. Jour. Med.*, Vol. 1 (1946), pp. 562–67.

Page 200

Jedlicka, V., "Paramyeloblastic Leukemia Appearing Simultaneously in Two Blood Cousins after Simultaneous Contact with Gammexane (Hexachlorcyclohexane)," *Acta Med. Scand.*, Vol. 161 (1958), pp. 447–51.

Page 200

Friberg, L., and J. Martensson, "Case of Panmyelopthisis after Exposure to Chlorophenothane and Benzene Hexachloride," (A.M.A.) *Archives Indus. Hygiene and Occupat. Med.*, Vol. 8 (1953), No. 2, pp. 166–69.

Pages 201–202

Warburg, Otto, "On the Origin of Cancer Cells," *Science*, Vol. 123, No. 3191 (Feb. 24, 1956), pp. 309–14.

Page 203

Sloan-Kettering Inst. for Cancer Research, *Biennial Report*, July 1, 1957–June 30, 1959, p. 72.

Page 203

Levan, Albert, and John J. Biesele, "Role of Chromosomes in Cancerogenesis, As Studied in Serial Tissue Culture of Mammalian Cells," *Annals New York Acad. Sci.*, Vol. 71 (1958), No. 6, pp. 1022–53.

Page 204

Hunter, F. T., "Chronic Exposure to Benzene (Benzol). II. The Clinical Effects," *Jour. Indus. Hygiene and Toxicol.*, Vol. 21 (1939), pp. 331–54.

Page 204

Mallory, T. B., et al., "Chronic Exposure to Benzene (Benzol). III. The Pathologic Results," *Jour. Indus. Hygiene and Toxicol.*, Vol. 21 (1939), pp. 355–93.

Page 204

Hueper, *Environmental and Occupational Cancer*, pp. 1–69.

Page 204

——, "Recent Developments in Environmental Cancer," *A.M.A. Archives Path.*, Vol. 58 (1954), pp. 475–523.

Page 204

Burnet, F. Macfarlane, "Leukemia As a Problem in Preventive Medicine," *New Eng. Jour. Med.*, Vol. 259 (1958), No. 9, pp. 423–31.

Page 204

Klein, Michael, "The Transplacental Effect of Urethan on Lung Tumorigenesis in Mice," *Jour. Natl. Cancer Inst.*, Vol. 12 (1952), pp. 1003–10.

Pages 205–206

Biskind, M. S., and G. R. Biskind, "Diminution in Ability of the Liver to Inactivate Estrone in Vitamin B Complex Deficiency," *Science*, Vol. 94, No. 2446 (Nov. 1941), p. 462.

Pages 205–206

Biskind, G. R., and M. S. Biskind, "The Nutritional Aspects of Certain Endocrine Disturbances," *Am. Jour. Clin. Path.*, Vol. 16 (1946), No. 12, pp. 737–45.

Pages 205–206

Biskind, M. S., and G. R. Biskind, "Effect of Vitamin B Complex Deficiency on Inactivation of Estrone in the Liver," *Endocrinology*, Vol. 31 (1942), No. 1, pp. 109–14.

Pages 205–206

Biskind, M. S., and M. C. Shelesnyak, "Effect of Vitamin B Complex Deficiency on Inactivation of Ovarian Estrogen in the Liver," *Endocrinology*, Vol. 30 (1942), No. 5, pp. 819–20.

Pages 205–206

Biskind, M. S., and G. R. Biskind, "Inactivation of Testosterone Propionate in the Liver During Vitamin B Complex Deficiency. Alteration of the Estrogen-Androgen Equilibrium," *Endocrinology*, Vol. 32 (1943), No. 1, pp. 97–102.

Page 205

Greene, H. S. N., "Uterine Adenomata in the Rabbit. III. Susceptibility As a Function of Constitutional Factors," *Jour. Exper. Med.*, Vol. 73 (1941), No. 2, pp. 273–92.

Page 205

Horning, E. S., and J. W. Whittick, "The Histogenesis of Stilboestrol-Induced Renal Tumours in the Male Golden Hamster," *Brit. Jour. Cancer*, Vol. 8 (1954), pp. 451–57.

Page 205

Kirkman, Hadley, *Estrogen-Induced Tumors of the Kidney in the Syrian Hamster.* U.S. Public Health Service, Natl. Cancer Inst. Monograph No. 1 (Dec. 1959).

Page 206

Ayre, J. E., and W. A. G. Bauld, "Thiamine Deficiency and High Estrogen Findings in Uterine Cancer and in Menorrhagia," *Science*, Vol. 103, No. 2676 (April 12, 1946), pp. 441–45.

Page 206

Rhoads, C. P., "Physiological Aspects of Vitamin Deficiency," *Proc. Inst. Med. Chicago*, Vol. 13 (1940), p. 198.

Page 206

Sugiura, K., and C. P. Rhoads, "Experimental Liver Cancer in Rats and Its Inhibition by Rice-Bran Extract, Yeast, and Yeast Extract," *Cancer Research*, Vol. 1 (1941), pp. 3–16.

Page 206

Martin, H., "The Precancerous Mouth Lesions of Avitaminosis B. Their Etiology, Response to Therapy and Relationship to Intraoral Cancer," *Am. Jour. Surgery*, Vol. 57 (1942), pp. 195–225.

Page 206

Tannenbaum, A., "Nutrition and Cancer," in Freddy Homburger, ed., *Physiopathology of Cancer.* New York: Harper, 1959. 2nd ed. A Paul B. Hoeber Book. P. 552.

Page 206

Symeonidis, A., "Post-starvation Gynecomastia and Its Relationship to Breast Cancer in Man," *Jour. Natl. Cancer Inst.*, Vol. 11 (1950), p. 656.

Page 206

Davies, J. N. P., "Sex Hormone Upset in Africans," *Brit. Med. Jour.*, Vol. 2 (1949), pp. 676–79.

Pages 206–207

Hueper, "Potential Role of Non-Nutritive Food Additives."

Page 207

VanEsch et al., "Production of Skin Tumours in Mice by Carbamates."

Page 207

Berenblum, I., and N. Trainin, "Possible Two-Stage Mechanism in Experimental Leukemogenesis," *Science*, Vol. 132 (July 1, 1960), pp. 40–41.

Pages 207–208

Hueper, W. C., "Cancer Hazards from Natural and Artificial Water Pollutants," *Proc.*, Conf. on Physiol. Aspects of Water Quality, Washington, D.C., Sept. 8–9, 1960, pp. 181–93. U.S. Public Health Service.

Page 208

Hueper and Payne, "Observations on Occurrence of Hepatomas in Rainbow Trout."

Pages 209–210

Sloan-Kettering Inst. for Cancer Research, *Biennial Report*, 1957–59.

Pages 209–211
    Hueper, W. C., To author.

CHAPTER 15: NATURE FIGHTS BACK

Page 213
    Briejèr, C. J., "The Growing Resistance of Insects to Insecti-
    cides," *Atlantic Naturalist*, Vol. 13 (1958), No. 3, pp. 149–55.
Pages 214–215
    Metcalf, Robert L., "The Impact of the Development of Or-
    ganophosphorus Insecticides upon Basic and Applied Sci-
    ence," *Bull. Entomol. Soc. Am.*, Vol. 5 (March 1959), pp. 3–15.
Page 215
    Ripper, W. E., "Effect of Pesticides on Balance of Arthro-
    pod Populations," *Annual Rev. Entomol.*, Vol. 1 (1956), pp.
    403–38.
Page 216
    Allen, Durward L., *Our Wildlife Legacy*. New York: Funk &
    Wagnalls, 1954. Pp. 234–36.
Page 216
    Sabrosky, Curtis W., "How Many Insects Are There?" *Year-
    book of Agric.*, U.S. Dept. of Agric., 1952, pp. 1–7.
Pages 216–217
    Bishopp, F. C., "Insect Friends of Man," *Yearbook of Agric.*,
    U.S. Dept. of Agric., 1952, pp. 79–87.
Page 217
    Klots, Alexander B., and Elsie B. Klots, "Beneficial Bees,
    Wasps, and Ants," *Handbook on Biological Control of Plant
    Pests*, pp. 44–46. Brooklyn Botanic Garden. Reprinted from
    Plants and Gardens, Vol. 16 (1960), No. 3.
Page 217
    Hagen, Kenneth S., "Biological Control with Lady Beetles,"
    Handbook on Biological Control of Plant Pests, pp. 28–35.
Page 217
    Schlinger, Evert I., "Natural Enemies of Aphids," *Handbook
    on Biological Control of Plant Pests*, pp. 36–42.
Page 217
    Bishopp, "Insect Friends of Man."
Page 218
    Ripper, "Effect of Pesticides on Arthropod Populations."

Page 219
> Davies, D. M., "A Study of the Black-fly Population of a Stream in Algonquin Park, Ontario," *Transactions*, Royal Canadian Inst., Vol. 59 (1950), pp. 121–59.

Page 219
> Ripper, "Effect of Pesticides on Arthropod Populations."

Page 219
> Johnson, Philip C., *Spruce Spider Mite Infestations in Northern Rocky Mountain Douglas-Fir Forests.* Research Paper 55, Intermountain Forest and Range Exper. Station, U.S. Forest Service, Ogden, Utah, 1958.

Page 219
> Davis, Donald W., "Some Effects of DDT on Spider Mites," *Jour. Econ. Entomol.*, Vol. 45 (1952), No. 6, pp. 1011–19.

Pages 220–221
> Gould, E., and E. O. Hamstead, "Control of the Red-banded Leaf Roller," *Jour. Econ. Entomol.*, Vol. 41 (1948), pp. 887–90.

Page 221
> Pickett, A. D., "A Critique on Insect Chemical Control Methods," *Canadian Entomologist*, Vol. 81 (1949), No. 3, pp. 1–10.

Page 221
> Joyce, R. J. V., "Large-Scale Spraying of Cotton in the Gash Delta in Eastern Sudan," *Bull. Entomol. Research*, Vol. 47 (1956), pp. 390–413.

Pages 221–222
> Long, W. H., et al., "Fire Ant Eradication Program Increases Damage by the Sugarcane Borer," *Sugar Bull.*, Vol. 37 (1958), No. 5, pp. 62–63.

Page 222
> Luckmann, William H., "Increase of European Corn Borers Following Soil Application of Large Amounts of Dieldrin," *Jour. Econ. Entomol.*, Vol. 53 (1960), No. 4, pp. 582–84.

Page 222
> Haeussler, G. J., "Losses Caused by Insects," *Yearbook of Agric.*, U.S. Dept. of Agric., 1952, pp. 141–46.

Page 222
> Clausen, C. P., "Parasites and Predators," *Yearbook of Agric.*, U.S. Dept. of Agric., 1952, pp. 380–88.

Page 223

——, *Biological Control of Insect Pests in the Continental United States.* U.S. Dept. of Agric. Technical Bulletin No. 1139 (June 1956), pp. 1–151.

Page 223

DeBach, Paul, "Application of Ecological Information to Control of Citrus Pests in California," *Proc.*, 10th Internatl. Congress of Entomologists (1956), Vol. 3 (1958), pp. 187–94.

Page 223

Laird, Marshall, "Biological Solutions to Problems Arising from the Use of Modern Insecticides in the Field of Public Health," *Acta Tropica*, Vol. 16 (1959), No. 4, pp. 331–55.

Page 223

Harrington, R. W., and W. L. Bidlingmayer, "Effects of Dieldrin on Fishes and Invertebrates of a Salt Marsh," *Jour. Wildlife Management*, Vol. 22 (1958), No. 1, pp. 76–82.

Page 224

Liver Flukes in Cattle. U.S. Dept. of Agric. Leaflet No. 493 (1961).

Page 224

Fisher, Theodore W., "What Is Biological Control?" *Handbook on Biological Control of Plant Pests*, pp. 6–18. Brooklyn Botanic Garden. Reprinted from *Plants and Gardens*, Vol. 16 (1960), No. 3.

Page 225

Jacob, F. H., "Some Modern Problems in Pest Control," *Science Progress*, No. 181 (1958), pp. 30–45.

Page 225

Pickett, A. D., and N. A. Patterson, "The Influence of Spray Programs on the Fauna of Apple Orchards in Nova Scotia. IV. A Review," *Canadian Entomologist*, Vol. 85 (1953), No. 12, pp. 472–78.

Page 226

Pickett, A. D., "Controlling Orchard Insects," *Agric. Inst. Rev.*, March–April 1953.

Page 226

——, "The Philosophy of Orchard Insect Control," 79th *Annual Report*, Entomol. Soc. of Ontario (1948), pp. 1–5.

Pages 226–227

——, "The Control of Apple Insects in Nova Scotia." Mimeo.

Page 227

Ullyett, G. C., "Insects, Man and the Environment," *Jour. Econ. Entomol.*, Vol. 44 (1951), No. 4, pp. 459–64.

CHAPTER 16: THE RUMBLINGS OF AN AVALANCHE

Pages 229–230

Babers, Frank H., *Development of Insect Resistance to Insecticides.* U.S. Dept. of Agric., E 776 (May 1949).

Pages 229–230

——, and J. J. Pratt, *Development of Insect Resistance to Insecticides. II. A Critical Review of the Literature up to 1951.* U.S. Dept. of Agric., E 818 (May 1951).

Pages 230–231

Brown, A. W. A., "The Challenge of Insecticide Resistance," *Bull. Entomol. Soc. Am.*, Vol. 7 (1961), No. 1, pp. 6–19.

Pages 230–231

——, "Development and Mechanism of Insect Resistance to Available Toxicants," *Soap and Chem. Specialties*, Jan. 1960.

Pages 230–231

Insect Resistance and Vector Control. World Health Organ. Technical Report Ser. No. 153 (Geneva, 1958), p. 5.

Pages 230–231

Elton, Charles S., *The Ecology of Invasions by Animals and Plants.* New York: Wiley, 1958. P. 181.

Pages 230–231

Babers and Pratt, *Development of Insect Resistance to Insecticides*, II.

Page 232

Brown, A. W. A., *Insecticide Resistance in Arthropods.* World Health Organ. Monograph Ser. No. 38 (1958), pp. 13, 11.

Page 232

Quarterman, K. D., and H. F. Schoof, "The Status of Insecticide Resistance in Arthropods of Public Health Importance in 1956," *Am. Jour. Trop. Med. and Hygiene*, Vol. 7 (1958), No. 1, pp. 74–83.

Pages 232–233
  Brown, *Insecticide Resistance in Arthropods.*
Pages 232–233
  Hess, Archie D., "The Significance of Insecticide Resistance in Vector Control Programs," *Am. Jour. Trop. Med. and Hygiene*, Vol. 1 (1952), No. 3, pp. 371–88.
Page 233
  Lindsay, Dale R., and H. I. Scudder, "Nonbiting Flies and Disease," *Annual Rev. Entomol.*, Vol. 1 (1956), pp. 323–46.
Page 233
  Schoof, H. F., and J. W. Kilpatrick, "House Fly Resistance to Organo-phosphorus Compounds in Arizona and Georgia," *Jour. Econ. Entomol.*, Vol. 51 (1958), No. 4, p. 546.
Page 233
  Brown, "Development and Mechanism of Insect Resistance."
Page 234
  ——, *Insecticide Resistance in Arthropods.*
Page 234
  ——, "Challenge of Insecticide Resistance."
Page 234
  ——, *Insecticide Resistance in Arthropods.*
Page 235
  ——, "Development and Mechanism of Insect Resistance."
Page 235
  ——, *Insecticide Resistance in Arthropods.*
Page 235
  ——, "Challenge of Insecticide Resistance."
Pages 235–236
  Anon., "Brown Dog Tick Develops Resistance to Chlordane," *New Jersey Agric.*, Vol. 37 (1955), No. 6, pp. 15–16.
Pages 235–236
  New York Herald Tribune, June 22, 1959; also J. C. Pallister, To author, Nov. 6, 1959.
Page 236
  Brown, "Challenge of Insecticide Resistance."
Pages 236–237
  Hoffmann, C. H., "Insect Resistance," *Soap*, Vol. 32 (1956), No. 8, pp. 129–32.

Page 237

Brown, A. W. A., *Insect Control by Chemicals.* New York: Wiley, 1951.

Page 237

Briejèr, C. J., "The Growing Resistance of Insects to Insecticides," *Atlantic Naturalist*, Vol. 13 (1958), No. 3, pp. 149–55.

Page 237

Laird, Marshall, "Biological Solutions to Problems Arising from the Use of Modern Insecticides in the Field of Public Health," *Acta Tropica*, Vol. 16 (1959), No. 4, pp. 331–55.

Page 238

Brown, *Insecticide Resistance in Arthropods.*

Page 238

——, "Development and Mechanism of Insect Resistance."

Page 239

Briejèr, "Growing Resistance of Insects to Insecticides."

Page 239

"Pesticides—1959," *Jour. Agric. and Food Chem.*, Vol. 7 (1959), No. 10, p. 680.

Page 239

Briejèr, "Growing Resistance of Insects to Insecticides."

CHAPTER 17:
THE OTHER ROAD

Page 242

Swanson, Carl P., *Cytology and Cytogenetics.* Englewood Cliffs, N.J.: Prentice-Hall, 1957.

Page 243

Knipling, E. F., "Control of Screw-Worm Fly by Atomic Radiation," *Sci. Monthly*, Vol. 85 (1957), No. 4, pp. 195–202.

Page 243

——, *Screwworm Eradication: Concepts and Research Leading to the Sterile-Male Method.* Smithsonian Inst. Annual Report, Publ. 4365 (1959).

Page 243

Bushland, R. C., et al., "Eradication of the Screw-Worm Fly by Releasing Gamma-Ray-Sterilized Males among the Natural Population," *Proc.*, Internatl. Conf. on Peaceful Uses of Atomic Energy, Geneva, Aug. 1955, Vol. 12, pp. 216–20.

Pages 243–244

Lindquist, Arthur W., "The Use of Gamma Radiation for Control or Eradication of the Screwworm," *Jour. Econ. Entomol.*, Vol. 48 (1955), No. 4, pp. 467–69.

Pages 244–245

——, "Research on the Use of Sexually Sterile Males for Eradication of Screw-Worms," *Proc.*, Inter-Am. Symposium on Peaceful Applications of Nuclear Energy, Buenos Aires, June 1959, pp. 229–39.

Pages 244–245

"Screwworm vs. Screwworm," *Agric. Research*, July 1958, p. 8. U.S. Dept. of Agric.

Pages 244–245

"Traps Indicate Screwworm May Still Exist in Southeast." U.S. Dept. of Agric. Release No. 1502-59 (June 3, 1959). Mimeo.

Pages 245–246

Potts, W. H., "Irradiation and the Control of Insect Pests," *Times* (London) Sci. Rev., Summer 1958, pp. 13–14.

Pages 245–246

Knipling, *Screwworm Eradication: Sterile-Male Method.*

Pages 245–246

Lindquist, Arthur W., "Entomological Uses of Radioisotopes," in *Radiation Biology and Medicine*. U.S. Atomic Energy Commission, 1958. Chap. 27, Pt. 8, pp. 688–710.

Pages 245–246

——, "Research on the Use of Sexually Sterile Males."

Page 246

"USDA May Have New Way to Control Insect Pests with Chemical Sterilants." U.S. Dept. of Agric. Release No. 3587-61 (Nov. 1, 1961). Mimeo.

Page 246

Lindquist, Arthur W., "Chemicals to Sterilize Insects," *Jour. Washington Acad. Sci.*, Nov. 1961, pp. 109–14.

Page 246

——, "New Ways to Control Insects," *Pest Control Mag.*, June 1961.

Page 246

LaBrecque, G. C., "Studies with Three Alkylating Agents As House Fly Sterilants," *Jour. Econ. Entomol.*, Vol. 54 (1961), No. 4, pp. 684–89.

Page 247
  Knipling, E. F., "Potentialities and Progress in the Devel-
  opment of Chemosterilants for Insect Control," paper pre-
  sented at Annual Meeting Entomol. Soc. of Am., Miami,
  1961.
Page 247
  ———, "Use of Insects for Their Own Destruction," *Jour.
  Econ. Entomol.*, Vol. 53 (1960), No. 3, pp. 415–20.
Page 247
  Mitlin, Norman, "Chemical Sterility and the Nucleic Acids,"
  paper presented Nov. 27, 1961, Symposium on Chemical Ste-
  rility, Entomol. Soc. of Am., Miami.
Page 248
  Alexander, Peter, To author, Feb. 19, 1962.
Page 248
  Eisner, T., "The Effectiveness of Arthropod Defensive Se-
  cretions," in Symposium 4 on "Chemical Defensive Mech-
  anisms," 11th Internatl. Congress of Entomologists, Vienna
  (1960), pp. 264–67. Offprint.
Page 248
  ———, "The Protective Role of the Spray Mechanism of the
  Bombardier Beetle, *Brachynus ballistarius* Lec.," *Jour. Insect
  Physiol.*, Vol. 2 (1958), No. 3, pp. 215–20.
Page 248
  ——— "Spray Mechanism of the Cockroach *Diploptera punc-
  tata*," *Science*, Vol. 128, No. 3316 (July 18, 1958), pp. 148–49.
Page 248
  Williams, Carroll M., "The Juvenile Hormone," *Sci. Ameri-
  can*, Vol. 198, No. 2 (Feb. 1958), p. 67.
Page 248
  "1957 Gypsy-Moth Eradication Program." U.S. Dept. of Ag-
  ric. Release 858-57-3. Mimeo.
Page 249
  Brown, William L., Jr., "Mass Insect Control Programs:
  Four Case Histories," *Psyche*, Vol. 68 (1961), Nos. 2–3, pp.
  75–111.
Page 249
  Jacobson, Martin, et al., "Isolation, Identification, and Syn-
  thesis of the Sex Attractant of Gypsy Moth," *Science*, Vol.
  132, No. 3433 (Oct. 14, 1960), p. 1011.

Pages 249–250

Christenson, L. D., "Recent Progress in the Development of Procedures for Eradicating or Controlling Tropical Fruit Flies," *Proc.*, 10th Internatl. Congress of Entomologists (1956), Vol. 3 (1958), pp. 11–16.

Page 250

Hoffmann, C. H., "New Concepts in Controlling Farm Insects," address to Internatl. Assn. Ice Cream Manuf. Conv., Oct. 27, 1961. Mimeo.

Page 250

Frings, Hubert, and Mable Frings, "Uses of Sounds by Insects," *Annual Rev. Entomol.*, Vol. 3 (1958), pp. 87–106.

Page 250

Research Report, 1956–1959. Entomol. Research Inst. for Biol. Control, Belleville, Ontario. Pp. 9–45.

Page 250

Kahn, M. C., and W. Offenhauser, Jr., "The First Field Tests of Recorded Mosquito Sounds Used for Mosquito Destruction," *Am. Jour. Trop. Med.*, Vol. 29 (1949), pp. 800–27.

Page 250

Wishart, George, To author, Aug. 10, 1961.

Page 250

Beirne, Bryan, To author, Feb. 7, 1962.

Page 250

Frings, Hubert, To author, Feb. 12, 1962.

Page 250

Wishart, George, To author, Aug. 10, 1961.

Page 250

Frings, Hubert, et al., "The Physical Effects of High Intensity Air-Borne Ultrasonic Waves on Animals," *Jour. Cellular and Compar. Physiol.*, Vol. 31 (1948), No. 3, pp. 339–58.

Page 251

Steinhaus, Edward A., "Microbial Control—The Emergence of an Idea," *Hilgardia*, Vol. 26, No. 2 (Oct. 1956), pp. 107–60.

Page 251

———, "Concerning the Harmlessness of Insect Pathogens and the Standardization of Microbial Control Products," *Jour. Econ. Entomol.*, Vol. 50, No. 6 (Dec. 1957), pp. 715–20.

Page 251
——, "Living Insecticides," *Sci. American*, Vol. 195, No. 2 (Aug. 1956), pp. 96–104.

Page 251
Angus, T. A., and A. E. Heimpel, "Microbial Insecticides," *Research for Farmers*, Spring 1959, pp. 12–13. Canada Dept. of Agric.

Page 251
Heimpel, A. M., and T. A. Angus, "Bacterial Insecticides," *Bacteriol. Rev.*, Vol. 24 (1960), No. 3, pp. 266–88.

Page 252
Briggs, John D., "Pathogens for the Control of Pests," *Biol. and Chem. Control of Plant and Animal Pests*. Washington, D.C., Am. Assn. Advancement Sci., 1960. Pp. 137–48.

Page 252
"Tests of a Microbial Insecticide against Forest Defoliators," *Bi-Monthly Progress Report*, Canada Dept. of Forestry, Vol. 17, No. 3 (May–June 1961).

Page 253
Steinhaus, "Living Insecticides."

Page 253
Tanada, Y., "Microbial Control of Insect Pests," *Annual Rev. Entomol.*, Vol. 4 (1959), pp. 277–302.

Page 253
Steinhaus, "Concerning the Harmlessness of Insect Pathogens."

Page 253
Clausen, C. P., *Biological Control of Insect Pests in the Continental United States*. U.S. Dept. of Agric. Technical Bulletin No. 1139 (June 1956), pp. 1–151.

Page 253
Hoffmann, C. H., "Biological Control of Noxious Insects, Weeds," *Agric. Chemicals*, March–April 1959.

Page 254
DeBach, Paul, "Biological Control of Insect Pests and Weeds," *Jour. Applied Nutrition*, Vol. 12 (1959), No. 3, pp. 120–34.

Page 255
Ruppertshofen, Heinz, "Forest-Hygiene," address to 5th

World Forestry Congress, Seattle, Wash. (Aug. 29–Sept. 10, 1960).

Page 255

—— To author, Feb. 25, 1962.

Page 255

Gösswald, Karl, *Die Rote Waldameise im Dienste der Waldhygiene.* Lüneburg: Metta Kinau Verlag, n.d.

Page 255

——, To author, Feb. 27, 1962.

Page 257

Balch, R. E., "Control of Forest Insects," *Annual Rev. Entomol.*, Vol. 3 (1958), pp. 449–68.

Page 257

Buckner, C. H., "Mammalian Predators of the Larch Sawfly in Eastern Manitoba," *Proc.*, 10th Internatl. Congress of Entomologists (1956), Vol. 4 (1958), pp. 353–61.

Page 257

Morris, R. F., "Differentiation by Small Mammal Predators between Sound and Empty Cocoons of the European Spruce Sawfly," *Canadian Entomologist*, Vol. 81 (1949), No. 5.

Page 257

MacLeod, C. F., "The Introduction of the Masked Shrew into Newfoundland," *Bi-Monthly Progress Report*, Canada Dept. of Agric., Vol. 16, No. 2 (March–April 1960).

Page 257

——, To author, Feb. 12, 1962.

Page 257

Carroll, W. J., To author, March 8, 1962.

OTHER WRITINGS
ON THE ENVIRONMENT

To Harold Lynch

9713 Sutherland Road
Silver Spring, Md.
July 15, 1945

Dear Mr. Lynch:

Many thanks for your letter of the 12th, and for your gracious settlement of our problems.

And now here is a query for your consideration—Practically at my backdoor here in Maryland, an experiment of more than ordinary interest and importance is going on. We have all heard a lot about what DDT will soon do for us by wiping out insect pests. The experiments at Patuxent have been planned to show what other effects DDT may have if applied to wide areas: what it will do to insects that are beneficial or even essential; how it may affect waterfowl, or birds that depend on insect food; whether it may upset the whole delicate balance of nature if unwisely used.

I believe there is a timely story in these tests. The incredible amount of painstaking work involved in setting up the test areas, the methods, results, and the interpretation from the biologist's point of view should add up to a pretty good article. It's something that really does affect everybody.

I am in position to cover the progress of the thing at first hand during the coming weeks, and with a little encouragement from you, I should do so with a view to turning out an article aimed for the pages of the *Digest*. The background is pretty well sketched in the enclosed release. Does the idea interest you?

Sincerely yours,

Mr. Harold Lynch
The Reader's Digest
Pleasantville, N. Y.

To Shirley Briggs

Saturday, Sept. 28

Dear Shirley:

My name must be written in large letters on your black list, for which I am very sorry, but honest I can't help it! As Kay can tell you, our days have been full of command cars, marshes, mud, sand dunes, mosquitoes, Audubonites, etc, etc, from dawn till dusk, and when at last we could crawl home, sunburned, black-and-blue, mosquito-bitten, and weary, it was time to change into respectable clothes (yes, we put on the disguise of respectable citizens for our evening appearance in Newburyport) and fare forth to our beloved Newburyport Manor for a juicy and tender steak. I have never eaten so much or so fast in my life; I am simply starved when dinner time finally arrives and wolf down everything they put before me, which is quite a lot. I don't know how we would have lived through this without the Manor. That is only for dinner; breakfast has to be eaten at a little place down town, not too bad, and mostly we have had sandwiches put up for lunch. Every day they look at us in amazement and ask if we are going on another picnic, and we smile and say yes, we think we will. Picnic, indeed.

Of course I can't deceive you; it has been a tremendous lot of fun and I will take on another one at the first opportunity. However, it has been strenuous from start to finish, and I'd like to curl up somewhere and hibernate a while. Unfortunately I am just beginning the tedious and relatively unattractive chore of library research which has to be done in order that we may publish some honest-to-goodness facts on Parker River for a change. Beginning Monday, I am going over all the old records filed in the library of the Mass. Audubon Society in Boston in order to get the records which will prove: (a) whether Parker River is a spot that has in the past been visited by large numbers and varieties of birds; (b) whether it is linked through migration paths with other areas to the south and so has more than local significance. It is amazing but true that we have never put these facts on record, and indeed do not have them in our possession. So I feel that this musty, dusty job, going over the local records, is really the heart of the whole problem.

Life has become more exciting around here the last few days, with rumors floating about that the enemy is threatening to sabotage the refuge. And more than rumors—several of the duck banding traps have been broken into and the ducks let out. The refuge manager maintains this is the work of a 4-legged animal, but some of the others aren't so sure. In fact, some are getting nervous about being shot at, but I think they have pretty good imaginations. However, if I don't come back——

I'm sorry you couldn't have experienced a command car tour of Plum Island; Chincoteague is like scooting along a super highway by comparison. All of the equipment, including the command car, regularly gets stuck and requires the services of all hands to pull it out; but luck was with us and it didn't happen when we were along. One trip, to the extreme southern tip of the refuge, was so wild that I still can't see how we got back without either getting stuck or turning over.

Kay has promised not to tell *all* the juiciest tidbits before I get back, so don't ask her too many questions. I expect to be most of the week in Boston, for I still have several people to see as well as all that library work. Then I expect to take enough leave to get around to P-town with Kitty. I'll be ready to relax for a few minutes by that time.

I expect to leave Boston for home Sunday night, so should see you Monday morning.

If you should be forgiving enough to write me, bear in mind the following: A letter mailed Monday or Tuesday should be sent to State Street; after that care of Kitty at 495 Commercial Street, Provincetown.

As ever,
Ray

1946

To Raymond J. Brown

9713 Sutherland Road
Silver Spring, Md.
October 15, 1946

Mr. Raymond J. Brown, Editor
Outdoor Life
353 Fourth Avenue
New York 10, N. Y.

Dear Mr. Brown:

Of course your letter of October 8 in regard to the Conservation Pledge competition interests me very much. I am sorry that my reply has been delayed, but I have just returned to Washington after an extended absence.

In complying with your request, let me introduce myself as follows: As long as I can remember, I have been interested in the world of nature, in wild creatures, and in natural, unspoiled beauty. I have always loved to watch birds, and to wander in the woods. Naturally enough, these interests led me to study the biological sciences in college and later to do graduate work in zoology at the Johns Hopkins University. For several years I assisted in the zoology laboratories at the Johns Hopkins Summer School and at the University of Maryland. Then I joined the biological staff of the former Bureau of Fisheries, now part of the Fish and Wildlife Service. Although first employed as a biologist, I was later transferred to the Information Division of the Service where my principal job is to prepare publications on conservation subjects for the public.

In spare time, I have also done some personal writing: one book, *Under the Sea Wind*, published by Simon and Schuster in 1941, and articles for various national magazines.

I am a native of Pittsburgh, but have spent the past 16 years in Baltimore and Washington. I belong to the District of Columbia Audubon Society, the Massachusetts Audubon Society, and the Hawk Mountain Sanctuary Association. My principal hobby is ornithology, in pursuit of which I am always willing to get up in the middle of the night or to get wet, cold, or dirty. I like to travel, especially in wild country, with a strong preference for the seacoast. I love boats, coffee, and cats. I suffer over writing, but do it anyway. That is about all there is to say, except that to me conservation is not an academic question for debate, but something that vitally and immediately concerns my whole way of life.

I should like to congratulate *Outdoor Life* on sponsoring this contest. I enjoyed taking part in it; my contribution was written while vacationing on the Maine coast.

Sincerely yours,

P.S. The photograph will follow in a few days.

To Edwin Way Teale

9713 Sutherland Road
Silver Spring, Md.
February 2, 1947

Dear Mr. Teale,

You will think I have forgotten my promise to write you about the eels, but I have not, and I hope I am not too late.

John Pearson, who has worked the South Atlantic coast pretty thoroughly, was in recently and I asked him where you should go. He was unable to suggest anything below North Carolina, but said he could take you to a small mill dam near the Beaufort Laboratory where he has seen the young eels concentrated below the dam. If you can look him up at the laboratory this might be your best bet. The Chesapeake, of course, has the largest eel population of any place on the coast, and you might come upon the youngsters in any of the small streams that empty into the Bay. However, Dr. Hildebrand, who has collected extensively for many years, says he does not remember ever having seen young eels in the water, although he has taken hundreds in his nets. Of course they are transparent except for scattered pigment cells, and the black little eyes. What with the early spring everywhere I should think the movement of the eels into the streams would take place by March in this locality, so perhaps you will be too late this year. But I thought you'd like to have these suggestions, discouraging as they are.

I have just returned from a few days at Lake Mattamuskeet. The mild winter there had sent most of the swans off early, much to my disappointment, but we were able to see a hundred or more at a time at close hand, and Canada geese by the thousand.

If you stop in Washington on your way north I hope you will come in. I'd be delighted if you and Mrs. Teale would have lunch or dinner with me. Did you find some one to leave your cats with? I do hope so. In any event, I envy you that trip, for I'm going to be shut up in the office for several months at least.

Sincerely,
Rachel L. Carson

To Maria Carson

Red Rock Lakes
Sunday, Sept 21, 1947

Dearest Mamma,

Well here we are in the most beautiful country I have ever seen or ever expect to see. I don't know where to begin to tell you about it, and I guess most of it will have to wait until I get home. My note book already has pages and pages in it, because for once I am trying to write down everything I possibly can. After a month I am afraid I'd get things mixed up if I didn't.

As I told you, we ran through snow all Thursday morning on the train, and it continued to snow most of the afternoon while we were in Butte. The wretched train to Monida was supposed to leave at about 6:40, I believe, but actually left a little after 8—waiting for a delayed Northern Pacific train from the west, and then had all the mail to transfer. We knew this would make us after midnight reaching Monida, and when we learned from the train men that there is nothing there but a couple of houses and a so-called hotel we began to wonder just what we would do if Dr. Sharp were not there. Fortunately, however, he was still waiting! He said we could all stay at the "hotel" if we were too tired to make the trip to the refuge, but we all seemed to feel we would rather get it over with. Of course Kay and I hadn't the slightest idea that it was a two-hour drive over muddy, snowy roads where a good many times there was considerable doubt whether the car would get through, or would leave us stuck in the middle of nowhere, in the middle of the night. But we did get through, reaching the Sharp home at 2:30!

They have a lovely, log-cabin home, very beautifully and comfortably furnished, and are treating us royally. Mrs. Sharp manages to serve delicious meals, although how she has provided all the fresh fruit and vegetables is beyond me. This place is only a little less isolated than Driftwood Valley, and their winter existence here, completely snowbound, is very like that described in the book. Today we had the most wonderful strawberry shortcake, this being the strawberry season in southern Montana!

We found they had had several days of snow before we came, but it was clear most of Friday, although cold enough for about the warmest gear I could muster. Yesterday was warmer, with the snow nearly all melted off, except on the highest peaks. Today has been a perfect day, with a deep blue, cloudless sky, and very comfortably warm in the sun. I am so glad the snow preceded us, for all the high peaks have been white, and indeed until today even the lower slopes were covered. The effect of the dark fir and pines, and the beautiful golden yellow of the aspens, against that background is simply beyond description. In this valley we are about 6,000 feet; the mountain tops hereabout are a little over 10,000 feet, their crests above timber line. The moose in the willow swamps around Upper Red Rock Lake have been putting on a regular show for us— cows, calves (twins) and big bulls with enormous antlers. The swans are seen only at a considerable distance, being very wary. I think I have gotten even more pleasure out of the small birds. The mountain bluebirds (remember the Peterson card) are the most heavenly color I have ever seen. The red-shafted flickers are in migration, and are brilliantly colored with red under the wings that flashes as they fly. My bird list also includes the following new to me: magpie, Clarke nutcracker, pink-sided junco, western meadowlark, Calaveras warbler, russet-backed thrush, rock wren, pipit, and mountain chickadee. I don't suppose any of our books will have these species.

I had wanted to see a porcupine (yellow-haired) and the Sharps said we might find one by going out around the house with a flash light. So last night we went out, and sure enough there was one. Dr. Sharp ran and headed him off, with the result that he turned and ran straight for the back door, where an outside light was on and Kay and Mrs. Sharp were waiting. Then he went down in the holly hocks beside the steps, and didn't seem at all frightened, just a little annoyed. I stood within two feet of him.

I hope the camera is working all right, although I sure wish I could see the results before I use all the film. Tell Robert that the first time I went to use it and tried to sight through the view-finder, I couldn't see a thing. Then I remembered the book said that there is a plate inside that drops over the view-finder when the camera is held upside down, and that when this happened one should reverse the position of the camera to open it. Well, I reversed it in every possible manner, and still that little plate stays right across the hole you are supposed to sight through. So the only thing I could do was sight through the little frames at the side, which Robert said doesn't work as well, but it's that or nothing. I cannot imagine what is the matter with it.

We are going to take the bus to Butte tomorrow about 11 in the morning, stay all night in Butte, and go in to Missoula Tuesday as planned. We decided that would work better all around than to go in and stay in that horrible little hotel in Monida tomorrow night and try to struggle onto that early train.

Your two air mail letters and one from Shirley came yesterday. They ordinarily get mail only three days a week, but one of the refuge men happened to go in. I was so glad to have it, and glad the weather is more decent. ARE YOU RESTING? Kiss the pussies for me—imagine poor little Teedie calling for me in the cellar. I do hope she doesn't miss me too much, and won't stop eating. But imagine, more than a week of the time will have gone when you get this. I would not be surprised if we could do the Columbia River more quickly than we had planned. This is by all odds the most isolated, and eats up more time getting around.

I am writing in the refuge office; now I must close and go up to dinner.

Loads of love,
Rachel

To Shirley Briggs

Sept. 29

Dear Shirley,

What I lack more than anything else is a typewriter—if I had one you'd have had a full account of the Bison Range long since. Now here we are right in the midst of the Columbia River—so to speak—and not only my correspondence but my official log of the trip are far behind. Tomorrow we set out in a resplendent new Ford station wagon, and each night will see us somewhere else, battling Wenatchee apple pickers and Grand Coulee sheep growers for a place (nice and sanitary, I hope) to lay our heads.

I see a place on the Oregon map named Starvation Pass, and our recent experiences make us regard such names with great understanding. You start out to see some remote dam or fish ladder and happily think you will lunch en route. But mid afternoon finds you wearily driving through country that apparently never gave a thought to feeding travelers. When you finally come to just *any* place that can produce something edible you rush in.

But I must tell you a little about our accommodation in Ronan, Mont., for therein lies the explanation of our doing the Bison Range in record time, despite its exceptionally nice animals. After being housed and fed so well at Red Rock Lakes, we assumed the Bison would do equally well by us, but it seems the Mushbachs had no spare room, and so had gotten us a hotel room "nearby." Nearby proved to be 20 miles away over rather jiggly, dusty road, in a typical (I suppose) wild-west hamlet known as Ronan (on the main road to Glacier Park). We were in an "upstairs" hotel, the downstairs being a hardware store, with a movie house next door. Our "windows"—a narrow slit next the ceiling—overlooked the movie, so we were entertained by the sound effects of "Bob, Son of Battle" and its unidentifiable co-feature, until midnight. The place was presided over by a gaunt, deep voiced specimen who became known to us as The Witch.

Ronan had two eating places and we tried them both. Where we ate dinner, two men had a violent argument all during their meal, and it got so bad that we began making plans to lie on the floor if and when shooting began.

I shall have to tell you about the Mushbach pets when we return, also about the buffalo, mule deer, elk, and Rocky Mt. sheep. My bird list has some glamorous additions, Oregon having furnished Canada and Steller's jays.

— Interruption, and great —
passage of time

This is now Wednesday, and we are spending the night in tourist cabins outside North Bonneville, having seen the sights of the dam today.

To go back—Mrs. Watkins certainly did royally by us. She had us out the first evening, then took us picnicking to Mt. Hood with her niece and nephew Sunday. We had a grand time, and thought it mighty nice of her.

We had a wild ride to Astoria yesterday with one of the river biologists and were thankful to return to Portland and the chaperonage of the Barnabys. Mr. B. drives considerately, and both he and his wife are good company. The fishery part of the Astoria trip was a flop, but we enjoyed side trips to the Seaside Aquarium and the Portland rose garden.

When are you going to Boston with Mr. Lemon? And are you and Kitty going to P-town?

We have enjoyed your letters even if we (or I) have not been able to do equally well by you. See you in a couple of weeks (if you're there).

As ever—
Ray

1947

Mr. Day's Dismissal

The dismissal of Mr. Albert M. Day as director of the Fish and Wildlife Service is the most recent of a series of events that should be deeply disturbing to every thoughtful citizen. The ominous pattern that is clearly being revealed is the elimination from the Government of career men of long experience and high professional competence and their replacement by political appointees. The firing of Mr. Marion Clawson, director of the Bureau of Land Management, is another example. There are widespread rumors that the head of the Park Service, who, like Mr. Day, has spent his entire professional life in the organization he heads, will also be replaced. These actions strongly suggest that the way is being cleared for a raid upon our natural resources that is without parallel within the present century.

The real wealth of the Nation lies in the resources of the earth—soil, water, forests, minerals, and wildlife. To utilize them for present needs while insuring their preservation for future generations requires a delicately balanced and continuing program, based on the most extensive research. Their administration is not properly, and cannot be, a matter of politics.

By long tradition, the agencies responsible for these resources have been directed by men of professional stature and experience, who have understood, respected, and been guided by the findings of their scientists. Mr. Day's career in wildlife conservation began 35 years ago, when, as a young biologist, he was appointed to the staff of the former Biological Survey, which later became part of the Fish and Wildlife Service. During the intervening years he rose through the ranks, occupying successively higher positions until he was appointed director in 1946. He achieved a reputation as an able and fair-minded administrator, with courage to stand firm against the minority groups who demanded that he relax wildlife conservation measures so that they might raid these public resources. Secretary McKay, whose own grasp of conservation problems is yet to be demonstrated, has now decreed that the Nation is to be deprived of these services.

These actions within the Interior Department fall into place beside the proposed giveaway of our offshore oil reserves and the threatened invasion of national parks, forests and other public lands.

For many years public-spirited citizens throughout the country have been working for the conservation of the natural resources, realizing their vital importance to the Nation. Apparently their hard-won progress is to be wiped out, as a politically minded Administration returns us to the dark ages of unrestrained exploitation and destruction.

It is one of the ironies of our time that, while concentrating on the defense of our country against enemies from without, we should be so heedless of those who would destroy it from within.

RACHEL L. CARSON.
Silver Spring.
Washington Post, April 22, 1953

To Fon Boardman

Dear Fon:

You have asked for my opinion of Aldo Leopold's *Round River*. Here it is, and I repeat my permission to quote, although I doubt that you will want to avail yourself of it.

This is a truly shocking book. By the time I had somehow gotten through the first 30 or 40 pages, I was in a state of cold anger that I haven't experienced in many a day. Notwithstanding the pious sentiments on conservation expressed elsewhere in the book (and in the light of the journal entries they become disgusting hypocrisy), Mr. Leopold was a completely brutal man. A wild animal was to him only something to kill and torture. Whoever wrote your jacket copy did a slick but misleading job in selecting the quotes to be used as reader bait. These would have been more truly representative of the spirit of the book:

> "We caught a wet cormorant by diving him down . . . Set him out for live bait at one of the cat sets."

> "The bobcat trap contained a small coon. He looked very wet and lonesome. . . . we skinned him out and reset."

> "Sighted a big bobcat fishing on the island . . . it was inconceivable that this cat could get away from us. But she did."

> "Carl knocked a squirrel out of a tree with his slingshot."

> "A snowshoe rabbit hopped across the trail. We all popped at him with our slingshots . . . It was such a funny performance to kill a rabbit with a rubber gun that we all roared with laughter."

This sort of thing is the more shocking because it is presented as the words of a "naturalist," a "humanist" and a "pioneer in modern conservation." To put it very mildly, someone at Oxford needs to consult a dictionary.

In fairness, let me say that until this book so rudely disillusioned me, I, too, believed in the legend of Aldo Leopold. I accepted without question that he was a true naturalist and conservationist. Perhaps, in a twisted sort of way, Oxford has done a service in revealing one of the things that is wrong with conservation—that so much of it is in the hands of men who smugly assume that the end of conservation is to provide fodder for their guns—and that anyone who feels otherwise is a sentimental fool.

The most disturbing thing of all, however, is the glorification of cruelty. There is no point preaching at you, who did not write the book and probably are not responsible for its publication, but I shall have to express a very deep conviction: that until we have courage to recognize cruelty for what it is—whether its victim is human or animal—we cannot expect things to be much better in the world. There can be no double standard. We cannot have peace among men whose hearts find delight in killing any living creature. By every act that glorifies or even tolerates such moronic delight in killing, we set back the progress of humanity. I am sorry that Oxford published this book.

c. September 1953

To Dorothy Freeman

West Southport, September 28

Dear Dorothy,

This is the letter I wasn't supposed to write (you said) but I'm writing it anyway for what seems to me a very good reason—I really want to!

The big September tides have come and gone, and every time I was down there exploring I wished you were there, too—you would have enjoyed it so. Of course wind and surf made each day quite unpredictable; what should have been the lowest tide wasn't, and one day the surf was so heavy that just being down within reach of it was an adventure. I'm sending you a snapshot taken from the "edge of the edge" on September 22, to show what I mean—you know it's not supposed to look like that at low tide! (By the way, I think I see Dogfish Head in this picture. The near points are, of course, the entrance to Deep Cove, but I thought the dimly seen bit on the horizon might be yours. If so, I can wave to you at low water of spring's, anyway.) I am so sorry we didn't have a good tide the day you were here. When it is really low, it is nothing to see as many as thirty anemones under one ledge—big things six to eight inches long. And there are so many urchins right out on the flat rock surfaces where the tide falls down below all the Irish moss, and there is nothing but the rose-colored coralline algae encrusting the rock. And then all the Laminarias and other deep-water plants begin to show themselves, and somehow everything seems so very different.

I think the things you might have enjoyed most (as I believe I did) were these. The first day, when the tide was only "o.o" and so not as exciting as the later ones promised to be, I was poking around a big rock that had thick crusts of the coralline algae that looked as though they could be broken off. I found they could be—because the pink coating was covering some very large barnacles, or rather barnacle shells, for they were empty. So I took a small mass of the stuff to the house, and spent the evening being entertained by all the creatures that were living in and on that little world that wasn't more than two inches across in any direction. Among other things, there were tiny anemones living inside the empty barnacle shells. And on the outside of the shells there was attached a whole new generation of baby barnacles. When they fed (as they did, madly, all the time I was watching) I could see that the inside of their shells, and their own little feathery appendages, were for some odd reason colored the same deep pink as the algae that were cementing their world together. The whole catalogue of creatures would be too long to list, but there were many different species and literally hundreds of individuals. (They all went home on the next low tide.)

The other especially choice thing (and I really got excited about this, because I didn't know it at all and have never seen any report of it) was the discovery that, where the pink crust of corallines over the rock has become thick and heavy enough that pieces can be chipped off, there is a whole community of creatures living in it and under it. That, again, was a whole evening's work and entertainment at the microscope, the high point of which was the discovery of an exquisitely beautiful worm (don't laugh, and don't shiver—it is the most beautiful worm in the world!) that I had never personally collected before, though of course I've seen it. The whole algal crust is riddled with the borings of things that have made a home in it, and with winding tunnels going off in all directions; sometimes one of those very tiny crustaceans that has a single, glowing eye would come up out of the darkness of such a tunnel, always reminding me of a miner with his head lamp. Well, you see I had a good time.

And the high tides, too, were something to enjoy, and on the moonlight nights we didn't go to bed till "all hours"—for we could sit by the living room window and enjoy the water without even being cold. On one of those nights there was re- ally a tremendous surf—I don't know why, but there were real, open-ocean rollers. And almost every night, after we'd gone to bed, we could hear the surf trampling in over our ledges, and then I would think about all the creatures I knew were down there, and wonder how they were getting along.

At one point, I thought I'd ration these good tides—maybe go to your shore for one, to Ocean Point, and to Pemaquid; but finally I decided it made more sense to get to know my own shore really well this year, at all the various tidal levels. Maybe we can do yours together, on a really good tide, next year. And that reminds me to say that when you decide when to take your 1954 vacation, you really must plan it with a tide table in hand. No coming on any old neap tides!

I always like fall better than summer, and we stay on and hate to think of going. One day there was a school of porpoises over on the far shore; we have had loons just offshore several different days; and yesterday a big seal put his head out several times and swam in toward the rocks in front of the Mahard place. And the *Register*'s society notes this week contain items about a whale, a moose, and three deer. I seriously doubt that we shall be able to send ourselves home until they come to turn the water off, October 20th. Besides, I have certain man- uscript goals that I want to reach before that break comes, and progress is—well, I may as well admit it—SLOW.

I was interested in your note about the movie, and if you go, and like it, I am quite prepared to forgive you. Of course I was terribly prejudiced, and perhaps a person going to see it just for itself, without particular reference to the book, would find it enjoyable. Anyway, please tell me. It has just opened at home (how glad I am that I'm not in Washington) and the *Washington Post*, while liking some of the photography, roasted the script, for which I was duly grateful.

May I tell you that every one of your letters has given me a great deal of pleasure. It seems as though I had known you for years instead of weeks, for time doesn't matter when two people think and feel in the same way about so many things. I, too, am looking forward to lazy hours when we can sit on our beach next summer and just talk! That will be one of the great pleasures of the summer. In the meantime, if I steal time from this tyrant manuscript now and then to write you, you must realize that I am doing it chiefly from selfish motives, because I like writing to friends almost as much as I like hearing from them!

And I'm glad that the books (and I) stood the test of rereading after you had met the author. Now that I know you, I value your devotion to what I've written all the more. I doubt that I could explain how any particular feeling or effect is achieved (you know the story of the centipede) but if there is any simple explanation I think it is that my sensory impressions of, and emotional response to, the world of nature date from earliest childhood, and that the factual knowledge was acquired much later. I had no formal training in biology, for example, until my second year of college, yet I had felt at home with wild creatures all my life. And I loved the ocean with a purely vicarious love long before I had seen it.

Thanks for the suggestion of a book for Roger; he is already showing great delight in being read to. He has four of the Beatrix Potter books (I grew up on them—did you?) and brings them to his mother one after another to have them read.

If I don't stop you will feel that you have read another book! I have just thought of about ten other things I meant to tell you but for once I shall take myself in hand and go back to the manuscript!

My best to all the Freemans, and to you my affectionate regard—

Sincerely,
Rachel

1953

To Dorothy Freeman

Thursday, January 21

Dearest,

After this one I'm going to behave—by at least trying to write on schedule, and no more! But your note written Tuesday morning needs a prompt reply, not so much because of your question as something else it suggests. Of course I don't mind if you want to inflict me on the Garden Club, though I'm sure they'd rather hear about Martha. Please do whatever you like about it, and know that if you do talk about our first adventure together, and read from the letter, I'll have not the slightest misgivings about the way you would handle it. Now for the other: the fact that you have mentioned this meeting next week makes me want to anticipate something that would have gone into my next "regular" letter—if there is such a thing. There is a critical situation developing about one of the National Monument areas in Colorado and Utah—a plan to build a dam that would submerge an area that is not only beautiful, but of great interest and importance as a geologic record. There seems to be no question that the dam could serve its intended purpose just as well if built somewhere else—so to put it in the Dinosaur Monument, submerging these wonderful formations and fossil beds, is nothing but blind and wilful destructiveness. The issue of the magazine *The Living Wilderness* that reached us yesterday says that the Secretary of the Interior—who is supposed to have such things in his care—has recommended the dam in Dinosaur Monument to the President, in whose hands the thing now is, or will be shortly. They urged everyone interested to write, or better wire, the President urging preservation of the National Parks and Monument, and specifically urging that this project be abandoned. We are wiring, and I felt you and Stan would feel as we do if you knew about it. I have asked that a copy of the magazine be sent you (see pages 26–28, and the hand-written note on p. 29) along with some other literature on the Dinosaur Monument. Remembering Stan's pictures, the feeling you both have for the West, and your appreciation of Mr. Cloos, I don't have any doubt how you'll feel about this. Now, however, I'm wondering if you won't carry it further and try to interest your Garden Club in the situation, if they aren't already. I know so many instances where Garden Clubs have become a strong constructive force for conservation. If this group saw Stan's Grand Canyon pictures, they ought to understand what this means.

This is something we shall talk of much in the years and months ahead, I know. It is part of the general problem that is so close to my heart—the saving of unspoiled, natural areas from senseless destruction. But in this particular case the time is short and perhaps you will find time and inclination to say something about it at your meeting next week.

I have been undergoing a new permanent this afternoon— writing this at possible intervals. Not the best conditions for concentration, but I want to mail this today.

Darling, I do plan to call you Sunday, and since I can tell you are going to hover around the phone, I won't make it person-to-person this time. But the next time (whenever that is) I won't let you be tied down in that way. I'll try to make this between 10:45 and 11:15.

Will you forgive me for laughing when I read your description of waiting for the mailman Monday to see if "It" would be there? Just as if my handwriting hadn't been appearing in your mail practically every day!! We are really awful! Aren't we worse than ever? Now, I guess, no more till—when? By this time I don't even know when it should be.

> But my dearest love, always
> Rachel

1954

The Real World Around Us

(Theta Sigma Phi Matrix Table Dinner, Columbus, Ohio, April 21, 1954)

I HAVE FOUND that the writing of books has many unexpected consequences. One of the things I didn't anticipate when I wrote *The Sea Around Us* was the number of times that I should have around me a sea of faces. I never expected, for example, that I should be your Matrix Table speaker tonight. Before I wrote the book that did so many surprising things to my life, I had never willingly spoken to more than ten people at once. I had seldom done it even unwillingly. And even tonight, despite certain experiences that have been thrust upon me since then, I am going to try to forget that there are 10 × 10 × 10 of you—for I want to speak quite simply and personally to each of you about some of the things that have come into one writer's life, and about some of the things she believes.

Some of the experiences writers have are very amusing. Some are delightful. A few are very moving. Tonight I'd like to share a few of each kind with you. I'd also like to take you to some of the places I have been, as a reporter and interpreter of the natural world, and, if I can, let you see those places as I have, and perhaps share my feelings about them.

One of the results I hadn't foreseen, when I wrote the story of part of our earth, was the fact that people would immediately become very curious about the author. That was very naive of me, I suppose, but it had never occurred to me, for example, that people would go to great length just to have a look at someone whose book was a best-seller.

A few months after *The Sea* was published, I was on a long southern field trip for my new book. In a strange town, I went into a beauty shop, and while I was sitting under the drier—which until then I had considered an inviolate sanctuary—the proprietor came over, turned off the drier, and said, "I hope you don't mind, but there is someone who wants to meet you." I admit I felt hardly at my best, with a towel around my neck and my hair in pin curls.

At another place during the same trip, a knock came at the door of our motor court early one morning. When my mother opened the door, a determined woman pushed past her and presented two books for autographing to an author who was still abed, and, if the truth must be told, very much annoyed.

When part of my book appeared as "A Profile of the Sea" in *The New Yorker*, a large proportion of the readers who wrote to the editors of the magazine demanded a profile of Rachel Carson. One wrote: "Please let me know in a hurry who Rachel Carson is. That girl keeps me awake night after night."

In publishing the book, Oxford University Press didn't foresee all this clamor and failed to put my picture on the jacket. So the field was open for all sorts of speculation—what did I look like, how old was I, how did I happen to discover the sea?

Among male readers there was a certain reluctance to acknowledge that a woman could have dealt with a scientific subject. Some, who apparently had never read the Bible enough to know that Rachel is a woman's name, wrote: "I assume from the author's knowledge that he must be a man." Another, addressing me properly as *Miss* Rachel Carson, nevertheless began his letter "Dear Sir:". He explained his salutation by saying that he had always been convinced that the males possess the supreme intellectual powers of the world, and he could not bring himself to reverse his conviction.

Then others assumed that, since I knew enough to write *The Sea*, I must be grey haired and venerable. One of my unknown correspondents wrote that I was probably just what he was looking for as a wife—except that to have learned all I put in the book must have taken a long time and perhaps I was too old for him. I think I could have qualified, but that was one of life's opportunities I passed by. I was a bit hesitant. One of *his* qualifications, he told me, was that he had spent most of his life in a mental institution.

Even in my publishing house, those who hadn't met me while I was writing the book seemed to expect something quite different from the reality. One of the editors said to me: "You are such a surprise to me. I thought you would be a very large and forbidding woman."

Now I am sure none of you had such misconceptions—but if you did, you see the truth before you now. You may have wondered, however, how I came to write *The Sea Around Us*, how I came to be a writer, and how I happen to be in a somewhat unusual field—the interpretation of part of the science of the earth and its life. At least, those are the questions that are most often asked me.

To answer them, I shall have to go back almost to my own beginnings. For I can remember no time, even in earliest childhood, when I didn't assume I was going to be a writer. I have no idea why. There were no writers in the family. I read a great deal almost from infancy, and I suppose I must have realized someone wrote the books, and thought it would be fun to make up stories, too.

Also, I can remember no time when I wasn't interested in the out-of-doors and the whole world of nature. Those interests, I know, I inherited from my mother and have always shared with her. I was rather a solitary child and spent a great deal of time in woods and beside streams, learning the birds and the insects and flowers.

There is another thing about my childhood that is interesting now, in the light of later happenings. I might have said, with Emily Dickinson:

> I never saw a moor,
> I never saw the sea;
> Yet know I how the heather looks,
> And what a wave must be.

For I never saw the ocean until I went from college to the marine laboratories at Woods Hole, on Cape Cod. Yet as a child I was fascinated by the thought of it. I dreamed about it and wondered what it would look like. I loved Swinburne and Masefield and all the other great sea poets.

I had my first prolonged contact with the sea at Woods Hole. I never tired of watching the tidal currents pouring through the Hole—that wonderful place of whirlpools and eddies and swiftly racing water. I loved to watch the waves breaking at Nobska Point after a storm. At Woods Hole, too, as a young biologist, I first discovered the rich scientific literature of the sea. But it is fair to say that my first impressions of the ocean were sensory and emotional, and that the intellectual response came later.

Before that meeting with the sea had been accomplished, however, I had a great decision to make. At least, I thought I had. I told you that I had always planned to be a writer; when I went to college, I thought the way to accomplish that was to major in English composition. Up to that time, despite my love for the world of nature, I'd had no training in biology. As a college sophomore, I was exposed to a fine introductory course in biology, and my allegiance began to waver. Perhaps I wanted to be a scientist. A year later the decision for science was made; the writing courses were abandoned. I had given up writing forever, I thought. It never occurred to me that I was merely getting something to write about. What surprises me now is that apparently it didn't occur to any of my advisors, either.

The merging of the two careers didn't begin until several years after I had left Johns Hopkins, where I had gone to do graduate work in zoology. Those were depression and post-depression years, and after a period of part-time teaching jobs, I supplemented them with another part-time assignment. The Bureau of Fisheries in Washington had undertaken to do a series of radio broadcasts. They were looking for someone to take over writing the scripts—someone who knew marine biology and who also could write. I happened in one morning when the chief of the biology division was feeling rather desperate—I think at that point he was having to write the scripts himself. He talked to me a few minutes and then said: "I've never seen a written word of yours, but I'm going to take a sporting chance."

That little job, which eventually led to a permanent appointment as a biologist, was in its way a turning point. One week I was told to produce something of a general sort about the sea. I set to work, but somehow the material rather took charge of the situation and turned into something that was, perhaps, unusual as a broadcast for the Commissioner of Fisheries. My chief read it and handed it back with a twinkle in his eye. "I don't think it will do," he said. "Better try again. But send this one to the *Atlantic*." Eventually I did, and the *Atlantic* accepted it. Since then I have told my chief of those days that he was really my first literary agent.

From those four *Atlantic* pages, titled "Undersea," everything else followed. Quincy Howe, then editor for Simon and Schuster, wrote to ask why I didn't do a book. So did Hendrik Willem van Loon. My mail had never contained anything so exciting as his first letter. It arrived in an envelope splashed with the green waves of a sea through which van Loon sharks and whales were poking inquiring snouts.

That was only the beginning of a wonderful correspondence, for it seemed Hendrik van Loon had always wanted to know what lay undersea, and he was determined I should tell the world in a book or books. His typing was amazing but his hand-written letters were almost illegible. Often he substituted a picture for a word, and that helped. After a few weeks of such correspondence, I spent a few days with the van Loons in their Connecticut home, during which I was properly introduced to my future publisher.

To a young and very tentative writer, it was a stimulating and wonderful thing to have the interest of this great man, so overwhelming in his person and his personality, but whose heart was pure gold. Through him, I had glimpses of a world that seemed exciting and fabulous, and I am sure his encouragement had a great deal to do with the fact that my first book, *Under the Sea Wind*, was eventually published.

When that happened, however, on the eve of Pearl Harbor, the world received the event with superb indifference. The reviewers were kind, but that rush to the book store that is the author's dream never materialized. There was a Braille edition, a German translation, and use of various chapters in anthologies. That was all. I was busy with war work, and when I thought at all about writing, it was in terms of magazine pieces; I doubted that I would ever write another book. But I did, and ten years after *Under the Sea Wind*, *The Sea Around Us* was published.

The fifteen years that I spent in fishery and wildlife conservation work with the Government have taken me into certain places where few other women have been. Perhaps you would like to hear about some of those.

While I was doing information work for Fish and Wildlife, the Service acquired a research vessel for work at sea, specifically on the famous fishing ground known as Georges Bank, that lies some 200 miles east of Boston and south of Nova Scotia. Some of the valuable commercial fishes are becoming scarce on the Bank, and the Service is trying to find the reason. The *Albatross III*, as this converted fishing trawler was called, operated out of Woods Hole, making repeated trips to

Georges. She was making a census of the fish population; this was done by fishing according to a systematic plan over a selected series of stations. Of course, various scientific data on water temperatures and other matters were collected, too.

It was decided finally—and I might have had something to do with originating the idea—that perhaps I could do a better job of handling publications about the *Albatross* if I had been out on her. But there was one great obstacle. No woman had ever been on the *Albatross*. Tradition is important in the Government, but fortunately I had conspirators who were willing to help me shatter precedent. But among my male colleagues who had to sign the papers, the thought of one woman on a ship with some fifty men was unthinkable. After much soul searching, it was decided that maybe *two* women would be all right, so I arranged with a friend, who was also a writer, to go with me. Marie thought she would write a piece about her experiences, and declared that her title would be: "I Was a Chaperone on a Fishing Boat."

And so one July day we sailed from Woods Hole into ten days of unusual adventure. This is not the place to tell about the scientific work that was done—but there was a lighter side, especially for us who were mere observers, and there were unforgettable impressions of fishing scenes; of fog on Georges, where the cold water and the warm air from the Gulf Stream are perpetually at war at that season of the year; and of the unutterable loneliness of the sea at night as seen from a small vessel.

As to the lighter side—a fishing trawler is not exactly a luxury liner, and both of us were on our mettle to prove that a woman could take it without complaining. Hardly had the coast of Massachusetts disappeared astern when some of the ship's officers began to give us a vivid picture of life aboard. The *Albatross*, they told us, was a very long and narrow ship and rolled like a canoe in a sea, so that everyone got violently seasick. They described some of the unpleasant accidents that sometimes occur in handling the heavy gear. They told us about the bad food. They made sure we understood that the fishing process went on night and day, and that it was very noisy.

Well—not all the things those Job's comforters predicted came true, but a great many of them did. However, we learned in those ten days that one gets used to almost anything.

We learned about the fishing the very first night. After steaming out through Nantucket Channel late in the afternoon, we were to reach our first fishing station about midnight. Marie and I had gone to bed and were sound asleep when we heard a crash, presumably against the very wall of our cabin, that brought us both upright in our bunks. Surely we had been rammed by another vessel. Then a series of the most appalling bangs, clunks, and rumbles began directly over our heads, a rhythmic thundering of machinery that would put any boiler factory to shame. Finally it dawned upon us that this was fishing! It also dawned on us that this was what we had to endure for the next ten nights. If there had been any way to get off the *Albatross* then I'm sure we would have taken it.

At breakfast the next morning there were grins on the faces of the men. "Hear anything last night?" they asked. Both of us wore our most demure expressions. "Well," said Marie, "once we thought we heard a mouse, but we were too sleepy to bother." They never asked us again. And after a night or two we really did sleep through the uproar like old salts.

One of the most vivid impressions I carried away from the *Albatross* was the sight of the net coming up with its load of fish. The big fishing trawlers such as this one drag a cone-shaped net on the floor of the ocean, scraping up anything lying on the bottom or swimming just above it. This means not only fish but also crabs, sponges, starfish and other life of the sea floor. Much of the fishing was done in depths of about 100 fathoms, or 600 feet. After a half hour of trawling the big winches would begin to haul in the cables, winding them on steel drums as they came aboard. There is a marker on every hundred fathoms of cable, so one can tell when to expect the big net to come into view, still far down in the green depths.

I think that first glimpse of the net, a shapeless form, ghostly white, gave me a sense of sea depths that I never had before. As the net rises, coming into sharper focus, there is a stir of excitement even among the experienced fishermen. What has it brought up?

No two hauls are quite alike. The most interesting ones came from the deeper slopes. Georges Bank is like a small mountain resting on the floor of a surrounding deeper sea—most of the fishing is done on its flat plateaus, but sometimes the net is dragged down on the slopes near the mountain's base. Then it brings up larger fish from these depths. There is a strange effect, caused by the sudden change of pressure. Some of the fish become enormously distended and float helplessly on their backs. They drift out of the net as it nears the surface but they are quite unable to swim down.

Then one sees the slender shapes of sharks moving in to the kill. There was something very beautiful about those sharks to me—and when some of the men got out rifles and killed them for "sport" it really hurt me.

In those deep net hauls, too, there were often the large and grotesque goosefish or angler fish. The angler has a triangular shape, and its enormous mouth occupies most of the base of the triangle. It lives on the floor of the sea, preying on other fish. The anglers always seemed to have been doing a little fishing of their own as the net came up, and sometimes the tails of two or three large cod would be protruding from their mouths.

Sometimes at night we would go up on the deck to watch the fishing. Then the white splash of electric light on the lower deck was the only illumination in a world of darkness and water. It was a colorful sight, with the men in their yellow oilskins and their bright flannel shirts, all intensified and made somehow dramatic by the blackness that surrounded them.

There is something deeply impressive about the night sea as one experiences it from a small vessel far from land. When I stood on the afterdeck on those dark nights, on a tiny man-made island of wood and steel, dimly seeing the great shapes of waves that rolled about us, I think I was conscious as never before that ours is a water world, dominated by the immensity of the sea.

However, it is a curious thing that one sometimes experiences a sense of the sea on land. A few years ago I had a wonderful opportunity to go far into the interior of the Everglades in Florida. Many people have crossed this great wilderness by way of the Tamiami Trail. That is better than not seeing it at all, but until one has penetrated far into the interior, into the trackless, roadless areas of the great swamp, one does not know the Everglades.

The difficulties of travel there are great, and no ordinary means of transportation will do. But a few pioneering individuals have developed wonderful vehicles called "glades buggies." They were first used, I believe, to prospect for oil in the interior of the Everglades. They are completely independent of roads; they can go through water, they can navigate the seas of "saw-grass" or even push through low-growing thickets of trees and shrubs; they can make their way—painfully but surely—over ground pitted with holes and strewn with jagged boulders.

I learned about the glades buggies when I was on a trip for my office to the area that is now the Everglades National Park. At that time the Fish and Wildlife Service had responsibility for protecting the wildlife of the area. Two of us were staying at a hotel in Miami Beach, visiting various wildlife areas in the vicinity. When we heard about Mr. Don Poppenhager and his wonderful glades buggy, we decided to try to arrange a trip.

Mr. Poppenhager had never taken a woman into the swamp and at first he was hesitant. He warned us that it was a very un-comfortable experience; we assured him we could take it and really wanted to go. So he agreed to meet us at a little store on the Tamiami Trail kept by a character known as Ma Szady.

I think our elegant Miami Beach hotel had been a little sus-picious of our comings and goings on strange errands and in strange costumes, but the morning we left for the Everglades trip was almost too much for them. One of the Fish and Wild-life men was to pick us up at 5 A.M. and take us over the trail. This was in the summer, and a tropical darkness still hung over Miami at that hour. Not wanting to arouse the hotel, Shirley and I crept down the stairs laden with all our strange gear. As we tiptoed through the lobby, the head of a very sleepy but thoroughly suspicious clerk rose above the desk. "Are you ladies checking out?" he asked. I don't think his estimate of us rose when a very noisy, two-ton Government truck roared down the street and stopped at the hotel for its passengers.

The glades buggy that was waiting for us was a wonderful conveyance. It was built something like a tractor, with six pairs of very large wheels. Its engine was completely naked and ex-posed, and during the trip blasted its heat on the three of us perched on the buggy's single seat. There were various tools—pliers, screwdrivers, etc.—in a little rack against the motor block, and from time to time Mr. Poppenhager leaned out as we jogged along and turned something or jabbed at the motor.

It seemed to be in a perpetual state of boiling over; and now and then Mr. P. would stop and get out with a tin can and dip up some water—there was water everywhere—and pour it into the radiator. Usually he would drink a little—"the best water in the world" he would say.

But as I said a while ago, there was a curious sense of the sea there in the heart of the Everglades. At first I couldn't analyze it but I felt it strongly. There is first of all a sense of immense space from the utter flatness of the land and the great expanse of sky. The feeling of space is almost the same as at sea. The cloud effects were beautiful and always changing; then rain came over the grass, making a beautiful soft play of changing color—all grey and soft green. And again I found myself remembering rain at sea, dimpling the soft grey sheet of water. And in the Everglades the coral rock is always cropping out—underlying the water and raised in jagged boulders among the grass. Once that rock was formed by coral animals, living in a shallow sea that covered this very place. There is today the feeling that the land has formed only the thinnest veneer over this underlying platform of the ancient sea—that at any time the relations of sea and land might again be reversed.

And as we traveled from one to another of the "hammocks" of palmetto and other trees that rise here and there in the great sea of grass, we thought irresistably of islands in the ocean. Except for scattered cypresses, all the trees of this part of the everglades are concentrated in the hammocks, which form where depressions in the rock accumulate a little soil. Everywhere else there is only rock, water, and grass. The hammocks are famous for their tree snails, which live on certain locust-like trees, feeding on mossy growths on the bark. The shells of the tree snails are brightly colored, with an amazing variety of patterns. They are so much sought by collectors that the more accessible hammocks have been stripped bare. On our steaming iron monster, we rode along through the hammocks, passing under the trees and picking off tree snails, as in childhood we used to snatch the iron rings on a merry-go-round.

During the day we went calling on several alligators known by Mr. Poppenhager to inhabit certain "holes." The first one was not at home; the second was. He was apparently out in his front yard, but at our approach he went crashing through the willows and into his pond. In the Everglades, a "'gator hole" is typically a water-filled depression in the middle of a small hammock. Usually there is a rocky cave in the floor of this pond to which the alligator can retreat.

The Everglades is, of course, the land of the Seminole Indians. Far in the interior of the Glades we visited the sites of two ancient Indian villages. Some of these are being studied by archaeologists who have found evidence of early tribes who antedated the Seminoles by several hundred years. Near one of the modern settlements, Mr. Poppenhager took us to visit an Indian grave. Because of the solid limestone floor of this whole region, there is no burial in the ordinary sense; the coffin is placed on the ground and the man is given his gun and other equipment he will need for his life in the next world.

To us the whole area seemed as trackless and as lacking in landmarks as the sea, but our guide knew exactly where he was going. Our only bad moments came late in the afternoon, when there began to be some question whether we had enough gas to get us back to the Trail. Mosquitoes had been with us all day, settling in clouds every time we stopped moving. So the thought of a night in the swamp wasn't pleasant. However, we made it about dusk, just as the Game Warden and the Fish and Wildlife patrolman were beginning to line up cars along the Trail to guide us back by their headlights.

That Fish and Wildlife patrolman was such an unforgettable character that I must tell you a little about him before we leave the Everglades. As Service patrolman for the area, it was his job to protect the birds and alligators and other wildlife from being molested. That meant he had to live far out in a wild part of the Everglades, where days went by without his seeing another person. The Service had had trouble filling the job. There were few men that would have taken it; and perhaps no one else as beautifully fitted for it as Mr. Finneran. He was tired of the northern cities where he had spent most of his life, and for about ten years he had known this wilderness of southern Florida. He had somehow gained the confidence of the Seminoles, who ordinarily have no love for the white man. But they admired and trusted Mr. Finneran—so much that they had given him a name and practically adopted him into their tribe. When the Service offered Mr. Finneran this lonely job, he took it gladly, and moved into the little shack that was to serve as home and headquarters. There he lived with a little dog, a few chickens, and a blue indigo snake named Chloe. He had five tree snails on a tree beside the house. He was very proud of them, and when we returned from our glades buggy trip, we brought him a few snails as a gift. He couldn't have been more pleased if they had been pure gold. I remember how feelingly he spoke to me of the beauty of the Everglades in the early morning, with dew on the grass and thousands of spider webs glistening. He spoke of the birds coming in such numbers they were like dark clouds in the sky. He told of the eerie silver light of the moon, and the red, glowing orbs of alligators in the ponds. His paradise had its flaws, as he acknowledged. He couldn't have a light in his shack at night because of the terrible Glades mosquito. Sometimes, on rainy nights, fire ants invaded the house and even swarmed into his bed. The Indians said ghosts haunted the place because it was built on an old Indian mound; but Mr. Finneran had heard no ghosts he couldn't explain. When city dwellers asked him how he stood the loneliness out there, he always asked how they endured sitting around in night clubs. "I wouldn't trade my life for anything," he told us.

* * * * * *

From what I have told you, you will know that a large part of my life has been concerned with some of the beauties and mysteries of this earth about us, and with the even greater mysteries of the life that inhabits it. No one can dwell long among such subjects without thinking rather deep thoughts, without asking himself searching and often unanswerable questions, and without achieving a certain philosophy.

There is one quality that characterizes all of us who deal with the sciences of the earth and its life—we are never bored. We can't be. There is always something new to be investigated. Every mystery solved brings us to the threshold of a greater one.

I like to remember the wonderful old Swedish oceanographer, Otto Pettersson. He died a few years ago at the age of 93, in full possession of his keen mental powers. His son, also a distinguished oceanographer, tells us in a recent book how intensely his father enjoyed every new experience, every new discovery concerning the world about him. "He was an incurable romantic," the son wrote, "intensely in love with life and with the mysteries of the Cosmos which, he was firmly convinced, he had been born to unravel." When, past 90, Otto Pettersson realized he had not much longer to enjoy the earthly scene, he said to his son: "What will sustain me in my last moments is an infinite curiosity as to what is to follow."

The pleasures, the values of contact with the natural world, are not reserved for the scientists. They are available to anyone who will place himself under the influence of a lonely mountain top—or the sea—or the stillness of a forest; or who will stop to think about so small a thing as the mystery of a growing seed.

I am not afraid of being thought a sentimentalist when I stand here tonight and tell you that I believe natural beauty has a necessary place in the spiritual development of any individual or any society. I believe that whenever we destroy beauty, or whenever we substitute something man-made and artificial for a natural feature of the earth, we have retarded some part of man's spiritual growth.

I believe this affinity of the human spirit for the earth and its beauties is deeply and logically rooted. As human beings, we are part of the whole stream of life. We have been human beings for perhaps a million years. But life itself—passes on something of itself to other life—that mysterious entity that moves and is aware of itself and its surroundings, and so is distinguished from rocks or senseless clay—life arose many hundreds of millions of years ago. Since then it has developed, struggled, adapted itself to its surroundings, evolved an infinite number of forms. But its living protoplasm is built of the same elements as air, water, and rock. To these the mysterious spark of life was added. Our origins are of the earth. And so there is in us a deeply seated response to the natural universe, which is part of our humanity.

Now why do I introduce such a subject tonight—a serious subject for a night when we are supposed to be having fun? First, because you have asked me to tell you something of myself—and I can't do that without telling you some of the things I believe in so intensely.

Also, I mention it because it is not often I have a chance to talk to a thousand women. I believe it is important for women to realize that the world of today threatens to destroy much of that beauty that has immense power to bring us a healing release from tension. Women have a greater intuitive understanding of such things. They want for their children not only physical health but mental and spiritual health as well. I bring these things to your attention tonight because I think your awareness of them will help, whether you are practicing journalists, or teachers, or librarians, or housewives and mothers.

What are these threats of which I speak? What is this destruction of beauty—this substitution of man-made ugliness—this trend toward a perilously artificial world? Unfortunately, that is a subject that could require a whole conference, extending over many days. So in the few minutes that I have to devote to it, I can only suggest the trend.

We see it in small ways in our own communities, and in larger ways in the community of the state of the nation. We see the destruction of beauty and the suppression of human individuality in hundreds of suburban real estate developments where the first act is to cut down all the trees and the next is to build an infinitude of little houses, each like its neighbor.

We see it in distressing form in the nation's capital, where I live. There in the heart of the city we have a small but beautiful woodland area—Rock Creek Park. It is a place where one can go, away from the noise of traffic and of man-made confusions, for a little interval of refreshing and restoring quiet—where one can hear the soft water sounds of a stream on its way to river and sea, where the wind flows through the trees, and a veery sings in the green twilight. Now they propose to run a six-lane arterial highway through the heart of that narrow woodland valley—destroying forever its true and immeasurable value to the city and the nation.

Those who place so great a value on a highway apparently do not think the thoughts of an editorial writer for the *New York Times* who said: "But a little lonesome space, where nature has her own way, where it is quiet enough at night to hear the patter of small paws on leaves and the murmuring of birds, can still be afforded. The gift of tranquillity, wherever found, is beyond price."

We see the destructive trend on a national scale in proposals to invade the national parks with commercial schemes such as the building of power dams. The parks were placed in trust for all the people, to preserve for them just such recreational and spiritual values as I have mentioned. Is it the right of this, our generation, in its selfish materialism, to destroy these things because we are blinded by the dollar sign? Beauty—and all the values that derive from beauty—are not measured and evaluated in terms of the dollar.

Years ago I discovered in the writings of the British naturalist Richard Jefferies a few lines that so impressed themselves upon my mind that I have never forgotten them. May I quote them for you now?

> "The exceeding beauty of the earth, in her splendor of life, yields a new thought with every petal. The hours when the mind is absorbed by beauty are the only hours when we really live. All else is illusion, or mere endurance."

Those lines are, in a way, a statement of the creed I have lived by, for, as perhaps you have seen tonight, a preoccupation with the wonder and beauty of the earth has strongly influenced the course of my life.

Since *The Sea Around Us* was published, I have had the privilege of receiving many letters from people who, like myself, have been steadied and reassured by contemplating the long history of the earth and sea, and the deeper meanings of the world of nature. These letters have come from all sorts of people. There have been hairdressers and fishermen and musicians; there have been classical scholars and scientists. So many of them have said, in one phrasing or another: "We have been troubled about the world, and had almost lost faith in man; it helps to think about the long history of the earth, and of how life came to be. And when we think in terms of millions of years, we are not so impatient that our own problems be solved tomorrow."

In contemplating "the exceeding beauty of the earth" these people have found calmness and courage. For there is symbolic as well as actual beauty in the migration of birds; in the ebb and flow of the tides; in the folded bud ready for the spring. There is something infinitely healing in these repeated refrains of nature—the assurance that dawn comes after night, and spring after winter.

Mankind has gone very far into an artificial world of his own creation. He has sought to insulate himself, with steel and concrete, from the realities of earth and water. Perhaps he is intoxicated with his own power, as he goes farther and farther into experiments for the destruction of himself and his world. For this unhappy trend there is no single remedy—no panacea. But I believe that the more clearly we can focus our attention on the wonders and realities of the universe about us, the less taste we shall have for destruction.

To Dorothy Freeman

Tuesday afternoon, June 1

Darling—

How I did wish you were with me last night! For we heard the veeries most beautifully in the very spot where I first heard them, in the same sort of green twilight, with almost the same magical effect. (When the first experience of anything beautiful has been exactly right, I don't suppose any repetition can quite equal it.) When we arrived at the spot, there was a great deal of other song in the woods—flycatchers, chickadees, water thrushes, ovenbirds, and even discordant cries of night herons flying over the creek. Finally the sound for which I'd been straining my ears—the short "Whew" that says veeries are about. And at last the songs began, the first so far off in the woods, or so faint, that I had to hold my breath to hear at all. But oh—unmistakably!—a veery. Soon others began, some quite close to us. As perhaps you found, they seem never to sing in chorus, but responsively, one voice quickly answering another. And all that unearthly quality—"of a spirit not to be discovered" as Mr. Halle said—was there. But one of the spirits revealed itself, for suddenly a slim little bird materialized on the branch of a dogwood, almost over our heads. Oh, how I wanted it to sing, then and there! But that was not to be. In a minute or so it flew off into a tangle of honeysuckle along the path. Then there began to be suspicious movements in the low growth at the very edge of the path. I was just beginning cautious investigations when three people came over the foot bridge. So we tried to look as if we'd never heard of a veery, for you can hardly imagine the rating a person must have with me before I'll reveal even a *possible* site of a bird's nest. "Have you heard anything?" they asked. "Oh, yes—there have been water thrushes—and ovenbirds." Being very cautious, you see. They looked disappointed. And after all, they each wore binoculars, so I relented a bit. "And the veeries have been singing." You should have seen their faces brighten. Obviously that was what they'd come for, so I was doubly glad when the spirits began calling again. As soon as we came home, I looked up Mr. Halle's description of the song—it could hardly be improved on.

I was going to write more but think I'll mail this now, as my greeting for today. Maybe there will be word from you tomorrow. And guess what? I love you.

Rachel

1954

To Edwin Way Teale

Tuesday, August 16th

Dear Edwin,

We are so sorry to hear that Nellie has had a seizure in the hospital with an operation and complications afterward. That is bad enough any time but it always seems worse to be laid up in hot weather. I do so hope all will go smoothly from now on.

All summer (and I think all winter, too) I have been planning to write you, but this book seemed to have no end. It really died hard, and kept me harassed one way or another, up to about a week ago. Just to make things really tough, page proof and index—all with split-second timing and the printer breathing down my neck—came at the same time as the most intensive work on the *New Yorker* condensation. That is out this week and next—August 20 and 27th. They left major sections of the book (including Florida Keys) untouched, which is a good thing. They have used some of Bob's illustrations, in a surprising break with *New Yorker* traditions. The book will be out around the first of November—exact date apparently depending on the production schedule.

I hope I did write long ago to thank you for your good letter of advice about cameras, but in the pressure of work maybe I didn't. Anyway, I got an Exakta in May—got a wonderful buy in a new, fully guaranteed camera at one of those freak reductions—about 40% off! I am learning by degrees, and am really delighted with the camera, for now and then even a rank amateur like me can get really lovely results. Such detail, brilliance, and depth of focus! The marine subjects are toughest for a beginner, but flowers, mosses, scenes, etc. are more rewarding. Nevertheless, that camera can look right down through 4 or 5 feet of water and see the bottom—as my eye can't!

You will both be glad to hear that I have extended my boundaries here by acquiring the vacant lot south of me. It has tempted me ever since I bought this one—in fact I originally considered it instead of this, but am so glad it all worked out as it did. I now have about 350 feet of shoreline, with the house well protected on both sides, and the new lot has such lovely woodland that it would have been a shame to build on it. Such wonderful ferns, mosses, lichens, glades full of bunchberry and Clintonia, wood lilies, Indian pipes, ladies slippers—real Maine woods.

This was the year I was going to do a lot outside, but only now can I begin to think about it. This week or next a man is going to move in some clumps of fern from the woods, and transplant some of that lovely mountain holly near the house. The berries are red now, and even those down on the slope overlooking the shore have berries, for the first time. Each year I love the place a little more—now I can't imagine I shall ever want to go anywhere else in the summer.

Is there any chance you might be driving this way for little vacation in the fall? I think you'd enjoy settling down in this coastal area for a bit, though I know you like the seclusion of the north woods. But people clear out fast after Labor Day and then lots of little places are available. It would be good to see you here again.

How wonderful that you still have Silver—and perhaps you may for some years to come. One does hear of cats living into their 20's. Marie Rodell has recently acquired as a client Margaret Cooper Gay, author of *How to Live with a Cat*, and finds her quite delightful. I have a telephone acquaintance with her, having discussed Jeffie's ailments with her last fall. Jeffie, I'm glad to say, is so much better than he has been since infancy. He dearly loves it here and I'm sure it does him good.

One of our little red squirrels has become so tame—largely through Marjie, who has been visiting us—that it sits on our laps or shoulders and follows us like a dog—even down over the weeds to the low-tide line!

We do hope to hear soon that Nellie is much better, and that you are holding up. Just now the thought of having to write makes me ill—so you know how deeply I feel for you, tied to an unfinished book! Of course I'm "tied" to one not even begun, but I'm resolutely not thinking about that!

Our very best to you both, and tell Nellie we do hope she will feel very much better soon.

As ever,
Rachel

Mrs. Newman does sound like an interesting person. In the three years here, I've never visited the opposite shore of the bay (about 40 miles each way) but if the weather clears we may take Marjie over to Reid State Park this week, and perhaps we might find her.

1955

Help Your Child to Wonder

ONE STORMY AUTUMN night when my nephew Roger was about 20 months old I wrapped him in a blanket and carried him down to the beach in the rainy darkness. Out there, just at the edge of where-we-couldn't-see, big waves were thundering in, dimly seen white shapes that boomed and shouted and threw great handfuls of froth at us. Together we laughed for pure joy—he a baby meeting for the first time the wild tumult of Oceanus, I with the salt of half a lifetime of sea love in me. But I think we felt the same spine-tingling response to the vast, roaring ocean and the wild night around us.

A night or two later the storm had blown itself out and I took Roger again to the beach, this time to carry him along the water's edge, piercing the darkness with the yellow cone of our flashlight. Although there was no rain the night was again noisy with breaking waves and the insistent wind. It was clearly a time and place where great and elemental things prevailed.

Our adventure on this particular night had to do with life, for we were searching for ghost crabs, those sand-colored, fleet-legged beings whom Roger had sometimes glimpsed briefly on the beaches in daytime. But the crabs are chiefly nocturnal, and when not roaming the night beaches they dig little pits near the surf line where they hide, seemingly watching and waiting for what the sea may bring them. For me the sight of these small living creatures, solitary and fragile against the brute force of the sea, had moving philosophic overtones, and I do not pretend that Roger and I reacted with similar emotions. But it was good to see his infant acceptance of a world of elemental things, fearing neither the song of the wind nor the darkness nor the roaring surf, entering with baby excitement into the search for a "ghos."

It was hardly a conventional way to entertain one so young, I suppose, but now, with Roger a little past his fourth birthday, we are continuing that sharing of adventures in the world of nature that we began in his babyhood, and I think the results are good. The sharing includes nature in storm as well as calm, by night as well as day, and is based on having fun together rather than on teaching.

I spend the summer months on the coast of Maine, where I have my own shoreline and my own small tract of woodland. Bayberry and juniper and huckleberry begin at the very edge of the granite rim of shore, and where the land slopes upward from the bay in a wooded knoll the air becomes fragrant with spruce and balsam. Underfoot there is the multi-patterned northern groundcover of blueberry, checkerberry, reindeer moss and bunchberry, and on a hillside of many spruces, with shaded ferny dells and rocky outcroppings—called the Wildwoods—there are ladyslippers and wood lilies and the slender wands of clintonia with its deep blue berries.

When Roger has visited me in Maine and we have walked in these woods I have made no conscious effort to name plants or animals nor to explain to him, but have just expressed my own pleasure in what we see, calling his attention to this or that but only as I would share discoveries with an older person. Later I have been amazed at the way names stick in his mind, for when I show color slides of my woods plants it is Roger who can identify them. "Oh, that's what Rachel likes—that's bunchberry!" Or, "That's jumer (juniper) but you can't eat those green berries—they are for the squirrels." I am sure no amount of drill would have implanted the names so firmly as just going through the woods in the spirit of two friends on an expedition of exciting discovery.

In the same way Roger learned the shells on my little triangle of sand that passes for a beach in rocky Maine. When he was only a year and a half old, they became known to him as winkies (periwinkles), weks (whelks) and mukkies (mussels) without my knowing quite how this came about, for I had not tried to teach him.

We have let Roger share our enjoyment of things people ordinarily deny children because they are inconvenient, interfering with bedtime or involving wet clothing that has to be changed or mud that has to be cleaned off the rug. We have let him join us in the dark living room before the big picture window to watch the full moon riding lower and lower toward the far shore of the bay, setting all the water ablaze with silver flames and finding a thousand diamonds in the rocks on the shore as the light strikes the flakes of mica embedded in them. I think we have felt that the memory of such a scene, photographed year after year by his child's mind, would mean more to him in manhood than the sleep he was losing. He told me it would in his own way, when we had a full moon the night after his arrival last summer. He sat quietly on my lap for some time, watching the moon and the water and all the night sky. Then he snuggled closer and whispered, "I'm glad we comed."

A rainy day is the perfect time for a walk in the woods. I always thought so myself; the Maine woods never seem so fresh and alive as in wet weather. Then all the needles on the evergreens wear a sheath of silver; ferns seem to have grown to almost tropical lushness and every leaf has its edging of crystal drops. Strangely colored fungi—mustard-yellow and apricot and scarlet—are pushing out of the leaf mold and all the lichens and the mosses have come alive with green and silver freshness.

Now I know that for children, too, nature reserves some of her choice rewards for days when her mood may appear to be somber. Roger reminded me of it on a long walk through rain-drenched woods last summer—not in words, of course, but by his responses. There had been rain and fog for days, rain beating on the big picture window, fog almost shutting out sight of the bay. No lobstermen coming in to tend their traps, no gulls on the shore, scarcely even a squirrel to watch. The cottage was fast becoming too small for a restless three-year-old.

"Let's go for a walk in the woods," I said. "Maybe we'll see a fox or a deer." So into yellow oilskin coat and sou'wester and outside in joyous anticipation.

Having always loved the lichens because they have a quality of fairyland—silver rings on a stone, odd little forms like bones or horns or the shell of a sea creature—I was glad to find Roger noticing and responding to the magic change in their appearance wrought by the rain. The wood path was carpeted with the so-called reindeer moss, in reality a lichen. Like an old-fashioned hall runner, it made a narrow strip of silvery gray through the green of the woods, here and there spreading out to cover a larger area. In dry weather the lichen carpet seems thin; it is brittle and crumbles underfoot. Now, saturated with rain which it absorbs like a sponge, it was deep and springy. Roger delighted in its texture, getting down on chubby knees to feel it, and running from one patch to another to jump up and down in the deep, resilient carpet with squeals of pleasure.

It was here that we first played our Christmas tree game. There is a fine crop of young spruces coming along and one can find seedlings of almost any size down to the length of Roger's finger. I began to point out the baby trees.

"This one must be a Christmas tree for the squirrels," I would say. "It's just the right height. On Christmas Eve the red squirrels come and hang little shells and cones and silver threads of lichen on it for ornaments, and then the snow falls and covers it with shining stars, and in the morning the squirrels have a beautiful Christmas tree . . . And this one is even tinier—it must be for little bugs of some kind—and maybe this bigger one is for the rabbits or woodchucks."

Once this game was started it had to be played on all woods walks, which from now on were punctuated by shouts of, "Don't step on the Christmas tree!"

A child's world is fresh and new and beautiful, full of wonder and excitement. It is our misfortune that for most of us that clear-eyed vision, that true instinct for what is beautiful and awe-inspiring, is dimmed and even lost before we reach adulthood. If I had influence with the good fairy who is supposed to preside over the christening of all children I should ask that her gift to each child in the world be a sense of wonder so indestructible that it would last throughout life, as an unfailing antidote against the boredom and disenchantments of later years, the sterile preoccupation with things that are artificial, the alienation from the sources of our strength.

If a child is to keep alive his inborn sense of wonder without any such gift from the fairies, he needs the companionship of at least one adult who can share it, rediscovering with him the joy, excitement and mystery of the world we live in. Parents often have a sense of inadequacy when confronted on the one hand with the eager, sensitive mind of a child and on the other with a world of complex physical nature, inhabited by a life so various and unfamiliar that it seems hopeless to reduce it to order and knowledge. In a mood of self-defeat, they exclaim, "How can I possibly teach my child about nature—why, I don't even know one bird from another!"

I sincerely believe that for the child, and for the parent seeking to guide him, it is not half so important to *know* as to *feel*.

If facts are the seeds that later produce knowledge and wisdom, then the emotions and the impressions of the senses are the fertile soil in which the seeds must grow. The years of early childhood are the time to prepare the soil. Once the emotions have been aroused—a sense of the beautiful, the excitement of the new and the unknown, a feeling of sympathy, pity, admiration or love—then we wish for knowledge about the object of our emotional response. Once found, it has lasting meaning. It is more important to pave the way for the child to want to know than to put him on a diet of facts he is not ready to assimilate.

If you are a parent who feels he has little nature lore at his disposal there is still much you can do for your child. With him, wherever you are and whatever your resources, you can still look up at the sky—its dawn and twilight beauties, its moving clouds, its stars by night. You can listen to the wind, whether it blows with majestic voice through a forest or sings a many-voiced chorus around the eaves of your house or the corners of your apartment building, and in the listening, you can gain magical release for your thoughts. You can still feel the rain on your face and think of its long journey, its many transmutations, from sea to air to earth. Even if you are a city dweller, you can find some place, perhaps a park or a golf course, where you can observe the mysterious migrations of the birds and the changing seasons. And with your child you can ponder the mystery of a growing seed, even if it be only one planted in a pot of earth in the kitchen window.

Exploring nature with your child is largely a matter of becoming receptive to what lies all around you. It is learning again to use your eyes, ears, nostrils and fingertips, opening up the disused channels of sensory impression.

For most of us, knowledge of our world comes largely through sight, yet we look about with such unseeing eyes that we are partially blind. One way to open your eyes to unnoticed beauty is to ask yourself, "What if I had never seen this before? What if I knew I would never see it again?"

I remember a summer night when such a thought came to me strongly. It was a clear night without a moon. With a friend, I went out on a flat headland that is almost a tiny island, being all but surrounded by the waters of the bay. There the horizons are remote and distant rims on the edge of space. We lay and looked up at the sky and the millions of stars that blazed in darkness. The night was so still that we could hear the buoy on the ledges out beyond the mouth of the bay. Once or twice a word spoken by someone on the far shore was carried across on the clear air. A few lights burned in cottages. Otherwise there was no reminder of other human life; my companion and I were alone with the stars. I have never seen them more beautiful: the misty river of the Milky Way flowing across the sky, the patterns of the constellations standing out bright and clear, a blazing planet low on the horizon. Once or twice a meteor burned its way into the earth's atmosphere.

It occurred to me that if this were a sight that could be seen only once in a century or even once in a human generation, this little headland would be thronged with spectators. But it can be seen many scores of nights in any year, and so the lights burned in the cottages and the inhabitants probably gave not a thought to the beauty overhead; and because they could see it almost any night perhaps they will never see it.

An experience like that, when one's thoughts are released to roam through the lonely spaces of the universe, can be shared with a child even if you don't know the name of a single star. You can still drink in the beauty, and think and wonder at the meaning of what you see.

And then there is the world of little things, seen all too seldom. Many children, perhaps because they themselves are small and closer to the ground than we, notice and delight in the small and inconspicuous. With this beginning, it is easy to share with them the beauties we usually miss because we look too hastily, seeing the whole and not its parts. Some of nature's most exquisite handiwork is on a miniature scale, as anyone knows who has applied a magnifying glass to a snowflake.

An investment of a few dollars in a good hand lens or magnifying glass will bring a new world into being. With your child, look at objects you take for granted as commonplace or uninteresting. A sprinkling of sand grains may appear as gleaming jewels of rose or crystal hue, or as glittering jet beads, or as a melange of Lilliputian rocks, spines of sea urchins and bits of snail shells.

A lens-aided view into a patch of moss reveals a dense tropical jungle, in which insects large as tigers prowl amid strangely formed, luxuriant trees. A bit of pond weed or seaweed put in a glass container and studied under a lens is found to be populated by hordes of strange beings, whose activities can entertain you for hours. Flowers (especially the composites), the early buds of leaf or flower from any tree, or any small creature reveal unexpected beauty and complexity when, aided by a lens, we can escape the limitations of the human size scale.

Senses other than sight can prove avenues of delight and discovery, storing up for us memories and impressions. Already Roger and I, out early in the morning, have enjoyed the sharp, clean smell of wood smoke coming from the cottage chimney. Down on the shore we have savored the smell of low tide—that marvelous evocation combined of many separate odors, of the world of seaweeds and fishes and creatures of bizarre shape and habit, of tides rising and falling on their appointed schedule, of exposed mud flats and salt rime drying on the rocks. I hope Roger will later experience, as I do, the rush of remembered delight that comes with the first breath of that scent, drawn into one's nostrils as one returns to the sea after a long absence. For the sense of smell, almost more than any other, has the power to recall memories and it is a pity that we use it so little.

Hearing can be a source of even more exquisite pleasure but it requires conscious cultivation. I have had people tell me they had never heard the song of a wood thrush, although I knew the bell-like phrases of this bird had been ringing in their back yards every spring. By suggestion and example, I believe children can be helped to hear the many voices about them. Take time to listen and talk about the voices of the earth and what they mean—the majestic voice of thunder, the winds, the sound of surf or flowing streams.

And the voices of living things: No child should grow up unaware of the dawn chorus of the birds in spring. He will never forget the experience of a specially planned early rising and going out in the predawn darkness. The first voices are heard before daybreak. It is easy to pick out these first, solitary singers. Perhaps a few cardinals are uttering their clear, rising whistles, like someone calling a dog. Then the song of a white-throat, pure and ethereal, with the dreamy quality of remembered joy. Off in some distant patch of woods a whippoorwill continues his monotonous night chant, rhythmic and insistent, sound that is felt almost more than heard. Robins, thrushes, song sparrows, jays, vireos add their voices. The chorus picks up volume as more and more robins join in, contributing a fierce rhythm of their own that soon becomes dominant in the wild medley of voices. In that dawn chorus one hears the throb of life itself.

There is other living music. I have already promised Roger that we'll take our flashlights this fall and go out into the garden to hunt for the insects that play little fiddles in the grass and among the shrubbery and flower borders. The sound of the insect orchestra swells and throbs night after night, from midsummer until autumn ends and the frosty nights make the tiny players stiff and numb, and finally the last note is stilled in the long cold. An hour of hunting out the small musicians by flashlight is an adventure any child would love. It gives him a sense of the night's mystery and beauty, and of how alive it is with watchful eyes and little, waiting forms.

The game is to listen, not so much to the full orchestra, as to the separate instruments, and to try to locate the players. Perhaps you are drawn, step by step, to a bush from which comes a sweet, high-pitched, endlessly repeated trill. Finally you trace it to a little creature of palest green, with wings as white and insubstantial as moonlight. Or from somewhere along the garden path comes a cheerful, rhythmic chirping, a sound as companionable and homely as a fire crackling on a hearth or a cat's purr. Shifting your light downward you find a black mole cricket disappearing into his grassy den.

Most haunting of all is one I call the fairy bell-ringer. I have never found him. I'm not sure I want to. His voice—and surely he himself—are so ethereal, so delicate, so otherworldly, that he should remain invisible, as he has through all the nights I have searched for him. It is exactly the sound that should come from a bell held in the hand of the tiniest elf, inexpressibly clear and silvery, so faint, so barely-to-be-heard that you hold your breath as you bend closer to the green glades from which the fairy chiming comes.

The night is a time, too, to listen for other voices, the calls of bird migrants hurrying northward in spring and southward in autumn. Take your child out on a still October night when there is little wind and find a quiet place away from traffic noises. Then stand very still and listen, projecting your consciousness up into the dark arch of the sky above you. Presently your ears will detect tiny wisps of sound—sharp chirps, sibilant lisps and call notes. They are the voices of bird migrants, apparently keeping in touch by their calls with others of their kind scattered through the sky. I never hear these calls without a wave of feeling that is compounded of many emotions—a sense of lonely distances, a compassionate awareness of small lives controlled and directed by forces beyond volition or denial, a surging wonder at the sure instinct for route and direction that so far has baffled human efforts to explain it.

If the moon is full and the night skies are alive with the calls of bird migrants, then the way is open for another adventure with your child, if he is old enough to use a telescope or a good pair of binoculars. The sport of watching migrating birds pass across the face of the moon has become popular and even scientifically important in recent years, and it is as good a way as I know to give an older child a sense of the mystery of migration.

Seat yourself comfortably and focus your glass on the moon. You must learn patience, for unless you are on a well-traveled highway of migration you may have to wait many minutes before you are rewarded. In the waiting periods you can study the topography of the moon, for even a glass of moderate power reveals enough detail to fascinate a space-conscious child. But sooner or later you should begin to see the birds, lonely travelers in space glimpsed as they pass from darkness into darkness.

In all this I have said little about identification of the birds, insects, rocks, stars or any other of the living and nonliving things that share this world with us. Of course it is always convenient to give a name to things that arouse our interest. But that is a separate problem, and one that can be solved by any parent who has a reasonably observant eye and the price of the various excellent handbooks that are available in quite inexpensive editions.

I think the value of the game of identification depends upon how you play it. If it becomes an end in itself I count it of little use. It is possible to compile extensive lists of creatures seen and identified without ever once having caught a breath-taking glimpse of the wonder of life. If a child asked me a question that suggested even a faint awareness of the mystery behind the arrival of a migrant sandpiper on the beach of an August morning I would be far more pleased than by the mere fact that he knew it was a sandpiper and not a plover.

What is the value of preserving and strengthening this sense of awe and wonder, this recognition of something beyond the boundaries of human existence? Is the exploration of the natural world just a pleasant way to pass the golden hours of childhood or is there something deeper?

I am sure there is something much deeper, something lasting and significant. Those who dwell, as scientists or laymen, among the beauties and mysteries of the earth are never alone or weary of life. Whatever the vexations or concerns of their personal lives, their thoughts can find paths that lead to inner contentment and to renewed excitement in living. Those who contemplate the beauty of the earth find reserves of strength that will endure as long as life lasts. There is symbolic as well as actual beauty in the migration of the birds, the ebb and flow of the tides, the folded bud ready for the spring. There is something infinitely healing in the repeated refrains of nature—the assurance that dawn comes after night, and spring after the winter.

I like to remember the distinguished Swedish oceanographer, Otto Pettersson, who died a few years ago at the age of 93, in full possession of his keen mental powers. His son, also world-famous in oceanography, has related in a recent book how intensely his father enjoyed every new experience, every new discovery concerning the world about him.

"He was an incurable romantic," the son wrote, "intensely in love with life and with the mysteries of the cosmos." When he realized he had not much longer to enjoy the earthly scene, Otto Pettersson said to his son: "What will sustain me in my last moments is an infinite curiosity as to what is to follow."

The lasting pleasures of contact with the natural world are not reserved for such scientists but are available to anyone who will place himself under the influence of earth, sea and sky and their amazing life. In my mail recently was a letter that bore eloquent testimony to the lifelong durability of a sense of wonder. It came from a reader who asked advice on choosing a seacoast spot for a vacation, a place wild enough that she might spend her days roaming beaches unspoiled by civilization, exploring that world that is old but ever new.

Regretfully she excluded the rugged northern shores. She had loved the shore all her life, she said, but climbing over the rocks of Maine might be difficult, for an eighty-ninth birthday would soon arrive. As I put down her letter I was warmed by the fires of wonder and amazement that still burned brightly in her youthful mind and spirit, just as they must have done fourscore years ago.

Woman's Home Companion, July 1956

To Stanley and Dorothy Freeman

Wednesday A.M., August 8

Dear Stan and Dorothy—

This morning I achieved the difficult feat of getting up without disturbing anyone but Jeffie, so maybe I can write a letter before breakfast.

Knowing you can't be at Southport as soon as you want to be, I'm always of two minds now about talking of the place or telling you anything special that happens—should I share it with you, or is it mean to talk about things you want so badly to see or do yourselves? That, in general, is my predicament, but this time I *have* to tell you about something strange and wonderful.

We are now having the spring tides of the new moon, you know, and they have traced their advance well over my beach the past several nights. Roger's raft has to be secured by a line to the old stump, so Marjie and I have an added excuse to go down at high tide. There had been lots of swell and surf and noise all day, so it was most exciting down there toward midnight—all my rocks crowned with foam, and long white crests running from my beach to Mahard's. To get the full wildness, we turned off our flashlights—and then the real excitement began. Of course, you can guess—the surf was full of diamonds and emeralds, and was throwing them on the wet sand by the dozen. Dorothy, dear—it was the night we were there all over, but with everything intensified: a wilder accompaniment of noise and movement, and a great deal more phosphorescence. The individual sparks were so large—we'd see them glowing in the sand, or sometimes, caught in the in-and-out play of water, just riding back and forth. And several times I was able to scoop one up in my hand in shells and gravel, and think surely it was big enough to see—but no such luck.

Now here is where my story becomes different. Once, glancing up, I said to Marjie jokingly, "Look—one of them has taken to the air!" A firefly was going by, his lamp blinking. We thought nothing special of it, but in a few minutes one of us said, "There's that firefly again." The next time he really got a reaction from us, for he was flying so low over the water that his light cast a long surface reflection, like a little headlight. Then the truth dawned on me. He "thought" the flashes in the water were other fireflies, signaling to him in the age-old manner of fireflies! Sure enough, he was soon in trouble and we saw his light flashing urgently as he was rolled around in the wet sand—no question this time which was insect and which the unidentified little sea will-o-the-wisps!

You can guess the rest: I waded in and rescued him (the will-o-the-wisps had already had me in icy water to my knees so a new wetting didn't matter) and put him in Roger's bucket to dry his wings. When we came up we brought him as far as the porch—out of reach of temptation, we hoped.

It was one of those experiences that gives an odd and hard-to-describe feeling, with so many overtones beyond the facts themselves. I have never seen any account scientifically, of fireflies responding to other phosphorescence. I suppose I should write it up briefly for some journal if it actually isn't known. Imagine putting that in scientific language! And I've already thought of a child's story based on it—but maybe that will never get written.

Then everyone got up, and the day began!

The tide was very low this morning (1.8′) but I didn't go out. I could see surf on Bull Ledge as I was writing.

Now I'm at the Witch's, about to have my hair fixed so I'll have to wear a hat all the time!

Dorothy, I'm going to call you—or try to, Thursday night. I hope I'll hear of lots of improvement.

I know it must be hard to be patient—but just remember those doctors have all the skill needed to get you back home as soon as possible. Our love to you both—

Rachel

1956

To Curtis and Nellie Lee Bok

December 12, 1956

Dear Curtis and Nellie Lee,

The purpose of this letter, which may become a long one, is to tell you of a dream of mine that quite suddenly and miraculously begins to seem possible. I tell you about it now, at this season when perhaps we are supposed to be thinking only of Christmas, for the practical reason that I shall soon need advice. So will you tuck the idea away somewhere and think about it in odd moments when you can?

I must try to be brief, but that is going to be hard.

To begin with, I think you understand this in me, even though we've had little chance to talk about it—my feeling for whatever beautiful and untouched oases of natural beauty remain in the world, my belief that such places can bring those who visit them the peace and spiritual refreshment that our "civilization" makes so difficult to achieve, and consequently my conviction that whenever and wherever possible, such places must be preserved.

When, a few years back and for the first time in my life, money somewhat beyond actual needs began to come to me through *The Sea Around Us*, I felt that, almost above all else, I wished some of the money might go, even in a modest way, to furthering these things I so deeply believe in. All I could reasonably do at that time was to provide in my will that, beyond my mother's needs and a few family bequests, the remainder should go to the Nature Conservancy to aid in the saving of unspoiled seashore areas. Rather soon after that, however, my family obligations, present and future, increased to such an extent that I realized I must remove even this provision from my will.

Just now a literary project has appeared on the horizon that has completely transformed my thinking, bringing the vision of something I might even attempt in my own lifetime. (And how much more satisfactory *that* is!) But first let me describe the dream itself as it has taken form in my mind during the past three years—at first a purely impersonal dream, now some thing I might undertake myself, at least on a small scale.

It concerns a tract of wooded Maine shore that lies a little to the north of us, between Deep Cove and Dogfish Head. I don't know its size, but I should think there might be several hundred acres. It is owned by Gustaf Tenggren, the artist, whose work in illustrating children's books you probably know. I have been told that when the Tenggrens bought this land some years ago, they paid $10,000 for the whole tract plus some additional land that has since been disposed of. I hesitate to think what it would bring now if sold as shore-front lots!

Its charm for me lies in its combination of rugged shore and unspoiled Maine woods. The rocks bordering the shore rise in rather steep cliffs for the most part, and are cut in several places by deep chasms where the storm surf must create a magnificent scene. Even the peaceful high tides explore them and leave a watermark of rockweeds, barnacles and periwinkles. There is one unexpected, tiny beach where the shore makes a sharp curve and there is a protective jutting out of rocks. At another place, something about the angle of the shore and the set of the currents must have produced just the right conditions to trap the driftwood that comes down the bay, and there is an exciting jumble of logs and treetrunks and stumps of fantastic shape. I suppose there is about a mile of shoreline. Behind this is the wonderful, deep, dark woodland—a cathedral of still-ness and peace. Spruce and fir, some hemlock, some pine, and hardwoods along the edges where a fire once destroyed what was there and set in motion the restorative forces of nature. It is a living museum of mosses and lichens, which in some places form a carpet many inches deep. Rocks jut out here and there, as a flat floor where only lichens may grow, or rising in shad-owed walls. For the most part the woods are dark and silent, but here and there one comes out into open areas of sunshine filled with woods smells. It is a treasure of a place to which I have lost my heart, completely.

Before I went to Southport to live, the Tenggrens had ap-parently entertained the horrid thought of subdividing this land into lots. About four years ago they bulldozed one road-way straight through to the shore, and another at right angles to this, penetrating part of the woods. Then nothing more was done. As matters stand now, there has been no serious harm. But the roads are a reminder and a threat.

I have had many precious moments in these woods, and this past fall as I walked there the feeling became overwhelming that something must be done. I had just played a small part in helping to organize a Maine chapter of the Nature Conser-vancy. My rather nebulous plans of last fall had to do with try-ing to enlist aid from that quarter. But while the Conservancy can help, the real job has to be pretty well provided for.

Now, at last, I come to what has happened to make me feel my dream might be realized.

First: *The Sea Around Us*, through a juvenile adaptation now being arranged for, is about to yield unexpected additional income that probably will be fairly substantial. I think this new income, added to what is already assured, will pretty well solve my family problems. This is important because of the way it has allowed me to think about a second new project that came up when I was in New York for the National Council Award. Briefly, it is the preparation and editing of an enormous, practically encyclopedic anthology of writings in the field of nature. I will tell you more of this later and now shall say only that in my pondering over it these past two weeks I have been able to visualize it as a really creative and worthwhile undertaking— something I could feel happy about doing. The publisher was moved to propose this to me because of the enormous success of another anthology he published a few months ago—in the rather improbable field of mathematics—of which there are already 100,000 sets in print! The publisher is confident my subject would be at least as popular. The prospective income to me, after rather considerable expenditures for library and stenographic help, *might* run around $100,000.

Sometimes I can believe things are meant-to-be. When Marie Rodell (my literary agent) told me quite out of a clear sky about this proposition, it instantly meant to me only one thing: my Lost Woods saved.

Now for the practical details on which I need advice: from you, Curtis, as a legal man, from both of you out of your experience with the Singing Tower and Sanctuary. Not that the two are strictly comparable; here the singing will be provided by the hermit thrushes, and their Tower will be the spruces. But some of the problems must be the same. And in this particular situation, the problems abound. If I let the income from the anthology come direct to me, taxes would take most of it and there would be no realization of The Dream. So I suppose I must set up some sort of foundation or whatever you would call it, just as simply as might be to comply with the law. Somehow, I would guess, the income must go direct to the Foundation. Just what must be provided in the way of public use in order to escape taxes I don't know. I can visualize wonderful use in which groups of children or adults would visit the place to receive an interpretation of its meaning and to be refreshed by it. But all that involves a staff and ideally a building that would provide library and exhibit facilities. This is hardly within the means I can reasonably expect to command. However, if I can save a substantial part of the property itself, I think the Nature Conservancy would solicit contributions so that operating funds might be built up.

So, I think my present question to you is how to proceed at the moment. The publisher is obviously eager. I rather think the Foundation, as I am calling it, must be established before there is a contract. On that, chiefly, I want your advice: how to do it as simply, inexpensively, and quickly as possible, and in such a way that its income would be tax-free.

Maurice Greenbaum, who is the lawyer at Greenbaum, Wolff, and Ernst who works most directly with me, has said in another connection that editorship of an anthology constitutes "services" rather than original creation and therefore cannot be considered appropriate for making a "Deed of Gift"—a device I have already used and am about to again to provide direct income to a dependent and so reduce taxes. So, presumably, I could not convey the rights in this anthology to the theoretical Foundation, and perhaps it, not I, must make the contract.

In all that I've said, I seem to have passed over one important point: will the owners sell? If they prefer to hold the property to sell, lot by lot, then there would be no choice but to forget it, for that would make the price astronomical. But if, by chance, the dream had some appeal for them as well, they might settle for a price—for all or a large block of the property—that made sense. Assuming they are willing to consider it, I suppose I would have to ask for a long-term option.

If all this sounds completely mad to you, you may feel reassured when I tell you that sometimes I think it is, too, and sometimes I can laugh at myself for even thinking of undertaking something so big on what is, I suppose, a mere shoestring. But some years ago I began to understand that one must dream greatly if one is ever to accomplish anything. And in this I am supported by a deep faith that it is the thing to attempt—the thing to do if it can possibly be done.

If you will, then, be turning all this over in your minds I shall be deeply grateful. I suppose the publisher will wish to complete a contract fairly soon. The Christmas season is upon us.

Sometime soon thereafter, I could arrange to spend part of a day in Philadelphia, if we could talk it over then. I know this is in line with all that both of you hold dear, so I do not hesitate to ask your thought and advice.

My warm and affectionate remembrance to all the family.

To DeWitt Wallace

January 27, 1958

Dear Mr. Wallace:

I have been told that the *Reader's Digest* has accepted for publication an article looking favorably upon the use of chemical poisons disseminated by aerial spraying for the control of insect pests. If I am correctly informed, your article refers especially to the control of the gypsy moth.

If this is true, I cannot refrain from calling to your attention the enormous danger—both to wildlife and, more frighteningly, to public health—in these rapidly growing projects for insect control by poisons, especially as widely and randomly distributed by airplanes.

Before I began to devote my time to writing, I spent some years in conservation work in the Federal Government. Many of my former colleagues, as well as representatives of other agencies directly concerned with public health as affected by food and water, tell me of their genuine alarm at what is happening. This alarm is shared by a large number of scientists, naturalists, physicians, and public health officials throughout the nation.

Let me pass on to you just a few of the reports that have recently come to my attention.

After widespread spraying for the control of the gypsy moth in New York, New Jersey, and Pennsylvania last year, the Laboratory of Industrial Hygiene of New York found milk from cows pastured on sprayed land to be unfit for human consumption and especially hazardous for children. The milk was condemned and dumped.

One of the leading manufacturers of baby foods discontinued production of one of its products because it could find no vegetables that had not been contaminated by spraying.

Members of the Department of Health, Education and

Welfare are deeply disturbed about the pollution of water supplies from aerial spraying. Some of the newly developed poisons are toxic to fish in the concentration of one part to one billion. What, the Department asks, about humans drinking this water?

The University of Iowa became so concerned about the problem that it established an Institute of Agricultural Medicine two years ago, one of its principal functions being the study of the effect of insecticides and fungicides on human populations. Typical of the as yet unanswered questions posed by the Institute is this: When plants are poisoned to kill herbivorous insects—and a chicken eats the poisoned insect—and a man eats the chicken, what happens to the man?

It has been proven beyond doubt that spraying with DDT induces sterility in various birds. Many responsible scientists are asking whether the same effect might not be induced, directly or indirectly, in human beings?

A new poison, dieldrin, now being prepared for widespread use in the southeast for control of the fire ant, has been shown by tests to be many times as toxic as DDT.

Many reputable biologists feel that other methods, especially in the field of biological control, can achieve good results in insect control without imperiling both wildlife and human populations.

This is only a small sampling of the facts that warrant your earnest consideration before publishing an article that gives direct or implied support to programs of the type now proposed. I am sure that a publication with the *Digest*'s enormous power to influence public thinking all over the country would not wish to put its seal of approval on something so potentially hazardous to public welfare.

Very sincerely yours,

Mr. DeWitt Wallace
The Reader's Digest
Pleasantville, New York

1. Carson at Hawk Mountain Sanctuary in eastern Pennsylvania, 1945.

2. Birding at Cobb Island, Virginia, with the Washington, D.C., Audubon Society, 1947.

3. Exploring the Everglades in a "glades buggy," 1950.

4. In the Florida Keys with artist Bob Hines, 1952.

5. At Woods Hole, Massachusetts, 1950.

6. Exploring tide pools, West Southport, Maine, Summer 1955.

7. Dusting sheep with DDT, Medford, Oregon, 1948.

8. Testing an insecticidal fogging machine, Jones Beach, New York, July 1945.

9. A farmer surveys the synthetic fertilizers and insecticides used over the course of an average year at his 78-acre farm in Lititz, Pennsylvania, April 1955.

10. Francis Uhler (*left*) and Eugene Surber (*right*) test the effects of DDT on
wildlife at the Patuxent Wildlife Research Center in Maryland, June 1945.
Carson offered to write an article about the Patuxent experiments for
Reader's Digest, but the magazine declined.

11. Carson read reports by bald eagle enthusiast Charles L. Broley, who attributed precipitous declines in Florida eagle populations to DDT spraying. Here, Broley holds up a newly banded eagle in Tampa, Florida, March 1948.

12. A mock-up of the dust jacket for the first edition of *Silent Spring*.

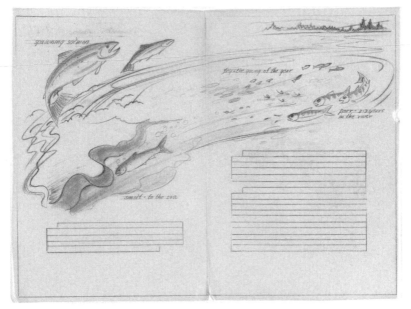

13. Preliminary illustration for *Silent Spring.*

14. Lois and Louis Darling, illustrators of *Silent Spring*,
with their golden retriever Tiggy, Fall 1968.

15. Carson with her grandnephew and adopted son Roger Christie, West Southport, Maine, 1960.

16. Photographed for *Life* magazine in Glover Archbold Park, Washington, D.C., September 1962.

17. After the publication of *Silent Spring*, Carson began a collection of the many cartoons in which she was mentioned. The *Peanuts* strip (*above*) by Charles M. Schulz appeared on November 12, 1962, and the *New Yorker* drawing, by James Stevenson, on May 18, 1963.

"Now, don't sell me anything Rachel Carson wouldn't buy."

18. Interviewed by Eric Sevareid for the *CBS Reports* program "The Silent Spring of Rachel Carson," which aired on April 3, 1963, to an audience estimated at 10–15 million.

19. Carson testifies before a U.S. Senate subcommittee in Washington, D.C., June 4, 1963.

To Dorothy Freeman

February 1

About the book: I'll see if I can make any sense about it briefly. The theme remains what I have felt for several years it would be: Life and the relations of Life to the physical environment. (The older ideas of dealing just with theories of the origin of life or with the course of evolution were discarded long ago.) But I have been mentally blocked for a long time, first because I didn't know just what it was I wanted to say about Life, and also for a reason more difficult to explain. Of course everyone knows by this time that the whole world of science has been revolutionized by events of the past decade or so. I suppose my thinking began to be affected soon after atomic science was firmly established. Some of the thoughts that came were so unattractive to me that I rejected them completely, for the old ideas die hard, especially when they are emotionally as well as intellectually dear to one. It was pleasant to believe, for example, that much of Nature was forever beyond the tampering reach of man—he might level the forests and dam the streams, but the clouds and the rain and the wind were God's—the God of your ice-crystal cathedral in that beautiful passage of a recent letter of yours.

It was comforting to suppose that the stream of life would flow on through time in whatever course that God had appointed for it—without interference by one of the drops of the stream—man. And to suppose that, however the physical environment might mold Life, that Life could never assume the power to change drastically—or even destroy—the physical world.

These beliefs have almost been part of me for as long as I have thought about such things. To have them even vaguely threatened was so shocking that, as I have said, I shut my mind—refused to acknowledge what I couldn't help seeing. But that does no good, and I have now opened my eyes and my mind. I may not like what I see, but it does no good to ignore it, and it's worse than useless to go on repeating the old "eternal verities" that are no more eternal than the hills of the poets. So it seems time someone wrote of Life in the light of the truth as it now appears to us. And I think that may be the book I am to write. Oh—a brief one, darling—suggesting the new ideas not treating them exhaustively. Probably no one could; certainly I couldn't.

I still feel there is a case to be made for my old belief that as man approaches the "new heaven and the new earth"—or the space-age universe, if you will, he must do so with humility rather than arrogance. (I was pleased to notice that word in the little editorial on snow I sent you. I think I wrote you of Frank Lloyd Wright's use of it.)

And along with humility I think there is still a place for wonder. (By the way, I hope you didn't think I was serious, some weeks back, in asserting a claim to the word "wonder.")

Well, darling, that's not an outline of the book, but at least indicates the approach.

Of course, in pre-*Sputnik* days, it was easy to dismiss so much as science-fiction fantasies. Now the most farfetched schemes seem entirely possible of achievement. And man seems actually likely to take into his hands—ill-prepared as he is psychologically—many of the functions of "God."

Glad you are reading Conrad. I don't remember ever studying him formally, but of course I've read a good deal of him. Does your book include selections from *Mirror of the Sea*—which I love. See Bennett Cerf's biographical note on Galsworthy for an incidental bit on Conrad.

Rain on the roof now—maybe snow tomorrow. Had an interruption a couple of pages back to hear the announcement of the launching (?) of an American satellite. Still a maybe as I write.

Perhaps I'll put this out in the morning for the mailman who comes early Saturday. If it snows I probably wouldn't go in later.

So I'll say goodnight, my dear one, in case there's no time to say more tomorrow. Now to sleep—perhaps to dream—Guess what?

I love you
Rachel

1958

To Marie Rodell

February 2

Dear Marie,

Following our telephone conversation, I am sending along a somewhat haphazard memorandum which nevertheless does incorporate a few of the horrifying facts about what is happening through the mass application of insecticides.

Having spent a substantial part of two weeks in phone calls, correspondence and searching through references, and having finally struck what appears to be "pay dirt"—I naturally feel I should like to do an article myself. In the course of all this, I have made certain valuable contacts and discovered many leads still to be followed up; this in its full value would be difficult to transfer to another writer. I don't see, however, why more than one article might not be written to good advantage—if you can find editors willing to listen.

If the *Digest* is about to publish an article on the horrors of clinical X-rays, I still don't see why this subject wouldn't be appealing—given the documentation, which I now know exists in plenty.

If you could get an expression of real interest from the *Ladies' Home Journal*, I might be willing to gamble, provided they would give a prompt decision.

I meant to ask if you would look in the phone book and see if Beechnut has a New York office. I want to verify the report, given me by the information man at Food and Drug, that Beechnut discontinued production of one specific vegetable (I don't know what one) because they could get no uncontaminated supplies.

Love

MEMORANDUM ON THE EFFECT OF COMMONLY
USED INSECTICIDES ON MAN AND ANIMALS:

Between the years 1940 and 1950, the volume of pesticides in use in the United States was multiplied by seven times. More than $500,000,000 worth of pesticides were produced for retail sale in the United States in 1956. Production by 1975 is expected to exceed 2 billion dollars' worth of chemicals annually.

The effect of these chemical poisons—some of great toxicity— is only beginning to be assessed in its relation to the lives and welfare, not only of the whole community of animals in the area subjected to ground and aerial sprays, but of the human population as well. There exists already, however, a large body of well documented evidence that these highly toxic poisons, as presently used, represent an alarming threat to human welfare, and also to the basic balance of nature on which human survival ultimately depends.

The following notes, here presented at random, are thoroughly documented by published reports of the U. S. Public Health Service, the Food and Drug Administration, the Fish and Wildlife Service, the Conservation Foundation, the National Wildlife Federation, and the National Audubon Society.

The program for the control of the fire ant, soon to be launched in southeastern United States, contemplates the use of the poison dieldrin in concentrations of 2 (or possibly in some areas 4) pounds per acre. This is to be spread in granular form from airplanes. The following facts about dieldrin are relevant:

DIELDRIN: Dieldrin is considered to be at least 20 times as toxic as DDT. Applied in concentrations of ¼ pound per acre (or one-eighth to one-sixteenth the concentrations now proposed) dieldrin resulted in heavy mortality of birds, fish, and mammals. The Public Health Service warns men engaged in control programs using dieldrin to use extreme care, wear protective gloves and clothing, and avoid breathing of dieldrin dusts or sprays. It cautions that dieldrin must not be applied in dairy barns, to food or food products or man or animals, nor directly to animals. Dieldrin is one of a group of toxic compounds that cannot be washed off foods, even with a hot bath of detergents. The Public Health Service states that at least 10% of the men using dieldrin in control programs have exhibited symptoms of poisoning, 3 to 24 months after exposure. These may include convulsions, coma, and a state indistinguishable from epilepsy.

A recent report of the Conservation Foundation and the New York Zoological Society states that Benzine Hexachloride, which has actually been used on human beings for the control of lice, has recently been found to produce brain tumors. Yet, according to the "Operational Memoranda" published by the U. S. Public Health Service, this product is an "agricultural insecticide" commonly used for the control of various pests. The Service warns that contamination of food products of man and animals should be avoided. Yet it cites examples of dissemination of this poison by airplanes for the control of mosquitoes.

One alarming aspect of the problem is the fact that many insects have developed strains resistant to DDT and other poisons in the general class of chlorinated hydrocarbons. Dangerous as these poisons are, there are others even worse. To combat resistant insect strains, recourse is now being had to compounds of organic phosphates which include some of the most dangerous of all poisons. An example is Parathion, concerning which some facts follow:

PARATHION: The Public Health Service described parathion as "extremely dangerous"—to be handled with "utmost care—no eating, chewing, or smoking while handling." Fatal human poisoning has followed ingestion, skin exposure alone, and sometimes inhalation. The Service also reports that "accidental exposure of children to open or even 'empty' containers has been a major and dramatic source of fatal poisonings." So dangerous is this substance that physicians or first aid workers handling victims of parathion poisoning are cautioned to wear rubber gloves while removing the patient's clothing. Yet it is reported in the Public Health Service's Operational Memoranda that parathion is "used extensively for protection of fruit, truck, and field crops from insect and mite damage." Some 3,000,000 pounds were used in the crop year 1952–53. In California, according to the Public Health Service, "ground and airplane treatments of parathion are being utilized" for the control of mosquitoes. In Arkansas, parathion has been introduced into irrigation waters for the control of insect pests in rice fields.

It is now well established that spraying with DDT and some other chemicals produces an effect of sterility a year or more after the spraying in various birds in the sprayed area. Adult bobwhite quail can stand considerable exposure to DDT, but 90% of their offspring are crippled and soon die. Many other forms of wildlife become incapable of reproduction after exposure to insecticides. The large colony of bald eagles in Florida has produced almost no young in recent years, following heavy spraying in the area. Similar results have been observed in pheasants and many other forms of wildlife. It should be noted especially that this is a delayed effect, not observable during the season of spraying. Many responsible biologists and conservationists are pointing out that the potential effect on human reproduction is not known, and might not make itself felt for some years. John Baker, President of the National Audubon Society, recently pointed out that "Insecticide hazards may well rank in seriousness of adverse effects with the dangers of radioactive fallout." He warned that "cumulative secondary poisoning of human beings and wildlife, which already exists to some extent, may become catastrophic."

Some recognition of these facts has already been made by segments of the very industry that manufactures these poisons for distribution. The president of the Thompson Chemical Corporation of Los Angeles and St. Louis recently said: "We have decided to withdraw entirely from the production, distribution and research of the presently known agricultural insecticides. . . . The ingestion of presently employed insecticide residues by human and other warm-blooded animals is a correlative problem of a highly serious nature."

The University of Iowa recently established an Institute of Agricultural Medicine, one of its functions being the study of these problems. One of the questions under serious consideration by this Institute is this: "Is there a potential human sterility problem that might not appear for several generations?"

The imperfect state of our knowledge of these insecticides was emphasized in the current report of the Conservation Foundation, which says: "It should be pointed out that there are some agents which we formerly thought were relatively non-toxic to humans but which were later found to be toxic . . . In the past, carcinogenic agents have been offered as insecticides without any knowledge of their cancer-producing effects."

The National Audubon Society, the National Wildlife Federation and many others have called on Secretary Benson to withhold further mass spraying of insecticides until the facts can be determined. An official of the Department of Agriculture said to me last week: "The automobile kills hundreds of people every year, but no one has suggested doing away with automobiles."

1958

To E. B. White

11701 Berwick Road, Silver Spring, Maryland
February 3, 1958

Dear Mr. White:

Several times in recent years our friends, especially Cass Canfield and Paul Brooks, have made beginning attempts to bring us together. Largely because of the infrequency of my own visits to New York, I suspect, these attempts have failed, to my regret.

Let it be said at once that I watch for, and devour, every word of yours that appears in *The New Yorker*!

The immediate purpose of this letter is to bring to your attention something that, it may well be, you are already well aware of. In recent weeks I have been reminded of my own former conviction that the mass spraying of DDT and other even more dangerous insecticides is a threat to the entire balance of nature and even more immediately to the welfare of the human population. An enormous literature on the subject exists. Should you wish, I can send you scores of references that establish the hazards beyond doubt.

However, my present concern is this. Dr. Robert Cushman Murphy and certain other residents of Long Island have brought suit against the Federal Government (seeking an injunction, I believe, against Secretary Benson) because of the disastrous results of spraying on Long Island last summer. The case is to come up in a Federal Court next week—February 10th, if my information is correct. A large amount of important testimony is certain to be presented. Should the injunction be granted, the results would be far-reaching. It is my hope that you might cover these court hearings for *The New Yorker*.

Dr. Murphy and others (including John Kieran and Edwin Way Teale) have organized a Committee Against Mass Poisoning, 3100 Chrysler Building, New York 17, where you could verify the time and place of the Court hearing. I am told that the Committee has on hand testimony in the form of affidavits on proven toxicity of foodstuffs resulting from DDT spraying, death of birds and fish, and cases in which the same area was sprayed as many as fifteen times.

As you may know, various conservation organizations, including the National Audubon Society and the National Wildlife Federation, have implored Benson to halt the spraying programs until more complete information is gained as to the effects on wildlife and the whole balance of nature. The Conservation Foundation and the New York Zoological Society have recently issued a publication surveying the problem of pesticides. The World Health Organization has published a 114-page survey of the hazards of pesticides to man. And publications of the U. S. Public Health Service that I have just received contain really horrifying facts—all expressed in most dispassionate language.

Following its 3,000,000-acre spraying for gypsy moth control in 1957, the Department of Agriculture plans a similar campaign for 1958. I have been told that Maine is involved. I have several acres near Boothbay Harbor. When I called the Department for information I was told the program had not yet been "firmed up." However, there are very firm plans for the spraying of large areas in the southeastern states for the control of the fire ant. Use of a relatively new and extremely dangerous poison, dieldrin, is planned, in concentrations of 2 or more pounds per acre. Dieldrin is at least 20 times as toxic as DDT, so this is equivalent to approximately 40 pounds of DDT per acre. I have alarming information on the effects of dieldrin on men and animals.

It would delight me beyond measure if you should be moved to take up your own pen against this nonsense—though that is far too mild a word! There is an enormous body of fact waiting to support anyone who will speak out to the public—and I shall be happy to supply the references.

Meanwhile, let me express again my delight in all you have written.

Sincerely yours,

Mr. E. B. White
The New Yorker
25 West 43rd Street
New York 36

To Dorothy Freeman

Thursday, June 12
On train

Darling,

Here we are—on noon train from N.Y., due Washington at 3:50. I've decided this should be waiting for you on your return home Sunday—I assume your plans haven't changed.

It has been a very good trip, with all major things accomplished and really a lot covered in the short time—*New Yorker*—Joan Daves—Fairfield Osborn—lawyers.

The visit with Mr. Shawn was really a delight and most satisfying. We talked for 2½ hours! He is completely fascinated with the theme and obviously happy and excited at the prospect of presenting it in the *New Yorker*. Best of all, I can (indeed, he *wants* me to) present it strictly from my point of view, pulling no punches. He says, "After all there are some things one doesn't have to be objective and unbiased about—one doesn't condone murder!" Besides the importance of the theme ("We don't usually think of the *New Yorker* as changing the world, but this is one time it might") he feels the material is just plain fascinating, and now thinks he'd like to have 50,000 words!! (This would mean 3 installments, I think—that alone will emphasize its importance in their eyes.) Since I had meant to keep the book down to about this length, it will mean the two will be about the same.

I'll probably save until I see you some of the "side issues" of our conversation—about E. B. White's awareness of this problem many years ago and his almost prophetic vision regarding it—he says Mr. White will be utterly horrified when he gets all the *facts* I present—about the profile of Dr. Newton Harvey, which did result from my suggestion some years ago—about my proposed Harper book, which seemed to impress him so greatly that I guess that, too, could well find space in his pages. Now I've suggested Vincent Schaefer as a profile subject, and also given him a little horror story on radioactivity which I got from Edwin Teale. It concerns Dick Pough—I planned to send you the letter after I've answered it. Mr. Shawn wrote it down at once, is going to call Dick Pough, and I imagine it will appear in "Talk of the Town."

From there I went to the New Weston and got established, and about six Joan Daves came over with our Scandinavian agent in tow—he just happened to be in her office. He stayed for "a drink"—then left and we had dinner and talked until eleven.

This morning I went to breakfast at the Osborns' at 8:15. We had about an hour to talk, then shared a cab as far as my hotel. From there I went immediately to Mr. Greenbaum—I have asked him to keep on the will, which has been a thorny problem, and still is, with tremendous decisions to be made.

And that took me up to train time! It was really a good trip. The only thing I couldn't carry out was seeing Elizabeth Lawrence, who left for a western trip yesterday, but that was chiefly a courtesy call so it didn't matter.

Now—enough about me. By this time your week in Maine is over. I do hope Stan came home minus any traces of his cold, and that you are refreshed by your time there. Your account of that first week-end sounded very full of people, but probably that changed Monday.

I laughed at Mrs. Pinkham's question. Did you say "Yes, *almost* every day"?

Darling, about your birthday, you know I'll want to call you but now that I've gotten organized enough to look at the calendar I realize it is Sunday. As you know, I can be more sure of peace and quiet to talk when Ida is there. Besides, you may go out. So how about Monday morning, about 10 o'clock?

How I'd love to hear that early June chorus of birds! And see the flowers. Maybe, sometime.

I don't think you'll want to read any more of this—and my arm is getting a cramp from trying to be legible.

How nice you have *The Return of the Native* before you! I confess I remember only two parts—the wonderful opening passage, and the scene about the glow worms.

All my love, dear.
Rachel

1958

To Edwin Way Teale

11701 Berwick Road, Silver Spring, Maryland
October 12, 1958

Dear Edwin,

It was good to hear from you, but I know I don't deserve to, for I have a sneaking suspicion I never answered your last letter.

Twenty-one chapters sounds like a very impressive achievement—how did you do it? With a great deal of luck, I might be through in time for publication in the fall of '59, too, but it will take some doing, for this job gets bigger and more complex all the time. I have already covered about as many references as for *The Sea Around Us*, which I thought was my limit as a research job. However, it has some fascinating aspects, and the correspondence involved is most rewarding. The right people are so glad I'm doing the job.

Your clipping gave me a laugh, though of course it really isn't funny that the public is being fed this sort of propaganda all the time. One of my delights in the book will be to take apart Dr. Wayland Hayes' much cited feeding experiment on "51" volunteers. It was "51" for only the first day of the experiment; thereafter the experimentees rapidly lost their taste for DDT and drifted away until only a mere handful finished the course of poisoned meals. Dr. Hayes thereupon got rid of them as fast as he could, and made no subsequent check on the welfare of any of them, although it is well known that the real damage takes many months or years to become evident. Dr. Hayes is just as slick a gentleman in his line as Admiral Strauss was in his!

However, I am personally convinced that the days of "squirt-gun control" of insects are numbered; there is a great deal of handwriting on the wall.

Are you, by any chance, coming to Washington for the Christmas meetings?

Along with 99 other lines of investigation, I am trying to collect all the specific data I can as to any actual declines in bird populations that have occurred—not just general impressions, but factual comparisons that will hold up under fire. I'm sure you know Mr. Broley's reports on the eagle. I now have some good material on the robin, especially from Michigan, with a beautiful tie-up to DDT via the earthworm (this from some Illinois studies). I have the promise of detailed data on about a score of other species from 2 biologists in Michigan. You probably saw also that Massachusetts Audubon reports that comparative records, based on its long-published "Records of New England Birds" show a disturbing decline, and the Society feels insecticides are probably responsible. Then the *Times* for Sept. 18 quotes various "naturalists" on Long Island (John J. Elliott of Seaford and Gilbert Raynor of Manorville) to the effect that certain warbler populations are off 50 to 90 percent, etc. etc. Do you know either of these men, or have you been in the field enough to have formed any impressions of your own, or have you heard any such reports from people whose judgment you respect?

Knowing you are "nip and tuck" I hate to ask this, and certainly don't want you to spend time asking anyone; but if some facts are just ready to spill out of you, perhaps some odd moment you could put them down. I thought I'd ask John Kieran, too.

Because of all my library research, we didn't get to Maine until mid-July, and had to leave a few days after Labor Day on account of Roger's school, but it was wonderful to be there. Did you and Nellie go anywhere—or did you take a vacation from traveling? I wish you'd both travel this way some time soon!

Mamma is well, though she gets about with more and more difficulty; Roger is just over a perfectly horrid case of ivy poisoning, that closed his eyes as well as doing other nasty things; and the cats and I are about as usual.

Our very best to both of you—
Rachel

To Marjorie Spock

Nov. 17, 1958

Dear Marjorie,

The record of bird song that came Friday gave Roger the greatest delight—first the thrill of receiving a package by mail—then the pleasure in the record itself. He has a very sweet feeling for all living things and loves to go out with me and look and listen to all that goes on. I know he will have great pleasure in recognizing the songs from this record—first the mockingbird that still pours his song down our chimney, then the cardinals that begin to whistle in January, and all the rest.

There was a special and unheralded thrill for me in the record. During the song of the wood pewee, I heard in the background the unmistakable voice of the veery—not once but several times. Of all bird song, that has the quality of purest magic for me!

It was very sweet of you to send this to him. In spite of the best I can do—and I'm afraid it's not always a very good best—I know he is often a lonely little boy, missing his own mommy terribly, and wishing for a father, although he does not remember his own. I really dread Christmas, for it was a hard time for him last year. But he loves people, and having friends, and to get mail from them is a special joy. When your "two-bears" card came, he turned on his brightest smile and said "Oh I know who that is—they're just pretending to be bears!"

Rain and fog today—still a few bright leaves here, and last week's warmth brought a quavering frog song from our swamp land!

Will write of other things soon. All our best to you both—

Rachel

To Marjorie Spock

December 4, 1958

Dear Marjorie,

You and Polly will grieve with us, I know, on hearing the sad news that my dear mother died early Monday morning, after a critical illness lasting about a week. She was stricken quite suddenly Saturday evening, November 22, with a slight stroke, followed by pneumonia and other complications. We kept her at home, knowing she would be so much happier, and since we had an oxygen tent, nurses, and every conceivable need, we know that no hospital could have done more. But in the end her wonderful heart could take no more.

I wish you could have known her better, but I'm happy you had the little exchange of correspondence recently. Yours to her, with the lovely enclosure of all the Christmas letters, came just a day or so before illness struck.

And your warm and understanding letter to me, chiefly about Roger, came after my mind had been filled with apprehension, but I was still free enough to read it with deep gratitude for your affectionate comprehension of the problem. Soon I shall want to return to it, and to hear more of your thoughts. So few understand. I wish I had known you sooner.

Poor little fellow—this is a new blow for him, for he loved his Grandma, and her love for him was overflowing. Also, it is so obviously recalling to him all the memories of the loss of his mother less than two years ago.

It is good for me, I'm sure, that I am forced to think of him and of how to ease the readjustment, which will be difficult, to say the least, for all of us.

Sometime I want to tell you more of her. Her love of life and of all living things was her outstanding quality, of which everyone speaks. More than anyone else I know, she embodied Albert Schweitzer's "reverence for life." And while gentle and compassionate, she could fight fiercely against anything she believed wrong, as in our present Crusade! Knowing how she felt about that will help me to return to it soon, and to carry it through to completion.

When I can, I shall write of that, but meanwhile I wanted to tell you of our sorrow.

As ever—
Rachel

To William Shawn

February 14, 1959

Mr. William Shawn
The New Yorker
25 W. 43rd Street
New York
New York

Dear Mr. Shawn:

I have wanted for a very long time to write you and bring you up-to-date on the progress of the article. I am sure it must seem a long time in the making, but in the end I believe you will feel, as I do, that my long and thorough preparation is indispensable to doing an effective job. I can see clearly now that an article I might have written last summer would have been half-baked, at best. Now it is as though all the pieces of an extremely complex jig-saw puzzle are at last falling into place.

As you know, it has always been my intention to give principal emphasis to the menace to human health, even though setting this within the general framework of disturbances of the basic ecology of all living things. As I look over my reference material now, I am impressed by the fact that the evidence on this particular point outweighs by far, in sheer bulk and also significance, any other aspect of the problem.

I have a comforting feeling that what I shall now be able to achieve is a synthesis of widely scattered facts, that have not heretofore been considered in relation to each other. It is now possible to build up, step by step, a really damning case against the use of these chemicals as they are now inflicted upon us.

It is an amusing fact that although the American Medical Association and the Public Health Service, when asked to take a stand, are rather on the fence, their various published statements constitute quite an indictment. Of course, I plan to quote them so as to take due advantage of the weight of their authority.

I shall be concerned less with acute poisoning, which occurs usually through accident or carelessness, than with the slow, cumulative and hard-to-identify long term effects. It is chiefly when life-span experiments are conducted with animals that the real damage shows up. No one now can honestly say what the effects of lifetime exposure in man will be, because not enough time has elapsed since these chemicals came into use. But we do know that every child born today carries his load of poison even at birth, for studies prove that these chemicals pass through the placenta. And after birth, whether breast-fed or bottle-fed, the child continues to accumulate poisons, for checks of mothers' milk, as well as of the dairy product, always show some content of DDT or other chlorinated hydrocarbons. There is also a body of evidence to show that young, rapidly growing animals are more seriously affected than adults.

Besides the effects on liver and nervous system, which are generally recognized by professional men if not by the public. I think I shall be able to support a claim to even more serious and insidious effects, which include the most basic functions of every living cell. This has been an extremely interesting line of research, and, to me, a terrifying one. One of its aspects has led me to the fundamental recent research of Otto Warburg, of the Max Planck Institute for Cell Physiology in Germany, on the physiological changes that may lead cells into the wild proliferation of cancer. Besides and beyond this vitally important fact, I shall be able to show that the chemicals used as insecticides interfere with many of the enzymes that control the most basic functions of the body. There is also scattered evidence, needing additional research, indicating that some of these chemicals interfere with normal cell division and may actually disturb the hereditary pattern.

If you receive the releases of the National Audubon Society and the Conservation Foundation, you are up to date on the far-reaching, inadvertent effects of the fire ant spraying in the south. This will probably be my "Exhibit A" among examples of ill-advised, irresponsible Governmental actions. I am told by friends in the National Wildlife Federation that the reaction in the south is so strong that there is hope funds may be shut off, at least for the fiscal year 1960!

On the positive side, there are many new lines of thought. It is encouraging to find that many different people are working on alternatives to chemical sprays. A great deal of publicity was given a few weeks ago (in part through an article in *Life*) to the possibility of controlling insects through hormones, which would interfere with their normal metamorphosis. I have corresponded with the scientists chiefly involved, and get the impression that this Utopia, if it ever materialized, is a long way off insofar as practical application is concerned. However, the very attempt is a fascinating story.

Personally, I am much more impressed with attempts at control through man-induced insect diseases. The Japanese beetle problem is already on the way to solution by this means. I gather the Dean of this whole school of research is Dr. Steinhaus of California, with whom students are now in charge of a Department of Agriculture unit concerned with such research. I am to see him soon and get the whole story. It is a very live thing, for recently it has been announced that two chemical firms are producing agents for microbial control—whether on a large scale or merely experimentally was not made clear. Also,

Food and Drug has given its approval of such attempts in a list of some sixteen states. The bacteria or viruses are specific for the insect and have further advantages, such as leaving no residues.

There is also a new Agriculture unit dealing with Insect Physiology. One of its staff told me that the real purpose is to develop alternatives to spraying. This, again, is a story I shall cover by visiting the laboratory. An example of practical application is the recent attempt to control screwworms and, even more recently, mosquitoes, by releasing males rendered sterile by radioactive cobalt.

I am really happy about these new developments. The older forms of biological control, depending on introduction of parasites and predators, had their values but also their limitations. One difficulty about pointing to them as the sole alternative was that they seemed to belong to an earlier age—whereas radioactive cobalt can firmly assert its place in 1959.

This will give you a general idea of the direction the article is taking. For myself, I feel happy about it, although I would not be so rash as to predict when you will have the manuscript. In another month, I should be better able to guess.

There is a bare possibility I may be in New York for the Wildlife Federation Meeting, February 27th. If I think I can, I'll call you several days in advance and hope to see you.

Sincerely,

"Vanishing Americans"

Your excellent March 30 editorial, "Vanishing Americans," is a timely reminder that in our modern world nothing may be taken for granted—not even the spring songs that herald the return of the birds. Snow, ice and cold, especially when visited upon usually temperate regions, leave destruction behind them, as was clearly brought out in the report of the National Audubon Society you quote.

But although the recent severe winters in the South have taken their toll of bird life, this is not the whole story, nor even the most important part of the story. Such severe winters are by no means rare in the long history of the earth. The natural resilience of birds and other forms of life allows them to take these adverse conditions in their stride and so to recover from temporary reduction of their populations.

It is not so with the second factor, of which you make passing mention—the spraying of poisonous insecticides and herbicides. Unlike climatic variations, spraying is now a continuing and unremitting factor.

During the past 15 years, the use of highly poisonous hydrocarbons and of organic phosphates allied to the nerve gases of chemical warfare has built up from small beginnings to what a noted British ecologist recently called "an amazing rain of death upon the surface of the earth." Most of these chemicals leave long-persisting residues on vegetation, in soils, and even in the bodies of earthworms and other organisms on which birds depend for food.

The key to the decimation of the robins, which in some parts of the country already amounts to virtual extinction, is their reliance on earthworms as food. The sprayed leaves with their load of poison eventually fall to become part of the leaf litter of the soil; earthworms acquire and store the poisons through feeding on the leaves; the following spring the returning robins feed on the worms. As few as 11 such earthworms are a lethal dose, a fact confirmed by careful research in Illinois.

The death of the robins is not mere speculation. The leading authority on this problem, Professor George Wallace of Michigan State University, has recently reported that "Dead and dying robins, the latter most often found in a state of violent convulsions, are most common in the spring, when warm rains bring up the earthworms, but birds that survive are apparently sterile or at least experience nearly complete reproductive failure."

The fact that doses that are sub-lethal may yet induce sterility is one of the most alarming aspects of the problem of insecticides. The evidence on this point, from many highly competent scientists, is too strong to question. It should be weighed by all who use the modern insecticides, or condone their use.

I do not wish to leave the impression that only birds that feed on earthworms are endangered. To quote Professor Wallace briefly: "Tree-top feeders are affected in an entirely different way, by insect shortages, or actual consumption of poisoned insects . . . Birds that forage on trunks and branches are also affected, perhaps mostly by the dormant sprays." About two-thirds of the bird species that were formerly summer residents in the area under Professor Wallace's observation have disappeared entirely or are sharply reduced.

To many of us, this sudden silencing of the song of birds, this obliteration of the color and beauty and interest of bird life, is sufficient cause for sharp regret. To those who have never known such rewarding enjoyment of nature, there should yet remain a nagging and insistent question: If this "rain of death" has produced so disastrous an effect on birds, what of other lives, including our own?

RACHEL CARSON.
Silver Spring.
Washington Post, April 10, 1959

To Beverly Knecht

April 12, 1959

Dear Beverly,

Not long ago you mentioned Housman's lovely poem about the cherry tree, which ended with his seeing the cherry hung with snow. Now I can see his poem just by looking out of the window, for this silly snowstorm has put an unwanted frosting not only on my lovely little weeping cherry, but on daffodils, hyacinths, and everything else. April 12th—I can't believe it. And after 90° just a few days ago!

I'm enclosing a copy of a letter I was moved to write to the *Washington Post* the other day, because it will give you an idea of the subject of my present book. However, my concern is considerably broader than this problem of bird life, which I discuss here. It deals with a quite incredible modern situation through which the whole earth, including the environment of all living things, is being alarmingly changed by the broadcast of chemical insecticides. This is the subject that brought me into contact with Dr. Biskind. Of all the medical men, he has been the most clear-thinking and the most articulate in pointing out the hazards to human beings through contamination of food, water, and even air—to say nothing of what is now considered commonplace household uses of these chlorinated hydrocarbons, which are deadly nerve and liver poisons. You will find references to the matter in some of the papers of his I sent you. Anyway, I became convinced somewhat more than a year ago that this was a problem that ranks second only to radioactive fall-out in its serious threat to all of us—so I put aside the writing that was already under way, and have ever since then been wading through the really enormous technical literature on the subject. The average person has no idea of the problem, and I decided this was more important than anything else I might deal with just now. I think I shall dedicate the book to Albert Schweitzer, who said of this very problem: "Modern man has lost the capacity to foresee or to forestall. He will end by destroying the earth—" I do think that in recent months there are signs in the wind that there is enough awakening to bring about a salutary change in methods, so Schweitzer's sad prediction may not come about through this means, but it is, I am convinced, one of the gravest problems of this age of many grave problems.

This present book is part of a broader topic I shall deal with later, in a book I've been promising to write for years. Its subject is the way life of all kinds changes and sometimes even creates its environment. It is a subject that intrigues me, for usually it is looked at the other way around.

The next book after the one I'm working on is to be a treatment of the same subject I dealt with in the magazine article I sent you—exploring nature with children. And the Sea Anthology—heavens, I must be counting on working a lot faster than I have in the past!

This letter seems to be all about me, but you were such a good listener when I told you about the anthology last night, that I decided it was about time to indicate what else is simmering in the back of my mind.

Your Schumann Piano Concerto in A Minor—yes, it is lovely. I dug it out of my collection and have played it several times since you mentioned it.

You know, we do seem to have so much to say to each other (and last night an hour sped by like ten minutes!) that I'm sure whatever time I can squeeze out between trains or planes will only be tantalizing. But at least it will be a delightful prelude to a longer visit I hope we can have somewhere before too long, when you are even better.

Dorothy was thinking of driving to Maine for the 6th birthday of her precious granddaughter, Martha, this week-end. I'm afraid their weather may be far worse than what we're getting, so I almost hope she didn't go. I don't know whether she told you, but her husband, Stan, had a very serious heart attack shortly after the Christmas holidays, and for a long time his recovery was most uncertain. He is progressing after a fashion, but I know she feels a constant, gnawing apprehension. So that was the big uncertainty in her thinking of going to Orono, where their son teaches in the University of Maine. Poor Stan also has a duodenal ulcer of the type that is subject to hemorrhage, so they have had an anxious time for several years.

Beverly dear—please don't rush your return to normal activities. You know the complete healing of that foot is the most important thing just now. I'm sure it must be wonderful, though, to be home and to do some of the old familiar things. I'll call you in a couple of days if it seems possible to see you Thursday or Friday.

Love,

To R. D. Radeleff

May 20, 1959

Dr. R. D. Radeleff
Agricultural Research Service
Animal Disease and Parasite Research Division
P. O. Box 232
Kerrville, Texas

Dear Dr. Radeleff:

Thank you for your letter of May 11 in reply to my inquiry as to the observed effect on domestic animals of chemicals used in Fire Ant control.

From my own background of scientific work in Government I realize that you may be restricted in your ability to release the complete data from your Field Laboratory. However, you will understand that in attempting to evaluate the known effects I shall need details. As you know, there have been numerous reports of the death of livestock in the areas treated with Heptachlor and other poisons. Were these deaths actually investigated by you or by other veterinarians whose judgment you accept? If so, what type of examination was made? What in your opinion was the cause of death in these cases that have been cited as examples of pesticide poisoning?

You state in your letter that you feel there is no hazard to domestic animals when "standard precautions" are taken. This leaves certain unanswered questions. Do you refer to precautions in the application of chemicals by control agencies or to precautions that should be taken by farmers and other land owners for the protection of their animals? If the latter, what were the official recommendations and to what extent, according to your observations, were these precautions heeded? What is actually known as to the amount of granular heptachlor that would be toxic to a chicken feeding in the treated area or that would accumulate potentially dangerous residues in such chickens intended for market?

My purpose in asking these numerous questions is to be better able to sum up the actual effect as part of a detailed study I am making of the problem of pesticide use. I shall sincerely appreciate, therefore, your early consideration of these questions.

If you are unable to release details of your study on the effect of the control chemicals on domestic animals, please refer me to the proper official in Washington for a further discussion of the problem.

Sincerely yours,

To J. I. McClurkin

September 28, 1959

Dr. J. I. McClurkin
Department of Biology
Memphis State College
Memphis, Tennessee

Dear Dr. McClurkin:

Before stating the questions I want to ask, I should like to explain my reasons for asking them. For many months I have been gathering information on the effects of chemicals on living things and on their environment. In the course of this study, which is the basis of a book I am writing, I have reviewed an enormous number of technical papers in various fields.

When I first saw references to the six-legged frogs that are said to have been produced in an artificial lake in Tennessee, I confess I was completely skeptical, dismissing the whole thing as a fanciful newspaper story. Recently, however, I have seen an article which quotes you in a way that leads me to think that you personally saw the frogs and had some direct knowledge of the situation. I shall therefore be very grateful to you if you will summarize for me just what your observations were. Is it true that there were actually a large number of frogs with this abnormality and that they were all produced in one pond, or at least in one general locality? Is there a definite record that the pond or the surrounding fields had been sprayed, and if so, when, and with what chemical?

I am familiar with many different investigations that show that some of the chemicals used as insecticides or as weed killers have the ability to cause gene mutations just as radiation does. One would think, however, that from such a cause the mutations would not necessarily have been identical, but would rather produce a variety of abnormalities. This is, however, an extremely interesting thing and I shall appreciate all you can tell me not only as to the observed facts, but also your own interpretation.

Sincerely,

To Grace Barstow Murphy

November 16, 1959

Dear Mrs. Murphy,

The enclosures are sent you in haste because I hope the Long Island group, through you, will express publicly its approval of Sec. Flemming's attempt to curb the sprayers in at least one small segment of their activity! The fine letters from Westwood and Tugwell (enclosed) seem to me just the sort of support Mr. Flemming desperately needs. If he is beaten down on this, I don't see how he can have the courage to take other needed actions.

Since some press reports have been garbled, I am sending you the official Food and Drug release—Flemming's original statement. From it you will learn that the cranberry has been clearly in violation of the law several times. They used the spray in 1957—a year before it had been registered by Agriculture for use at all. Then in 1958 it was registered for post harvest use *only*—but they, or at least some of the growers, used it before harvest. When this was discovered, F&D asked the industry not to ship from the suspect areas until berries could be tested. This they agreed to—then broke the agreement and shipped anyway! This is what broke the story open.

There is to be a public hearing in Washington Wednesday. From my conversations with F&D officials, I know any expressions of support would be most welcome. Knowing that you will understand as I do that this particular incident is just a skirmish in a battle of far wider implications, I hope you will feel it appropriate to send a telegram to Flemming. I think I am getting such action from Audubon, National Wildlife, and others. It will all help.

Heavens—a new crisis every day, isn't there?

My warm regards, always

To Dorothy Freeman

Thursday, November 19

Dearest,

So much I'd like to write, but I won't attempt to cover much today. Perhaps over the week-end. Of course I was tremendously interested in all you wrote of the Audubon Society meeting. I had known about it, and in fact had reasons for interest in it on the basis of my conversations here with Allen Morgan, and later correspondence with him. He is the new Russ Mason, you know—though I doubt he is quite a Russ Mason. I think I shall write him now and ask for a copy of Dr. Sears's paper. Yes, I imagine—or at least I surely hope—I know most of the facts that would be presented, but it is interesting to see the emphasis and interpretations given.

Yesterday I attended the big Cranberry Meeting, which was most interesting. Flemming certainly went way up in my estimation; he handled the whole thing with such quiet dignity and courtesy, but they didn't ever put anything over on him! At the conclusion of a speaker's statement, he would gently pick up the very thing I'd been hoping he would demolish. I think he's going to stand firm. I was delighted by the support he was given from various independent organizations that had asked to be heard. And I was impressed by the fact that the tone of industry spokesmen was on the whole mild and conciliatory. I had been told privately that industry heads had been shocked when they really began to look into the situation, and their attitude yesterday bore that out. A Congressman from Oregon—whom I had expected to be full of pleas for the industry, took quite a different line. He said that, sad as this might be for the industry, it has served to point up the whole enormous problem of contamination of food by pesticides and other chemicals, and called for better laws and more rigid enforcement. I was so impressed that I called his office later for a copy of the statement. You will laugh, as I did, at the response. When it became necessary to give my name for the mailing of the document, there was a sudden brief silence, then—"You frighten me—could you have written *The Sea Around Us?*" I assured her that, though I had, I was quite harmless! As you may have heard, the industry's prize exhibit was a doctor from Tufts who says he uses the chemical clinically in treatment of the thyroid and it could not be harmful. Oh dear—his testimony can be shot so full of holes as to be absolutely worthless, and the disheartening thing is that he must know this full well, if he is the great specialist they say he is. For those who might not see the holes, however, it was good that Consumer's Union had sent a man who gave the opinion of a leading cancer specialist of the U. of Chicago—exactly the opposite of the man from Tufts. And my files are full of things that refute the Tufts man!

I believe Roger Peterson showed his flamingo pictures here— either that or he's going to this winter. Of course we are his "home" group. But I never have time or opportunity to go now so I'm not sure. Of course I would love to have seen the slides, too. Perhaps I should ask Mr. Morgan for a list of the photographers represented, so that when I ever get around to collecting the actual pictures for the book I can investigate them. Up to now, as you know, we have been concerned only in equivalent samples as a basis of cost estimates.

If I'm to send this with Ida, as I planned, I must stop now. It seems to me the weeks now are about 2 or 3 days long— no more, ever. I wish I could find some way to salvage the week-ends for work, but I never accomplish much then.

All love, dear, to the others and to *you*.

Rachel

1959

To Morton S. Biskind

December 3, 1959

Dear Dr. Biskind:

Thank you for your note and for passing on the informative letter from Mrs. Hunter. And I am late in thanking you for the material on Food and Drug's recent and somewhat astonishing action on heptachlor. There must be a great deal going on behind the scenes. Mr. Rankin, the Deputy Commissioner in charge of Pesticides, had told me about a week in advance that the heptachlor action was to be taken. When the announcement was made, I was perplexed by the comment that the action was taken on the basis of "new evidence" re the conversion to the epoxide. I called Mr. Rankin and mentioned the fact that of course this evidence was by no means new (antedating the setting of the original tolerance). His reply was somewhat fuzzily to the effect that of course they had to confirm the work that had been done! Now there seems to be a mighty, though belated, stirring of activity about stilbestrol, residues of penicillin and pesticides in milk, etc. Marjorie Spock says you told her of the appointment of a toxicologist to the Public Health staff—a retired man who is fully aware of all these hazards and may be having something to do with all this. She could not remember his name. I wonder if you would drop me a note very soon on this—perhaps I might very profitably be in touch with him, or at least be on the alert for results of his activity.

Recently some of my thinking on all this has begun to fit together like the pieces of a jig-saw puzzle. As a result, I now plan to devote an entire chapter to the cancer hazards related to pesticides. At one point I thought the evidence rather shaky; now I find it is extremely strong. Through your guidance into this field, I have been pursuing all the evidence about the disturbance of enzyme systems, uncoupling of phosphorylation, etc. I have been fascinated to discover recently that, according to Greenstein, von Euler, and others, defective operation of the cytochrome system is "the main characteristic of the respiratory mechanism of the cancerous cell." Now that I have tabulated all the pesticidal chemicals that disturb cytochrome oxidase, a great light is breaking in my mind. I have never seen this connection fully pursued. Perhaps I shall be "a fool rushing in" where angels would not tread, but I think this possible mechanism should be suggested as one means by which the cell may be converted to the cancerous way of life. Also, of course, Warburg leads one to realize that uncoupling phosphorylation from respiration may lead to the same end. Here again, the list of chemicals disturbing phosphorylation is impressive. And

Weinbach's work on pentachlorophenol (to which you led me) seems to tie up with a paper presented at the American Chemical Society's recent meetings, in which, among other things, it was pointed out that fermentation takes place outside the mitochondria, normal oxidation inside. Weinbach, you know, found damage to the mitochondria by pentachlorophenol. To me, it all seems to fit together. When I finish putting down all of this evidence, I would be tremendously grateful if you would look it over and see if it seems to make sense from the professional viewpoint.

As to the references in testicular atrophy, I have several: Draize et al, 1944; Woodard et al, 1944; Nelson et al, 1944; as well as Burlington and Lindemann on chickens. I shall be pleased to add to these to the extent that you can supply other references.

No, I missed that particular *Times* editorial. I shall look it up. With kindest regards, always.

To Dorothy Freeman

Monday night

Dear,

Yesterday's talk was such a treat—perhaps all the more because nowadays it is rare. It was all so satisfying and good. I've had a big day and am weary—not trying to work tonight. Was up until one last night preparing to get the most out of my interviews—then up early so I could be dressed and breakfasted too before Ida came, for I needed time in the library with Mrs. D. before my 10 o'clock appointment. That was *wonderful*! I had expected what I wanted to ask Dr. Weinbach would take no more than half an hour; instead I was with him more than 2 hours! He is the kind whose eyes shine with the wonder and excitement of what he is doing, and—he gave me a 2-hour seminar lecture on biological oxidations! The entire subject of energy-production through phosphorylation is something developed long after my Hopkins days—since this really basic process is disrupted by some of these chemicals, I have to bring to life for my readers what goes on inside the cell and even inside the mitochondria! (We used to see them, knew their function was "unknown," but I can't remember we even wondered. Now they're known to be the basis of practically everything!) Well, I decided after all that to skip lunch, so Mrs. D. got me a package of cheese crackers and a chocolate bar while I ran through all the volumes she had assembled and marked passages for her to copy. Then an hour with Dr. Hueper, mostly picking up loose ends and clarifying, but very good! I must say I'll be glad to have a day at home, working in solitude except for Ida's ministrations, tomorrow. Dr. H. says he thinks the time now is right for the book for people are beginning to want the facts—sooner would have been premature, he thinks.

I can't remember the dates of the family's visit, though I do know your Christmas is the 19th. I do hope it can be a happy time for you, dear. You'll have fun with the children, I know.

Thought you might like Miss Worrell's 50-year reminiscences now. No hurry about returning any of this, of course.

All for now, dear. Very much love—

Rachel

c. December 7, 1959

To Marjorie Spock and Polly Richards

April 12

Dear Marjorie and Polly,

What seething activity! It makes me breathless even at this distance. But it is good you have the interest and joy of the new place to occupy your minds now.

I'm not sure the new address is in effect now, so will send this to the old, sure you will be picking up mail there for some time.

My hospital adventure turned into another set-back of some magnitude, wrecking my tight work schedule for the spring. Viruses—my own and Roger's—delayed my operation until last Monday, April 4. There were two tumors in the left breast. One was benign—the other suspicious enough to require a radical mastectomy. However, I made amazing progress and talked my way out of the hospital on Saturday—of course coming home to bed and good care, but I was desperately eager to get home to Roger. So far, I'm just concentrating on getting well, and the only possible attitude seems to be to feel thankful this discovery was made so early. I think there need be no apprehension for the future.

I am giving details to special friends like you—not to others, but I suppose it's a futile effort to keep one's private affairs private. Somehow I have no wish to read of my ailments in literary gossip columns. Too much comfort to the chemical companies!

You know my feelings about the Court action without my expressing them!

Love to you both—

Rachel

1960

To C. Girard Davidson

June 8, 1960

Mr. C. Girard Davidson
Democratic National Committeeman for Oregon
Equitable Building
Portland 4, Oregon

Dear Mr. Davidson:

This is in reply to your letter of May 17 enclosing an outline of topics for consideration by the Natural Resources Committee.

First let me say that I consider it a privilege to serve as a member of this Committee. As I explained to Mr. Butler on receiving his invitation to membership, my participation will have to be limited, because of temporary health problems and especially because of very heavy writing commitments. However, I do sincerely wish to do all I can.

You will understand that it is difficult to tell from an outline alone how fully or in what direction the various topics would be developed. My principal comments at this time relate to amplifications under the heading Pollution Control and to additional topics under the heading Conservation and Recreation.

Pollution Control: This seems to me one of the most important points and I hope it will serve as the basis of a strong plank in the platform. It is now believed by leading authorities in the field of medicine and public health that the diseases that now threaten mankind most seriously are of environmental origin, arising from elements introduced into air, water, food as by-products of industrial and agricultural operations. The problem of air pollution and its relation to cancer and other diseases has gained widespread recognition only within the past few years, and for that matter some of the most significant findings have come to light only within recent months. The problem of water pollution is growing in seriousness and is now complicated by the contamination of water supplies by chemical pesticides, detergents, etc. But federal neglect of the problem increases.

In discussing radioactive contamination—certainly a major point—I hope we can deal specifically with the problem of such contamination of the sea. Almost all oceanographers view present procedures as highly dangerous and irresponsible, based on the most fragmentary knowledge of the circulation of deep waters (and flatly disregarding what we do know) and ignoring the concentration of radioactive elements by marine organisms and their consequent redistribution. This whole problem is one of extreme urgency, requiring the establishment of a policy guided by the facts that are now known to oceanographers, and (perhaps even more importantly) recognizing that much remains to be learned. Once the widespread contamination of the sea has been accomplished—and we seem to be well on the way—it will be too late to apply the findings of research.

The problem of chemical contamination of the environment is comparable to that of radioactivity and in my opinion is equally serious. The chief source of such contamination is chemical insecticides and weed killers, being used in annually increasing quantities and applied on a massive scale (often by air) that distributes the poisons over areas not even intended as targets. Every major river system now contains traces of the chlorinated hydrocarbon insecticides. Residues in soil are so longlasting that a leading specialist on soil biology has warned of future impairment of agricultural soils. All commonly consumed foods now carry residues of pesticides. The record of the present Administration is particularly bad, especially as reflected in the Agriculture Department's sponsoring of huge programs which are being carried out in the absence of precise knowledge of the effect of the poisons on wildlife, livestock, or human beings, and often over the protests of the States concerned and of private and public conservation organizations. We should take a positive stand for (a) consultation with all agencies or interests concerned before a program is undertaken, (b) suspension of any program unless adequate research has been conducted into the effects of the chemicals under the proposed conditions of use, (c) change of emphasis, away from research designed to develop new and more dangerous poisons, to research aimed at alternative methods as substitutes for chemical control. Several promising methods of this sort have already been developed, but under this Administration such research projects are starved for funds.

Under Conservation and Recreation I should like to see a strong statement of the need to preserve natural areas, with direct support of the principles embodied in the Wilderness Bill now awaiting action by the Senate Committee on Interior and Insular Affairs. My own impression of the present outline on Conservation is that the emphasis is too strongly on recreation. I don't mean to underestimate the importance of setting aside adequate areas for recreation, but I think this must be balanced by clear recognition of the fundamental need to preserve the undisturbed balance of forests, waters and life in areas secure from invasion by conflicting interests. I feel very strongly on this point and should like to write you further on it.

If you are to be in Washington before June 20, I hope there may be an opportunity to see you. I leave for the summer about that time.

Sincerely yours,

To Marjorie Spock

September 27, 1960

Dear Marjorie:

I am sure you have interpreted my long silence correctly. I have just been keeping correspondence to a minimum so that I might devote every possible moment to work. Everything in your last letter (dated August 30th) was extremely interesting. That item about the article in the Rochester paper stating that 200 agricultural workers are cancer patients at the Roswell Memorial Institute is certainly extraordinary and I am writing Mr.

King to get further leads that I may follow. I meant to ask Dr. Hueper what he knew of this when I saw him recently, but we had so much else to talk about that I forgot. Of course if you have any further details, I should be glad to hear them.

You also mentioned something about an electric pesticide detector developed at Stanford which you thought you had already told me about. I am sure this is the first time I have heard of it and if you have any idea of the right person or department to write to at Stanford, I should like to follow it up.

Thank you for the information about Ryanicide. From what you tell me, I suppose the drug from which it is derived is related to rotenone, or at least is similar in its action. If so, I would like to suggest in passing that you exercise reasonable caution in handling it. These botanical insecticides are of course greatly to be preferred to the chlorinated hydrocarbons and as far as I know, do not produce chronic poisoning. However, they are poisons and as such, should be handled with proper respect. Just think what a scandal it would be if you should be hospitalized as a result of using an insecticide of this type!

You will remember that I wrote you during the summer in regard to having some work done by the Laboratory of Industrial Hygiene. I wrote to Colonel Johnson as you suggested, and outlined the problem, which was to take a sample of heptachlor epoxide and perform what apparently is a very simple test to determine whether it alkylates readily. Colonel Johnson thought my intention was that he should first prepare a sample of epoxide from the parent heptachlor and said that this would be quite a long and involved process, and he would need certain further information. I told him this was not my intention; that I was sure he could obtain a reference sample of the epoxide from one of the manufacturers of heptachlor, and suggested several possible sources. This was in July, and I have not heard from him again and do not believe he is interested in doing the work. If you are in touch with Dr. Pfeiffer I wonder if you would ask him if this could be done rather easily and inexpensively? All that the test involves is to take a sample of heptachlor epoxide and shake it up with some sodium thiosulfate. If the resulting solution becomes alkaline within ½ hour or so, this establishes the fact that heptachlor epoxide is not stable but alkylates readily. If this is true, the fact has some significance in relation to a potential cancer hazard.

I was sorry that the summer came to an end before I could find time and opportunity to get over to Macmahan to see your Mother. Do give her my love when next you write to her. Perhaps in the meantime, you and Polly have been in Maine, or perhaps you are going soon. We did hate so dreadfully to leave. I have never had less time to enjoy the out of doors there, but I don't think I have ever loved it more in any other summer.

We came back here about the 9th of September and Roger has now been in school for a couple of weeks. I am working furiously, and while I can see progress and feel that the writing now has real momentum, I still do not see the end.

I am belatedly returning this letter from the chairman of the Illinois Citizens Committee on Pesticides. Perhaps I shall write her and find out a little more about her organization. She certainly sounds alert. I am keeping the letter from Margaret Louise Hill from Austin, since it was marked a copy, but please let me know if you would like to have it back.

Roger and I send our love to both of you,

As ever,

To Dorothy Freeman

Wednesday, October 12

My dearest one,

This life is so different from anything I've known before in all the years since we met. Just no time for letters—no time to read—no time for anything, it seems, unless it is somehow related to the great projects that are uncompleted. I suppose as I grow older and become more aware that life is not only uncertain but short at best, the sense of urgency grows to press on with the things I need to say—things that may be less important than I think, but to me at least it is necessary that they be said. But I do think back with a bit of nostalgia, darling, to the times when this urgency seemed less. I like to *remember* at least, days when I spent a whole morning (or more!) writing a leisurely letter to you. As I addressed your letter this morning I wondered with a smile how many times I've written that address—and how many times my heart has leaped (as it still does) to the sight of the familiar handwriting in return.

And right now I do long for even a day with you, when you could say the things you want to say, and I could tell you some of the little things that never get written, and that I'd rather talk about, anyway.

A mockingbird is pouring out *his* heart to the whole world just now, as I sit in the study, writing. One day recently I recorded some minutes of song coming down the chimney.

Darling, I continue to be so relieved that this long ordeal is over. All those weeks I was so aware that each visit to your mother must just tear your heart. And how I know the feeling of dreading to waken each morning to that leaden weight. There will be sadness now, I know, but of such a different kind. Now, for the first time since Stan's retirement, you should begin to have some of the enjoyment of leisure that you deserve. You must feel so much better about him than you did even at the beginning of the summer. Now begin to take care of yourself, too, darling. Because I love you—remember.

Rachel

1960

To Clarence Cottam

11701 Berwick Road, Silver Spring, Maryland
January 4, 1961

Dr. Clarence Cottam
Rob and Bessie Welder Wildlife Foundation
Sinton, Texas

Dear Clarence:

I had hoped to attend the Water Pollution Conference and to see you there, but that week I was laid low with an attack of flu. My secretary, Mrs. Davis, attended part of it for me and brought me many of the papers, but it was a disappointment to miss seeing you and others.

My general impression of Dr. Decker (through publications —I have never met him, I believe) is that his efforts are always devoted to promoting further use of pesticides and to decrying any attempts to demonstrate possible harm from their use. I remember one bitter attack (in the *Journal of Agriculture and Food Chemistry* about a year ago) on the F.D.A. for maintaining its position that milk should be pesticide-free. I think this may have been what I mentioned to you. I'll try to look up this and perhaps other papers of his soon and perhaps find quotes to document what I mean. I'm skeptical enough to doubt you will convert him but it's a worthy effort.

I am glad to have the copy of your paper and in a quick scanning of it (it came only today) found one thing I'd like to ask you about. In discussing Clear Lake you mention that plankton organisms had accumulated DDD (5.3 ppm). My last information on Clear Lake is about a year old, but at that time I was told that the plankton studies had not been made. Evidently this gap has now been filled in. Can you give me the source of this statement?

Also, you refer to the situation at Tule and Klamath Lakes. Where can I find out more about this?

It was such a great pleasure to have you here when you were in Washington last fall and I hope we may soon have a chance for another visit—even though you might then discover that my culinary skill is really pretty feeble.

My warm good wishes for the New Year to you and your family.

Sincerely
Rachel

To Paul Brooks

11701 Berwick Road, Silver Spring, Maryland
June 26, 1961

Mr. Paul Brooks
Houghton Mifflin Co.
2 Park Street
Boston, Mass.

Dear Paul:
You have asked my opinion about chemical spraying on a community basis for the control of mosquitoes. I can say

without hesitation that such spraying operations are not only dangerous but they are also largely self-defeating. For the latter reason (as well as for others) they make very little sense.

The reason I say they are self-defeating is that insect populations very quickly develop resistance to chemical sprays. This nullifies all we are trying to accomplish by spraying. In the end we find we have gone to considerable expense and trouble for nothing. We have killed off or endangered fish, birds, and other wildlife, contaminated vegetables and fruits, damaged shrubbery and flowers, and introduced poisonous chemicals into soil and water supplies. After all this we find the mosquito has the last laugh, for while we have been progressively poisoning our own environment the mosquito has been breeding a superior race composed of individuals that are immune to chemical attack.

This phenomenon is exactly comparable to what is happening among bacterial populations subjected to antibiotics; it represents the typical ability of living organisms to adapt to unfavorable conditions. We are now confronted with the frightening fact that 100 species of insects have become completely resistant to one or more types of insecticides. These are divided about equally between agricultural pests and carriers of disease. Efforts to keep ahead of the insects by developing new and more toxic chemicals have gotten us nowhere. It is like a treadmill, for each new product is effective for only a few years—some times only a few months. Connecticut entomologists report that on at least one or two insect pests the *last available material* is now being used. The World Health Organization considers insect resistance to chemicals as one of our most serious world health problems, and public health officials generally fear that because of this situation we may lose control of insect-borne disease.

The common mosquitoes of the genus *Culex* develop resistance very quickly. DDT resistance among common house mosquitoes was detected in Boston and its suburbs several years ago. From DDT resistance it is only a step to resistance to dieldrin, chlordane, and all the other available materials. The irony of the situation is that the harder we spray, the more rapidly we bring on the day when nothing we can use will reduce the mosquito populations.

Spraying, if it is to be done at all, must be kept to an absolute minimum if we are to buy time in which to work out non-chemical methods of control. Again the analogy with human medicine is appropriate. The clamor of wonder drugs blinded hospital personnel for years to the prosaic fact that there is no substitute for the drudgery of meticulous aseptic procedures. The resistant strains of bacteria that now run rampant in many hospitals are the result. In the field of mosquito control the myth of the supposedly quick and easy method of wonder chemicals leads to equally dire results. Again there is no substitute for the unglamorous methods of good community housekeeping: eliminating standing water in ditches, under porches, in basements; turning tubs, pails, boats, or wheelbarrows upside-down, draining clogged gutters, stocking ornamental pools with fish.

By these methods we can certainly keep mosquito populations within tolerable bounds for the present. Looking toward the future, a number of highly promising non-chemical methods of insect control are now in the stage of development and testing. These promise safer, more efficient control without the hazards and drawbacks of chemicals. When public opinion becomes sufficiently informed and aroused to demand increased research in this field, we may look toward a completely new era in insect control.

As ever,
Rachel Carson

To Dorothy Freeman

Saturday night, January 6

Darling,

It seems a very long time since I have written an apple—opportunities for writing at all have just been so limited that double letters have been difficult to manage. I'm sure you understand, but I miss it and perhaps you do, too.

I hope you have been safely at home today. If the news is correct, Boston is having an ice storm with very bad driving conditions. Really we haven't had anything too bad so far this year—oh, several bad nights but nothing lasting very long.

Yes I do remember that you came to me a year ago this month and that it was a precious and happy visit in spite of the "undertones." And the premonitions there might have been of what lay ahead! Yes, there is quite a story behind *Silent Spring*, isn't there? Such a catalogue of illnesses! If one were superstitious it would be easy to believe in some malevolent influence at work, determined by some means to keep the book from being finished. Some of the earlier things have been more serious, but I don't think anything has been frustrating and maddening in quite the same way as this iritis. And of course having the end in sight when it struck makes it, in a way, all the worse. I just creep along, a few hours a day. And I know that before I can happily let it go to the printer, there is a tremendous lot of work that only my eyes can do. I have always known I am visual minded, and I've certainly been reminded of it now. Having Jeanne read is of such limited help. I have to see it, and on revision I have to keep going over and over a page—with my eyes!

Of course, darling, your Christmas letter made me very happy. I'm glad that before publication you have come to understand not only why it is important to me, but to the world. And I'm glad all these other things are happening to emphasize that people are ready for the book, and need it now. I, too, think a couple of years ago would have been too soon. But now I know that there are many, many people who are eager to do something and long to be given the facts to fight with.

Last year as I flew home from Cleveland I thought rather deeply, as you might suppose, and I knew then that if my time were to be limited, the thing I wanted above all else was to finish this book. Doing so, not swiftly and easily, but draggingly with the impediments of the arthritis and now the iritis, has been rather like those dreams where one tries to run and can't, or to drive a car and it won't go. But now that it seems I shall somehow make this goal, of course I'm not satisfied—now I want time for the *Help Your Child to Wonder* book, and for the big "Man and Nature" book. Then I suppose I'll have others— if I live to be 90 still wanting to say something!

There is rain on the roof and now and then strong gusts of wind. A cozy night, really, and now I'm going to turn out the light.

Another day, and I'll just finish this briefly so I can mail your letter. Rereading what I wrote last night, I'm afraid it sounds a little in a minor key. It shouldn't. Really, in spite of my complaints, I'm optimistic that the eye will let me stumble along somehow until I "come out into the light."

See the quotation from Camus in Lois's letter. I guess that's what I should strive for now.

One more thing. I'm so happy, darling, that you are feeling again the stirrings of your own urge to write. You know—or I hope you do—how much I hope you will. I hesitate to say much about it, for in this area it has always been my unhappy fate to say the wrong thing so that the effect isn't at all what I mean. Don't let that make you unhappy, dear—it's just a fact which I wish could be changed. But please know that I very much want you to set down in permanent form the lovely things that are in your heart and mind so that others may share them. It will make me deeply happy if you do.

> All my love, darling
> Rachel

1962

To Dorothy Freeman

> Tuesday, January 23
> Under drier without glasses

Dearest,

I'm going to call you tonight and will hope to find you at home. As you know already, last week was rather a momentous one, for I achieved the goal of sending the 15 chapters to Marie—like reaching the last station before the summit of Everest. I also sent her duplicate copies for Mr. Shawn, and learned on Sunday that she had sent them along to him Friday. Last night about 9 o'clock the phone rang and a mild voice said, "This is William Shawn." If I talk to you tonight you will know what he said and I'm sure you can understand what it meant to me. *Shamelessly*, I'll repeat some of his words—"a brilliant achievement"—"you have made it literature" "full of beauty and loveliness and depth of feeling."

And with his remark about publishing in the spring I suddenly feel full of what Lois once called "a happy turbulence"

—aware, of course, of how very much is to be done with last minute checking, etc., but so excited that the time is so close. Of course, this may well be June, and for Paul's sake I hope so—he will be frantic. But all the people who are so eager to have it to work with will snap up copies of the *New Yorker*—and it is needed for the spring and summer season.

I'm to see Dr. Wilber later today, but I can tell you her news tonight. The eye quieted down rapidly after the treatment last week. I have used it sparingly.

> More soon. Now much love—
> Rachel

Darling,

I longed so for you last night to share my thoughts and feelings. It was odd—I really had not been waiting breathlessly for Mr. Shawn's reaction, yet once I had it I knew how very much it meant to me. You know I have the highest regard for his judgment, and suddenly I knew from his reaction that my message would get across. After Roger was asleep I took Jeffie into the study and played the Beethoven violin concerto—one of my favorites, you know. And suddenly the tensions of four years were broken and I got down and put my arms around Jeffie and let the tears come. With his little warm, rough tongue he told me that he understood. I think I let you see last summer what my deeper feelings are about this when I said I could never again listen happily to a thrush song if I had not done all I could. And last night the thoughts of all the birds and other creatures and all the loveliness that is in nature came to me with such a surge of deep happiness, that now I *had* done what I could—I had been able to complete it—now it had its own life!

And those are the thoughts I would have shared had you been here. I wish you were!

> All my love,
> Rachel

1962

To Frank E. Egler

January 29, 1962

Dr. Frank E. Egler
Norfolk, Connecticut

Dear Dr. Egler:

I am quite overwhelmed at the thought of the hours you must have spent in reading and commenting on my manuscript. I would never have dreamed of asking you for so much time and I am more grateful than I can say for the very substantial help you have given me. I feel vastly more secure now that I know you found no actual errors and the chapter will be much better for many of the changes of wording or qualifications you have suggested. In addition, you have mentioned several things that are not now in the chapter, but which I feel need to be added. I am going to list some of these and if you have references easily available, will ask you to send them to me in order to expedite my search for material.

Do you, for example, know of a good account of the ecological history (if that is a reasonable term) of the sage lands—something that will amplify your paragraph stating that the range of sage was less extensive in the time of the Indians and that the grazing of cattle has extended the area of sagelands. One point in your paragraph is a little different from my understanding gained from reading or correspondence with such people as Olaus Murie. They seem to feel that the sage by holding moisture in the soil and providing a degree of shelter, actually favors the development of grass under and around the sage. In other words, if there were no sage there the grass would have a hard time establishing and maintaining itself. Does this sound reasonable? This is more or less what I am driving at in the early part of the chapter in saying or implying that the sprayers are trying to turn these lands into something they never were and probably could not be because of the factors of climate, etc.

Your suggestions about the spraying of timber lands certainly bring up a very important omission. It is odd, but none of this came to my attention until after the chapter was written. Now I am very anxious to get in a few paragraphs about it. Perhaps

I can get on the track of the information by calling the Forest Service, and if you think of any individuals who could give me the picture of what is going on, I expect you could save me some time. If I understand you correctly, they are spraying mixed forests to kill out the hardwoods and preserve the conifers. There was some reference to this from the standpoint of destruction of wildlife food in a paper by Goodrum and Reid at the 1956 Wildlife Conference.

Will you tell me more about the destruction of blueberries. This, I should think, would be a very sore point in Maine and certainly would arouse the ire of all those, both natives and summer visitors, who wander along the roads picking blueberries. What was the Sullivan County incident and what sort of notice did you publish? I shall not refer to this directly if you would rather I wouldn't, but it would help my own understanding to know. Do you mean it is 2,4-D that kills the blueberries? I have had correspondence with a woman who owns blueberry land in Maine and she tells me that 2,4-D is among the chemicals recommended by the Agricultural Extension Service for use on blueberry lands. Can you clarify what appears to be a contradiction?

This is quite a challenge to find a name for your method so that I can avoid a label that you find objectionable. I am sure you do not mind my attributing the method to you and once that is done, perhaps I can just say "selective spraying" instead of the "Egler method."

I should like to see statements of the roadside policies that are being drawn up by the Nature Conservancies and the Boy Scout Association. I should be able to get the former by a call to the headquarters here, but can you suggest an individual whom I might write to in the Boy Scouts?

In regard to your point about the increase in the graminoid weeds following the use of 2,4-D in agriculture, could you suggest one or two key references without spending too much time looking them up? I think this is important.

You have given me a number of new thoughts that I shall want to work into the chapter because I feel they will strengthen it enormously. I can't begin to tell you how grateful I am for your very generous help.

It may be that I shall want to telephone you to check on a few final points before we go to press. Is an evening hour likely to be convenient?

Yours sincerely,

To Lois Crisler

Feb. 8

Dear Lois—

It was good to have your letter yesterday—though it left me still hungry for news of *you*. Don't hesitate to write on account of my eye—I read all the time anyway and a letter is no strain at all.

A note from me probably crossed yours—telling you briefly that I've almost come to the end of the long road. I'm glad Dorothy filled in details I was able to give her by phone. I am, of course, deeply happy about Mr. Shawn's reaction. Paul, too, seems excited about the book. His editorial suggestions, which I now have, are most helpful, as always. Marie is coming this week end. We'll go over it page by page—then I'll make final changes. The tentative time for the *New Yorker* is April. I shall be very busy until after that—indeed, until the book itself is in page proof. By the time I go to Southport—this time without my load of books and papers—I should be indeed free!

I still haven't finished the last chapter!

No, I myself never thought the ugly facts would dominate, and I hope they don't. The beauty of the living world I was trying to save has always been uppermost in my mind—that, and anger at the senseless, brutish things that were being done. I have felt bound by a solemn obligation to do what I could—if I didn't at least try I could never again be happy in nature. But now I can believe I have at least helped a little. It would be unrealistic to believe one book could bring a complete change.

So much I want to write you but there is never time. But please do fill me in on events in your world.

Oh yes—the trip to Scripps! Poor Dr. Hard has been trying so patiently for 7 or 8 years—I couldn't say no again, especially since I shall be through. But "the things I have to say" are really part of the big book to come—the basic theme of man and nature. I hope I'll find enough time to think about it before June. It is a bad time in a way—just at the end of Roger's school, and wanting to get off to Maine, so I shall have to make it a very quick trip.

Now I must close. Write soon—

Love
Rachel

1962

To Dorothy Freeman

Thursday, April 5

Dearest,

I hope you are having as lovely a day as this, though far more wintry, I'm sure. Today I brought in the flowers of the red maple to marvel at as the microscope shows them. You know the two big trees off the corner of the house that intermingle so they appear as one? As one could tell only at this time of year, one is male, one female. It is the female trees that make so vivid a showing now, for the male flowers are delicate by comparison. I also brought in flowers of spice bush—so lovely, too. Yesterday I photographed the maple flowers against a blue sky. How I wish I could do it through the microscope.

Your note written yesterday A.M. is here. I marvel at you—to have preparations for your trip so in order you could write a letter at 6:20 A.M.!!

I thought you might like to read the enclosed—one of the finishing touches! Paul thinks we should use both quotations (as a sort of "motto" on one of the front pages, you know). The E. B. White is too perfect to pass up, and the Keats helps explain the title. We have worked all day on the bibliography and it seems to me we hardly got started! Miss Phillips called today (H.M. editor) and said text will go to printer today or tomorrow. Galleys in 2 or 3 weeks then! It is wonderful not to be going to hosp. for a few days. But the effects of radiation linger on so I'm not quite able to forget it. I wish we had your goldfinches! Have ordered a bluebird house—how I'd love to have it occupied. I don't really *realize* the book is done—it

can't be possible, it seems! This rambling note is to say welcome Home & to carry *much* love—

Rachel

1962

To Dorothy Freeman

Tuesday, May 20
Under drier

Dearest,

Another milestone (or do I mean mill stone?)—I'm mailing
tax data when I leave here. Now on to speech no. 1.

An unexpected diversion last night, but very pleasant and
worthwhile. Clarence Cottam phoned yesterday A.M. on arrival in Washington & invited me to a dinner of Natl. Parks
Assoc. trustees, and a later meeting of the Conservation Forum to be addressed by Justice Douglas. The Justice and Mrs.
D. were at the dinner. He came to me and said "Your book
is tremendous." She said "I'm selling it every place I go." In
his talk at the Forum, which was about Potomac River dams
and the Army Engineers, he digressed to speak of the havoc
wrought by "experts" in other fields, and added, "Everyone
should read Rachel Carson's forthcoming book, *Silent Spring*,
to learn what the chemical engineers are doing to our world."

An odd feeling, darling. It's out on its own now, my fourth
brainchild, and it's beginning to move.

I must hurry on. Oh—a perfectly HUGE box came from
Filene's—what in the world have you sent me? I want to keep
it for Sunday, but am tempted to open it before you call.

Home again—will just add a little and put out for mailman.
Darling, I've been thinking about something I want to mention. You are going back to Maine soon where you'll be seeing people who know me and will perhaps ask about me—the
Bennetts, Mahards, Pinkham family, etc. You know something
of how I feel about this, but probably not the depth of that
feeling. There is no reason even to say I have not been well. If
you want or think you need give any negative report, say I had a
bad time with iritis that delayed my work, but it has cleared up
nicely. And that you *never saw me look better*. Please say that. If

you look at my picture you will know you can say it truthfully. It is what everyone says. I know what happens when even an inkling of the other situation gets out. As last night, scraps of dinner table conversation about poor Senator Neuberger: "You know she had a cancer operation." . . . "They say she's down to 85 pounds" . . . "If you'd see her on the Senate floor you'd know she can't last." That's the sort of thing I couldn't bear, and the reason I have told so few people. Whispers about a private individual might not go far; about an author-in-the-news they go like wildfire. So let people think I am as well as I look. As to those few you have felt it necessary to tell, will you please try to impress on them how I feel about it?

Almost every day Roger says he can hardly wait to get to Maine. For me it will surely seem a haven of escape. But really, although at the end of each task I'm so exhausted I think I can never get up again, I'm fresh and all right by morning.

Last night, finishing my taxes after my return from dinner, it was one o'clock before I turned out the light. Before I slept a whippoorwill began singing so close I went to the window and looked out in the bright moonlight. Could see nothing but felt as though I could touch it. So I put on robe and slippers and went out. It was on the study roof—of course I couldn't see it, but stood right below it, hearing a funny little preliminary note one never hears at a distance. So that's my bird story.

Now my Wayside irises are a glory to behold—I never saw such thrilling ones. And the little Dutch irises are like orchids.

Perhaps you will have this before we talk. Darling, you *are* going to my house, aren't you? It would truly distress me if you didn't.

All my love
Rachel

1962

Of Man and the Stream of Time
(Scripps College, Claremont, California, June 12, 1962)

As I was carried here so swiftly across the continent by a jet airliner, it occurred to me that I have really been on the way for ten years, for it was that long ago that your President first invited me to come to Scripps College. Through the intervening years, he has renewed that invitation with infinite patience and courtesy. Now at last circumstances have allowed me to accept, and I am very happy to be here.

Had I come ten years ago, I am not certain what I would have talked about. But as I have lived and, I hope, learned, as I have reflected upon the problems that crowd in upon us today, one stands out in my mind as having such vast importance that I want to discuss it with you now.

I wish to speak today of man's relation to nature and more specifically of man's attitude toward nature. A generation ago this would perhaps have been an academic subject of little interest to any but philosophers. Today it is a subject of immediate and sometimes terrifying relevance.

The word Nature has many and varied connotations, but for the present theme I like this definition: "Nature is the part of the world that man did not make." You who have spent your undergraduate years here at Scripps have been exceptionally fortunate, living in the midst of beauty and comforts and conveniences that *are* creations of man—yet always in the background having the majestic and beautiful mountains to remind you of an older and vaster world—a world that man did not make.

Man has long talked somewhat arrogantly about the conquest of nature; now he has the power to achieve his boast. It is our misfortune—it may well be our final tragedy—that this power has not been tempered with wisdom, but has been marked by irresponsibility; that there is all too little awareness that man is *part* of nature, and that the price of conquest may well be the destruction of man himself.

Measured against the vast backdrop of geologic time, the whole era of man seems but a moment—but how portentous a moment! It was only within the past million years or so that the race of man arose. Who could have foretold that this being, who walked upright and no longer lived in trees, who lurked in caves, hiding in fear from the great beasts who shared his world—who could have guessed that he would one day have in his hands the power to change the very nature of the earth—the power of life and death over so many of its creatures? Who could have foretold that the brain that was developing behind those heavy brow ridges would allow him to accomplish things no other creature had achieved—but would not at the same time endow him with wisdom so to control his activities that he would not bring destruction upon himself?

I like the way E. B. White has summed it up in his usual inimitable style. "I am pessimistic about the human race," said Mr. White, "because it is too ingenious for its own good. Our approach to nature is to beat it into submission. We would stand a better chance of survival if we accommodated ourselves to this planet and viewed it appreciatively instead of skeptically and dictatorially."

Our attitude toward nature has changed with time, in ways that I can only suggest here. Primitive men, confronted with the awesome forces of nature, reacted in fear of what they did not understand. They peopled the dark and brooding forests with supernatural beings. Looking out on the sea that extended to an unknown horizon, they imagined a dreadful brink lying beneath fog and gathering darkness; they pictured vast abysses waiting to suck the traveler down into a bottomless gulf.

Only a few centuries have passed since those pre-Columbian days, yet today our whole earth has become only another shore from which we look out across the dark ocean of space, uncertain what we shall find when we sail out among the stars, but like the Norsemen and the Polynesians of old, lured by the very challenge of the unknown.

Between the time of those early voyages into unknown seas and the present we can trace an enormous and fateful change. It is good that fear and superstition have largely been replaced by knowledge, but we would be on safer ground today if the knowledge had been accompanied by humility instead of arrogance.

In the western world our thinking has for many centuries been dominated by the Jewish-Christian concept of man's relation to nature, in which man is regarded as the master of all the earth's inhabitants. Out of this there easily grew the thought that everything on earth—animate or inanimate, animal, vegetable, or mineral—and indeed the earth itself—had been created expressly for man.

John Muir, who knew and loved the California mountains, has described this naive view of nature with biting wit: "A numerous class of men are painfully astonished whenever they find anything, living or dead, in all God's universe, which they cannot eat or render in some way what they call useful to themselves. . . . Whales are storehouses of oil for us, to help out the stars in lighting our dark ways until the discovery of the Pennsylvania oil wells. Among plants, hemp is a case of evident destination for ships' rigging, wrapping packages, and hanging the wicked."

So Muir, with his pen dipped in acid, many years ago pointed out the incredible absurdity of such views. But I am not certain that in spite of all our modern learning and sophistication, we have actually progressed far beyond the self-oriented philosophy of the Victorians. I fear that these ideas still lurk about, showing themselves boldly and openly at times, at others skulking about in the shadows of the subconscious.

I have met them frequently, as I have pointed out some exquisite creature of the tide pools to a chance companion. "What is it for?" he may ask, and he is obviously disappointed if I can't assure him that it can be eaten or at least made into some bauble to be sold in a shop.

But how is one to assign a value to the exquisite flower-like hydroids reflected in the still mirror of a tide pool? Who can place in one pan of some cosmic scales the trinkets of modern civilization and in the other the song of a thrush in the windless twilight?

Now I have dwelt at some length on the fallacious idea of a world arranged for man's use and convenience, but I have done so because I am convinced that these notions—the legacy of an earlier day—are at the root of some of our most critical problems. We still talk in terms of "conquest"—whether it be of the insect world or of the mysterious world of space. We still have not become mature enough to see ourselves as a very tiny part of a vast and incredible universe, a universe that is distinguished above all else by a mysterious and wonderful unity that we flout at our peril.

Poets often have a perception that gives their words the validity of science. So the English poet Francis Thompson said nearly a century ago,

> Thou canst not stir a flower
> Without troubling of a star.

But the poet's insight has not become part of general knowledge.

Man's attitude toward nature is today critically important, simply because of his new-found power to destroy it. For a good many years there has been an excellent organization known as The International Union for the Protection of Nature. I clearly remember that in the days before Hiroshima I used to wonder whether nature—nature in the broadest context of the word—actually needed protection from man. Surely the sea was inviolate and forever beyond man's power to change it. Surely the vast cycles by which water is drawn up into the clouds to return again to the earth could never be touched. And just as surely the vast tides of life—the migrating birds—would continue to ebb and flow over the continents, marking the passage of the seasons.

But I was wrong. Even these things, that seemed to belong to the eternal verities, are not only threatened but have already felt the destroying hand of man.

Today we use the sea as a dumping ground for radioactive wastes, which then enter into the vast and uncontrollable movements of ocean waters through the deep basins, to turn up no one knows where. . . .

The once beneficent rains are now an instrument to bring down from the atmosphere the deadly products of nuclear explosions. Water, perhaps our most precious natural resource, is used and misused at a reckless rate. Our streams are fouled with an incredible assortment of wastes—domestic, chemical, radioactive, so that our planet, though dominated by seas that envelop three-fourths of its surface, is rapidly becoming a thirsty world.

We now wage war on other organisms, turning against them all the terrible armaments of modern chemistry, and we assume a right to push whole species over the brink of extinction. This is a far cry from the philosophy of that man of peace, Albert Schweitzer—the philosophy of "reverence for life." Although all the world honors Dr. Schweitzer, I am afraid we do not follow him.

So nature does indeed need protection from man; but man, too, needs protection from his own acts, for he is part of the living world. His war against nature is inevitably a war against himself. His heedless and destructive acts enter into the vast cycles of the earth, and in time return to him.

Through all this problem there runs a constant theme, and the theme is the flowing stream of time, unhurried, unmindful of man's restless and feverish pace. It is made up of geologic events, that have created mountains and worn them away, that have brought the seas out of their basins, to flood the continents and then retreat. But even more importantly it is made up of biological events, that represent that all-important adjustment of living protoplasm to the conditions of the external world. What we are today represents an adjustment achieved over the millions and hundreds of millions of years. There have always been elements in the environment that were hostile to living things—extremes of temperature, background radiation in rocks and atmosphere, toxic elements in the earth and sea. But over the long ages of time, life has reached an accommodation, a balance.

Now we are far on the way to upsetting this balance by creating an artificial environment—an environment consisting to an ever increasing extent of things that "man has made." The radiation to which we must adjust if we are to survive is no longer simply the natural background radiation of rocks and sunlight, it is the result of our tampering with the atom. In the same way, wholly new chemicals are emerging from the laboratories—an astounding, bewildering array of them. All of these things are being introduced into our environment at a rapid rate. There simply is no time for living protoplasm to adjust to them.

In 1955 a group of 70 scientists met at Princeton University to consider man's role in changing the face of the earth. They produced a volume of nearly 1200 pages devoted to changes that range from the first use of fire to urban sprawl. It is an astounding record. This is not to say, of course, that all the changes have been undesirable. But the distinguishing feature of man's activities is that they have almost always been undertaken from the narrow viewpoint of short-range gain, without considering either their impact on the earth or their long-range effect upon ourselves.

They have been distinguished, also, by a curious unwillingness to be guided by the knowledge that is available in certain areas of science. I mean especially the knowledge of biologists, of ecologists, of geneticists, all of whom have special areas of competence that should allow them to predict the effect of our actions on living creatures, including, of course, man himself.

This is an age that has produced floods of how-to-do-it books, and it is also an age of how-to-do-it science. It is, in other words, the age of technology, in which if we know *how* to do something, we do it without pausing to inquire whether we *should*. We know how to split the atom, and how to use its energy in peace and war, and so we proceed with preparations to do so, as if acting under some blind compulsion; even though the geneticists tell us that by our actions in this atomic age we are endangering not only ourselves but the integrity of the human germ plasm.

Instead of always trying to impose our will on Nature we should sometimes be quiet and listen to what she has to tell us. If we did so I am sure we would gain a new perspective on our own feverish lives. We might even see the folly and the madness of a world in which half of mankind is busily preparing to destroy the other half and to reduce our whole planet to radioactive ashes in the doing. We might gain what the English essayist Tomlinson called "a hint of a reality, hitherto fabulous, of a truth that may be everlasting, yet is contrary to all our experience," for "our earth may be a far better place than we have yet discovered."

I wish I could stand before you and say that my own generation had brought strength and meaning to man's relation to nature, that we had looked upon the majesty and beauty and terror of the earth we inhabit and learned wisdom and humility. Alas, this cannot be said, for it is we who have brought into being a fateful and destructive power.

But the stream of time moves forward and mankind moves with it. Your generation must come to terms with the environment. Your generation must face realities instead of taking refuge in ignorance and evasion of truth. Yours is a grave and a sobering responsibility, but it is also a shining opportunity. You go out into a world where mankind is challenged, as it has never been challenged before, to prove its maturity and its mastery—not of nature, but of itself. Therein lies our hope and our destiny. "In today already walks tomorrow."

Scripps College Bulletin, July 1962

Form Letter to Correspondents

August 2, 1962

Dear   :

I am glad to know you enjoyed my series in *The New Yorker*. I have read your thoughtful letter with much interest.

You ask what you can do. At the moment I think the most helpful thing is for people to express their concern in letters to the appropriate government departments and to their congressmen. Judging by letters and clippings I receive, there is already a strong tide of public protest and I hope this may lead to some constructive action.

In addition to this, I think that people must be alert to ill planned and unnecessary spraying activities in their own communities and should demand thorough justification of all such programs.

Thank you very much for writing to me.

Sincerely yours,

On the Reception of Silent Spring

By what may be the fortunate chance that I am a slow writer, enjoying the stimulating pursuit of research far more than the drudgery of turning out manuscript, *Silent Spring* was published about two years later than my original plans called for. For many reasons, the climate of 1962 seems to have been far more favorable for its reception than that of 1960, or indeed of any earlier year. This is, I think, because a variety of events in recent months have brought people to an uneasy realization that mankind is in many respects doing very badly in his self-imposed role of master of this planet. They have been shocked into skepticism of our eagerness to use new products without full knowledge of their potential for harm, and of our urge to adopt quick and easy solutions (which may be temporary and quite unsound) for whatever problem besets us, be it inability to sleep or dandelions in the lawn.

The problem I dealt with in *Silent Spring* is not an isolated one. The excessive and ill-advised use of chemical pesticides is merely one part of a sorry whole—the reckless pollution of our living world with harmful and dangerous substances. Until very recently, the average citizen assumed that "someone" was looking after these matters and that some little understood but confidently relied upon safeguards stood like shields between his person and any harm. Now he has experienced, from several different directions, a rather rude shattering of these beliefs.

If burning nostrils and bronchial distress do not inform him directly of the menace of air pollution, he has only to read the daily newspaper to learn of the rate at which mankind is using the once-pure air as a dumping place for radiological and chemical refuse. When white froth issues from his kitchen faucet instead of clear water he is reminded of the alarming facts of water pollution by detergents, poisons, and wastes of all kinds.

As weapons testing continues, even the citizen with no scientific training whatever has discovered that fallout levels, now reported routinely in the morning newspaper along with the weather, have an ominous meaning not only for himself but for his children and generations to come.

Almost simultaneously with the publication of *Silent Spring*, the problem of drug safety and drug control, which had been simmering in the press for many months, reached its shocking culmination in the thalidomide tragedy.

The freewheeling use of highly dangerous chemicals for situations ranging all the way from a roach on the kitchen floor to an infestation of budworms in a vast forest is another piece in this jig-saw puzzle, related to and affected by all the others. Judging by the strong and quite wonderful response that is coming to me in letters from all over the country, the public understands this relationship very well. It suddenly understands, also, that these problems do not correct themselves. What I read in these letters is not any reaction of panic, but rather a firm determination to bring under control the abuses I reported in *Silent Spring*.

December 2, 1962

Speech to the Women's National Press Club
(Washington, D.C., December 5, 1962)

My TEXT THIS afternoon is taken from the *Globe Times* of Bethlehem, Pa., a news item in the issue of October 12th. After describing in detail the adverse reactions to *Silent Spring* of the farm bureaus in two Pennsylvania counties, the reporter continued: "No one in either county farm office who was talked to today had read the book, but all disapproved of it heartily."

This sums up very neatly the background of much of the noisier comment that has been heard in this unquiet autumn following the publication of *Silent Spring*. In the words of an editorial in the *Bennington Banner*, "The anguished reaction . . . to *Silent Spring* has been to refute statements that were never made." Whether this kind of refutation comes from people who actually have not read the book or from those who find it convenient to misrepresent my position I leave it to others to judge.

Early in the summer—as soon as the first installment of the book appeared in the *New Yorker*—public reaction to *Silent Spring* was reflected in a tidal wave of letters—letters to Congressmen, to newspapers, to Government agencies, to the author. These letters continue to come and I am sure represent the most important and lasting reaction.

Even before the book was published, editorials and columns by the hundred had discussed it all over the country. Early reaction in the chemical press was somewhat moderate, and in fact I have had fine support from some segments of both chemical and agricultural press. But in general, as was to be expected, the industry press was not happy. By late summer the printing presses of the pesticide industry and their trade associations had begun to pour out the first of a growing stream of booklets designed to protect and repair the somewhat battered image of pesticides. Plans are announced for quarterly mailings to opinion leaders and for monthly news stories to newspapers, magazines, radio, and television. Speakers are addressing audiences everywhere. It is clear that we are all to receive heavy doses of tranquilizing information, designed to lull the public into the sleep from which *Silent Spring* so rudely awakened it. Some definite gains toward a saner policy of pest control have been made in recent months. The important issue now is whether we are to hold and extend those gains.

The attack is now falling into a definite pattern and all the well known devices are being used. One obvious way to try to weaken a cause is to discredit the person who champions it. So—the masters of invective and insinuation have been busy: I am a "bird lover—a cat lover—a fish lover"—a priestess of nature—a devotee of a mystical cult having to do with laws of the universe which my critics consider themselves immune to.

Another well known, and much used, device is to misrepresent my position and attack the things I have never said. I shall not belabor the obvious. Anyone who has really read the book knows that I favor insect control in appropriate situations— that I *do not* advocate complete abandonment of chemical control—and that I criticize the modern chemical method not because it *controls* harmful insects, but because it controls them *badly* and *inefficiently* and creates many dangerous side effects in doing so. I criticize the present methods because they are based on a rather low level of scientific thinking. We are capable of much greater sophistication in our solution of the problem.

Another piece in the pattern of attack largely ignores *Silent Spring* and concentrates on what I suppose would be called the soft sell—the soothing reassurances to the public. Some of these acknowledge the correctness of my facts, but say that the incidents I reported occurred some time in the past, that industry and Government are well aware of them and have long since taken steps to prevent their recurrence. It must be assured that the people who read these comforting reports read nothing else in their newspapers. Actually, pesticides have figured rather prominently in the news in recent months: Some items trivial, some almost humorous, some definitely serious.

These reports do not differ in any important way from the examples I cited in *Silent Spring*—so if the situation is under better control there is little evidence of it.

What are some of the ways pesticides have made recent news?

1. The *New York Post* of October 12 reported the seizure by the Food and Drug Administration of more than a quarter of a million pounds of potatoes—346,000 pounds to be exact—in

the Pacific Northwest. Agents said they contained about 4 times the permitted residues of aldrin and dieldrin.

2. In September, Federal investigators had to look into the charge that vineyards near the Erie County thruway had been damaged by weed-killer chemicals sprayed along the highway. Similar reports came from Iowa.

3. In California, fumes from lawns to which a chemical had been applied were so obnoxious that the fire department was called to drench the lawns with water. Thereupon the fumes increased so greatly that 11 firemen were hospitalized.

4. Last summer the newspapers widely reported the story of some 5000 Turkish children suffering from an affliction called *porphyria*, characterized by severe liver damage and the growth of hair on face, hands and arms, giving a monkey-like appearance to victims. This was traced to the consumption of wheat treated with a chemical fungicide. The wheat had been intended for planting, rather than for direct consumption. But the people were hungry and perhaps did not understand the restriction. This was an unplanned occurrence in a far part of the world but it is well to remember that large quantities of seed are similarly treated here.

5. You will remember that the bald eagle, our national emblem, is seriously declining in numbers. The Fish and Wildlife Service recently reported significant facts that may explain why this is so. The Service has determined experimentally how much DDT is required to kill an eagle. It has also discovered that eagles found dead in the wild have lethal doses of DDT stored in their tissues.

6. This fall also, Canadian papers carried a warning that woodcock being shot during the hunting season in New Brunswick were carrying residues of heptachlor and might be dangerous if used as food. Woodcock are migratory birds. Those that nest in New Brunswick winter in the southern United States, where heptachlor has been used extensively in the campaign against the fire ant. The residues in the birds were 3 to 3.5 ppm. The legal tolerance for heptachlor is ZERO.

7. Biologists of the Massachusetts Fish and Game Department have recently reported that fish in the Framingham Reservoir on the outskirts of Boston contain DDT in amounts as high as 75 ppm, or more than 10 times the legal tolerance.

This is, of course, a public water supply for a large number of people.

8. One more item—an Associated Press dispatch of November 16th: a sad commentary on technology gone wrong. A Federal Court Jury awarded a New York State farmer $12,360 for damages to his potato crop. The damage was done by a chemical that was supposed to halt sprouting. Instead, the sprouts grew inward.

We are told also that chemicals are never used unless tests have shown them to be safe. This, of course, is not an accurate statement. I am happy to see that the Department of Agriculture plans to ask the Congress to amend the FIFRA to do away with the provision that now permits a company to register a pesticide under protest, even though a question of health or safety has been raised by the Department.

We have other reminders that unsafe chemicals get into use—County Agents frequently have to amend or rescind earlier advices on the use of pesticides. For example, a letter was recently sent out to farmers recalling stocks of a chemical in use as a cattle spray. In September, "unexplained losses" occurred following its use. Several suspected production lots were recalled but the losses continued. All outstanding lots of the chemical have now had to be recalled.

Inaccurate statements in reviews of *Silent Spring* are a dime a dozen, and I shall only mention one or two examples. *Time*, in its discussion of *Silent Spring*, described accidental poisonings from pesticides as *very rare*. Let's look at a few figures. California, the only state that keeps accurate and complete records, reports from 900 to 1000 cases of poisoning from agricultural chemicals per year. About 200 of these are from parathion alone. Florida has experienced so many poisonings recently that this state has attempted to control the use of the more dangerous chemicals in residential areas. As a sample of conditions in other countries, parathion was responsible for 100 deaths in India in 1958 and takes an average of 336 deaths a year in Japan.

It is also worthy of note that during the years 1959, 1960, and 1961, airplane crashes involving crop-dusting planes totaled 873. In these accidents 135 pilots lost their lives.

This very fact has led to some significant research by the Federal Aviation Agency through its Civil Aeromedical

Unit—research designed to find out why so many of these planes crashed. These medical investigators took as their basic premise the assumption that spray poisons accumulate in the pilot's body—inside the cells, where they are difficult to detect.

These researchers recently reported that they had confirmed two very significant facts:

1. That there is a causal relation between the build-up of toxins in the cell and the onset of sugar diabetes.

2. That the build-up of poisons within the cell interferes with the rate of energy production in the human body.

I am, of course, happy to have this confirmation that cellular processes are not so "irrelevant" as a certain scientific reviewer of *Silent Spring* has declared them to be.

This same reviewer, writing in a chemical journal, was much annoyed with me for giving the sources of my information. To identify the person whose views you are quoting is, according to this reviewer, name-dropping.

Well, times have certainly changed since I received my training in the scientific method at Johns Hopkins! My critic also profoundly disapproved of my bibliography. The very fact that it gave complete and specific references for each important statement was extremely distasteful to him. This was padding to impress the uninitiated with its length.

Now I would like to say that in *Silent Spring* I have never asked the reader to take my word. I have given him a very clear indication of my sources. I make it possible for him—indeed I invite him—to go beyond what I report and get the full picture. This is the reason for the 55 pages of references. You cannot do this if you are trying to conceal or distort or to present half truths.

Another reviewer was offended because I made the statement that it is customary for pesticide manufacturers to support research on chemicals in the universities. Now—this is just common knowledge and I can scarcely believe the reviewer is unaware of it, because his own university is among those receiving such grants.

But since my statement has been challenged, I suggest that any of you who are interested make a few inquiries from representative universities. I am sure you will find out that the practice is very widespread. Actually, a visit to a good scientific library will quickly establish the fact, for it is still generally the custom for authors of technical papers to acknowledge the source of funds for the investigation. For example, a few gleaned at random from the *Journal of Economic Entomology* are as follows:

1. In a paper from Kansas State University, a footnote states: Partial cost of publication of this paper was met by the Chemagro Corporation.

2. From the University of California Citrus Experimental Station: The authors thank the Diamond Black-Leaf Co., Richmond, Virginia for grants-in-aid.

3. University of Wisconsin: Research was also supported in part by grants from the Shell Chemical Co., Velsicol Chemical Corporation & Wisconsin Canners Association.

4. Illinois Nat. Hist. Survey: This investigation was sponsored by the Monsanto Chem. Co. of St. Louis, Mo.

A penetrating observer of social problems has pointed out recently that whereas wealthy families once were the chief benefactors of the universities, now industry has taken over this role. Support of education is something no one quarrels with—but this need not blind us to the fact that research supported by pesticide manufacturers is not likely to be directed at discovering facts indicating unfavorable effects of pesticides.

Such a liaison between science and industry is a growing phenomenon, seen in other areas as well. The AMA, through its newspaper, has just referred physicians to a pesticide trade association for information to help them answer patients' questions about the effects of pesticides on man. I am sure physicians have a need for information on this subject. But I would like to see them referred to authoritative scientific or medical literature—not to a trade organization whose business it is to promote the sale of pesticides.

We see scientific societies acknowledging as "sustaining associates" a dozen or more giants of a related industry. When the scientific organization speaks, whose voice do we hear—that of science? or of the sustaining industry? It might be a less serious situation if this voice were always clearly identified, but the public assumes it is hearing the voice of science.

What does it mean when we see a committee set up to make a supposedly impartial review of a situation, and then discover that the committee is affiliated with the very industry whose profits are at stake?

I have this week read two reviews of the recent reports of a NAS committee on the relations of pesticides to wildlife. These reviews raise disturbing questions. It is important to understand just what this committee is. The two sections of its report that have now been published are frequently cited by the pesticide industry in attempts to refute my statements. The public, I believe, assumes that the Committee is actually part of the Academy. Although appointed by the Academy, its members come from outside. Some are scientists of distinction in their fields. One would suppose the way to get an impartial evaluation of the impact of pesticides on wildlife would be to set up a committee of completely disinterested individuals. But the review appearing this week in *The Atlantic Naturalist* described the composition of the Committee as follows: "A very significant role in this committee is played by the Liaison Representatives. These are of three categories. A.) Supporting Agencies. B.) Government Agencies. C.) Scientific Societies. The supporting agencies are presumably those who supply the hard cash. Forty-three such agencies are listed, including 19 chemical companies comprising the massed might of the chemical industry. In addition, there are at least 4 trade organizations such as the National Agricultural Chemical Association and the National Aviation Trades Association."

The committee reports begin with a firm statement in support of the use of chemical pesticides. From this predetermined position, it is not surprising to find it mentioning only some damage to some wildlife. Since, in the modern manner, there is no documentation, one can neither confirm or deny its findings. *The Atlantic Naturalist* reviewer described the reports as "written in the style of a trained public relations official of industry out to placate some segments of the public that are causing trouble."

All of these things raise the question of the communication of scientific knowledge to the public. Is industry becoming a screen through which facts must be filtered, so that the hard, uncomfortable truths are kept back and only the harmless morsels allowed to filter through?

I know that many thoughful scientists are deeply disturbed that their organizations are becoming fronts for industry.

More than one scientist has raised a disturbing question—whether a spirit of Lysenkoism may be developing in America today—the philosophy that perverted and destroyed the science of genetics in Russia and even infiltrated all of that nation's agricultural sciences. But here the tailoring—the screening of basic truth—is done, not to suit a party line—but to accommodate to the short-term gain—to serve the gods of profit and production.

These are matters of the most serious importance to society. I commend their study to you, as professionals in the field of communication.

Speech Accepting the Schweitzer Medal

(Animal Welfare Institute, Washington, D.C., January 7, 1963)

I CAN THINK of no award that would have more meaning for me or that would touch me more deeply than this one, coupled as it is with the name of Albert Schweitzer. To me, Dr. Schweitzer is the one truly great individuals our modern times have produced. If, during the coming years, we are to find our way through the problems that beset us, it will surely be in large part through a wider understanding and application of his principles.

I often reread his own account of the day when there suddenly dawned in his mind the concept of Reverence for Life. In few words, yet so vividly, he describes that scene on a remote river in Africa. He had traveled laboriously upstream for three days in a small river steamer, traveling 160 miles to treat the ailing wife of a missionary. On the way he had been deep in thought, struggling to formulate that universal concept he had been unable to find in any philosophy.

At sunset on the third day the streamer came upon a herd of hippopotami. Suddenly there flashed into his mind the phrase, "Reverence for Life," which all the world now knows.

He gives us few details—just that sand-choked river at sunset, the herd of great beasts—but there it was, that flash of deep insight, that sudden awareness.

In his various writings, we may read Dr. Schweitzer's philosophical interpretation of that phrase. But to many of us, the truest understanding of Reverence for Life comes, as it did to him, from some personal experience, perhaps the sudden, unexpected sight of a wild creature, perhaps some experience with a pet. Whatever it may be, it is something that takes us out of ourselves, that makes us aware of other life.

From my own store of memories, I think of the sight of a small crab alone on a dark beach at night, a small and fragile being waiting at the edge of the roaring surf, yet so perfectly at home in its world. To me it seemed a symbol of life, and of the way life has adjusted to the forces of its physical environment. Or I think of a morning when I stood in a North Carolina marsh at sunrise, watching flock after flock of Canada geese rise from resting places at the edge of a lake and pass low overhead. In that orange light, their plumage was like brown velvet. Or I have found that deep awareness of life and its meaning in the eyes of a beloved cat.

Dr. Schweitzer has told us that we are not being truly civilized if we concern ourselves only with the relation of man to man. What is important is the relation of man to all life. This has never been so tragically overlooked as in our present age, when through our technology we are waging war against the natural world. It is a valid question whether any civilization can do this and retain the right to be called civilized. By acquiescing in needless destruction and suffering, our stature as human beings is diminished.

All the world pays tribute to Dr. Schweitzer, but all too seldom do we put his philosophy into practice. The Schweitzer Medal is one means of disseminating the thoughts and ideas of this great man. I am very proud, and also very humble, to be a recipient of this award.

Speech to the Garden Club of America

(New York, January 8, 1963)

I AM PARTICULARLY glad to have this opportunity to speak to you. Ever since, ten years ago, you honored me with your Frances Hutchinson medal, I have felt very close to the Garden Club of America. And even as you have graciously honored me, I should like to pay tribute to you for the quality of your work and for the aims and aspirations of your organization. Certainly you have your place among the vital and affirmative forces of the world. Through your interest in plant life, your fostering of beauty, your alignment with constructive conservation causes, you promote that onward flow of life that is the essence of our world.

This is a time when forces of a very different nature too often prevail—forces careless of life or deliberately destructive of it and of the essential web of living relationships.

My particular concern, as you know, is with the reckless use of chemicals so unselective in their action that they should more appropriately be called biocides rather than pesticides. Not even their most partisan defenders can claim that their toxic effect is limited to insects or rodents or weeds or whatever the target may be.

The battle for a sane policy for controlling unwanted species will be a long and difficult one. The publication of *Silent Spring* was neither the beginning nor the end of that struggle. I think, however, that it is moving into a new phase, and I would like to assess with you some of the progress that has been made and take a look at the nature of the struggle that lies before us.

In the beginning we should be very clear about what our cause is. What do we oppose? What do we stand for? If you read some of my industry-oriented and industry-sponsored reviewers you will think that I am opposed to any efforts to control insects or other organisms. This, of course, is *not* my position and I am sure it is not that of the Garden Club of America. We differ from the promotors of biocides chiefly in the means we advocate, rather than the end to be attained.

It is my conviction that if we automatically call in the spray planes or reach for the aerosol bomb when we have an insect problem we are resorting to crude methods of a rather low scientific order. We are being particularly unscientific when we fail to press forward with research that will give us the new kind of weapons we need when we must control the numbers of some organism. Some such weapons now exist—brilliant and imaginative prototypes of what I trust will be the insect control methods of the future. But we need many more, and we need to make better use of those we have. Research men of the Department of Agriculture have told me privately that some of the measures they have developed and tested and turned over to the insect control branch have been quietly put on the shelf—it is easier, no doubt, just to go along with the same old methods.

I criticize the present heavy reliance upon biocides on several grounds: First, on the grounds of their inefficiency. This may surprise you in view of all we have been hearing recently about the indispensable nature of chemical controls. But I have here some comparative figures on the toll taken of our crops by insects before and after the DDT era. During the first half of this century, crop loss due to insect attack has been estimated by a leading entomologist at 10 percent a year. It is startling to find, then, that the National Academy of Science last year placed the present crop loss at 25 percent a year. If the percentage of crop loss is increasing at this rate, even as the use of modern insecticides increases, surely something is wrong with the methods used! We had better turn the job over to the kind of scientist who can produce better results. As an indication of what could be done, I would remind you that a non-chemical method gave 100% control of the screwworm fly—a degree of success no chemical has ever achieved.

Chemical controls are inefficient also because as now used they promote resistance among insects. The number of insect species resistant to one or more groups of insecticides has risen from about a dozen in pre-DDT days to nearly 150 today. This is a very serious problem, threatening, as it does, greatly impaired control.

Another measure of inefficiency is the fact that chemicals often provoke flarebacks or resurgences of the very insect they seek to control, because they have killed off its natural controls. Or they cause some other organism suddenly to rise to nuisance status, for the same reason. Every gardener knows that spider mites, once relatively innocuous, have become a world-wide pest since the advent of DDT.

My other reasons for believing we must turn to other methods of controlling insects have been set forth in detail in *Silent Spring* and I shall not take time to discuss them now. They have to do, of course, with the destruction of wildlife and wildlife habitat, with the contamination of water, air, soil and vegetation, and with the known and potential hazard to ourselves.

Obviously, it will take time to revolutionize our methods of insect and weed control to the point where dangerous chemicals are minimized. Meanwhile, there is much that can be done—much that needs to be done, to bring about some immediate improvement in the situation through better procedures and controls.

In looking at the pesticide situation today, it is possible to be optimistic about some developments and pessimistic about others. Let me take some of the encouraging things first.

The most hopeful sign is an awakening of strong public interest and concern. People are beginning to ask questions and to insist upon proper answers instead of meekly acquiescing in whatever spraying programs are proposed. This in itself is a wholesome thing.

There is increasing demand for better legislative control of pesticides. The State of Massachusetts has already made a beginning by setting up a Pesticide Board with actual authority. This Board has taken a very necessary step by requiring the licensing of anyone proposing to carry out aerial spraying. Incredible though it may seem, before this was done anyone who had money to hire an airplane could spray where and when he pleased. I am told that the State of Connecticut is now planning an official investigation of spraying practices. And of course on a national scale, the President last summer directed his science advisor to set up a committee of scientists to review the whole matter of the government's activities in this field.

Citizens' groups, too, are becoming active. For example, the Pennsylvania Federation of Women's Clubs recently set up a program to protect the public from the menace of poisons in the environment,—a program based on education and promotion of legislation. The National Audubon Society has advocated a 5-point action program involving both state and federal agencies. The North American Wildlife Conference this year will devote an important part of its program to the problem of pesticides. All these developments will serve to keep public attention focused on the problem, as it should be.

I was amused recently to read a bit of wishful thinking in one of the trade magazines. Industry "can take heart," it said, "from the fact that the main impact of the book, (i.e., *Silent Spring*) will occur in the late fall and winter—seasons when consumers are not normally active buyers of insecticides . . . it is fairly safe to hope that by March or April *Silent Spring* no longer will be an interesting conversational subject."

If the tone of my mail from readers is any guide, and if the movements that have already been launched gain the expected momentum, this is one prediction that will not come true.

This is not to say that we can afford to be complacent. Although the attitude of the public is showing a refreshing change, there is very little evidence of any reform in spraying practices. The same old patterns are repeating themselves with monotonous regularity. Very toxic materials are being applied with solemn official assurances that they will harm neither man nor beast. When wildlife losses are later reported, the same officials deny the evidence or declare the animals must have died from "something else." These statements are becoming cliches.

Exactly this pattern of events is occurring in a number of areas now. For example, clippings from a newspaper in East St. Louis, Illinois, describe the death of several hundred rabbits, quail and songbirds in areas treated with pellets of the insecticide dieldrin. One area involved was, ironically, a "game preserve." This was part of a program of Japanese beetle control.

The procedures seem to be the same as those I described in *Silent Spring*, referring to another Illinois community, Sheldon. At Sheldon the destruction of many birds and small mammals amounted almost to annihilation. Yet an Illinois Agriculture official is now quoted as saying dieldrin has no serious effect on animal life.

What looks like a significant case history is shaping up now in Norfolk, Virginia. Again the chemical is the very toxic dieldrin, but here the target is the white fringed beetle, which attacks some farm crops. This situation has several especially interesting features. One is the evident desire of the state agriculture officials to carry out the program with as little advance discussion as possible. When the Outdoor Editor of the Norfolk *Virginian-Pilot* "broke" the story on December 11, he reported that officials refused comment on their plans. The Norfolk health officer offered reassuring statements to the public on the grounds that the method of application guaranteed safety. The method he described was use of a machine that drills holes in the soil, the poison then being injected into the holes. "A child would have to eat the roots of the grass to get the poison" he is quoted as saying.

Two weeks later, however, when alert reporters dug out more facts, these assurances proved to be without foundation. The actual method of application is to be by seeders, blowers and helicopters, distributing dieldrin granules on the surface of the ground. This is the same type of procedure that in Illinois wiped out robins, brown thrashers and meadowlarks, killed sheep in the pastures, and contaminated the forage so that cows gave milk containing poison.

It is no wonder the Norfolk citizens are concerned and are demanding to be heard. Yet at a hearing of sorts granted on Wednesday of last week, they were told merely that the State's Department of Agriculture was committed to the program and that it would therefore be carried out.

I think this case fairly represents much of what is wrong with the handling of many problems of insect control. The fundamental wrong is the authoritarian control that has been vested in the agricultural agencies.

There are, after all, many different interests involved in such a case. There are problems of water pollution, of soil pollution, of wildlife protection, of public health. Yet the matter is approached as if the agricultural interest were the supreme, or indeed the only one.

It seems to me clear that this problem of the white fringed beetle in Norfolk—and all similar problems—should be resolved by a conference of representatives of all the interests involved.

I wonder whether citizens in such a situation would not do well to be guided by the strong hint given by the Court of Appeals reviewing the so-called DDT case of the Long Island citizens a few years ago.

You will no doubt remember that this group sought an injunction to protect them from a repetition of the gypsy moth spraying. The lower court refused the injunction and the United States Court of Appeals sustained this ruling on the grounds that the spraying had already taken place and could not be enjoined. However, the court made a very significant comment that seems to have been largely overlooked. Regarding the possibility of a repetition of the Long Island spraying, the judges pointed out that the district court should then re-examine the proposed practices and procedures. They then made this significant general comment: ". . . it would seem well to point out the advisability for a district court, faced with a claim concerning aerial spraying or any other program which may cause inconvenience and damage as widespread as this 1957 spraying appears to have caused, to inquire closely into the methods and safeguards of any proposed procedures so that incidents of the seemingly unnecessary and unfortunate nature here disclosed, may be reduced to a minimum, assuming, of course, that the government will have shown such a program to be required in the public interest."

Here the United States Court of Appeals spelled out a procedure whereby citizens may seek relief in the courts from unnecessary, unwise or carelessly executed programs. I hope it will be put to the test in as many situations as possible.

If we are ever to find our way out of the present deplorable situation, we must remain vigilant, we must continue to challenge and to question, we must insist that the burden of proof is on those who would use these chemicals to prove the procedures are safe.

Above all, we must not be deceived by the enormous stream of propaganda that is issuing from the pesticide manufacturers and from industry-related—although ostensibly independent organizations. If you read the trade magazines, you know that the announced strategy of the industry to concentrate on repairing and building up the somewhat battered image of pesticides—to present them as both essential to our welfare and safe. There is already a large volume of handouts openly sponsored by the manufacturers. There are other packets of material being issued by some of the state agricultural colleges, as well as by certain organizations whose industry connections are concealed behind a scientific front. This material, in enormous volume, is going to writers, editors, professional people, and other leaders of opinion.

It is characteristic of this material that it deals in generalities, unsupported by documentation. In its claims for safety to human beings, it ignores the fact that we are engaged in a grim experiment never before attempted. We are subjecting whole populations to exposure to chemicals which animal experiments have proved to be extremely poisonous and in many cases cumulative in their effect. These exposures now begin at or before birth and—unless we change our methods—will continue through the lifetime of those now living. No one knows what the result will be, because we have no previous experience to guide us.

Let us hope it will not take the equivalent of another thalidomide tragedy to shock us into full awareness of the hazard. Indeed, something almost as shocking has already occurred—the tragedy of the Turkish children who have developed a horrid disease through use of an agricultural chemical. To be sure, the use was unintended, but this does not lessen the human tragedy. And, as I shall explain, it does not guarantee against a repetition. A few months ago we were all shocked by newspaper accounts of this poisoning of thousands of Turkish children. The poisoning had been continuing over a period of some 7 years, unknown to most of us. What made it newsworthy in 1962 was the fact that a scientist gave a public report on it.

This was a disease known as toxic porphyria. It has turned some 5000 Turkish children into hairy, monkey-faced beings. The skin becomes sensitive to light and is blotched and blistered. Thick hair covers much of the face and arms. The victims have also suffered severe liver damage. Several hundred such cases were noticed in 1955. Five years later, when a South African physician visited Turkey to study the disease, he found 5000 victims. The cause was traced to seed wheat which had been treated with a chemical fungicide called hexachlorobenzene. The seed, which had been intended for planting, had instead been ground into flour for bread by the hungry people. Recovery of the victims is slow, and indeed worse may be in store for them. Dr. W. C. Hueper, a specialist on environmental cancer, tells me there is a strong likelihood these unfortunate children may ultimately develop liver cancer.

You may wonder why I take time to talk about something that happened in a distant part of the world—something that was the result of obvious misunderstanding by illiterate people and an act of desperation by hungry people. "This could not happen here" you might easily think.

It would surprise you, then, to know that the use of poisoned seed in our own country is a matter of present concern by the Food and Drug Administration. In recent years there has been a sharp increase in the treatment of seed with chemical fungicides and insecticides of a highly poisonous nature. Two years ago an official of the Food and Drug Administration told me of that agency's fear that treated grain left over at the end of a growing season was finding its way into food channels. He explained that this was particularly hard to detect because of the lack of any simple laboratory test for a fungicide.

Now, on last October 27, the Food and Drug Administration proposed that all treated food grain seeds be brightly colored so as to be easily distinguishable from untreated seeds or grain intended as food for human beings or livestock. I should like to read a paragraph from the Food and Drug Administration's report on this matter: "FDA has encountered many shipments of wheat, corn, oats, rye, barley, sorghum, and alfalfa seed in which stocks of treated seed left over after the planting seasons have been mixed with grains and sent to market for food or feed use. Injury to livestock is known to have occurred.

"Numerous Federal court seizure actions have been taken against lots of such mixed grains on charges they were adulterated with a poisonous substance. Criminal cases have been brought against some of the shipping firms and individuals.

"Most buyers and users of grains do not have the facilities or scientific equipment to detect the presence of small amounts of treated seed grains if the treated seed is not colored. The FDA proposal would require that all treated seed be colored in sharp contrast to the natural color of the seed, and that the color be so applied that it could not be readily removed. The buyer could then easily detect a mixture containing treated seed grain, and reject the lot." I understand, however, that objection has been made by some segments of the industry and that this very desirable and necessary requirement may be delayed.

This is a specific example of the kind of situation requiring public vigilance and public demand for correction of abuses.

The way is not made easy for those who would defend the public interest. In fact, a new obstacle has recently been created, and a new advantage has been given to those who seek to block remedial legislation. I refer to the income tax bill passed by the 87th Congress, a bill which becomes effective this year. This bill contains a little known provision which permits certain lobbying expenses to be considered a business expense deduction. This means that the lobbyists may deduct expenses incurred in appearing before legislative committees or submitting statements on proposed legislation. It means, to cite a specific example, that the chemical industry may now work at bargain rates to thwart future attempts at regulation.

But what of the non-profit organizations such as the Garden Clubs, the Audubon Societies and all other such tax-exempt groups? Their status is not changed. Under existing laws they stand to lose their tax-exempt status if they devote any "substantial" part of their activities to attempts to influence legislation. The word "substantial" needs to be defined. In practice, even an effort involving less than 5 percent of an organization's activity has been ruled sufficient to cause loss of the tax-exempt status.

What happens, then, when the public interest is pitted against large commerical interests? Those organizations wishing to plead for protection of the public interest do so under the peril of losing the tax-exempt status so necessary to their existence. The industry wishing to pursue its course without legal restraint is now actually subsidized in its efforts.

This is a situation which the Garden Club, and similar organizations, within their legal limitations, might well attempt to remedy.

There are other disturbing factors which I can only suggest today. One is the growing interrelations between professional organizations and industry, and between science and industry. For example, the American Medical Association, through its newspaper, has just referred physicians to a pesticide trade association for information to help them answer patients' questions about the effects of pesticides on man. I am sure physicians have a need for information on this subject. But I would like to see them referred to authoritative scientific or medical literature—not to a trade organization whose business it is to promote the sale of pesticides.

We see scientific societies acknowledging as "sustaining associates" a dozen or more giants of a related industry. When the scientific organization speaks, whose voice do we hear—that of science? or of the sustaining industry? It might be a less serious situation if this voice were always clearly identified, but the public assumes it is hearing the voice of science.

Another cause of concern is the increasing size and number of industry grants to the universities. On first thought, much support of education seems desirable, but on reflection we see that this does not make for unbiased research—it does not promote a truly scientific spirit. To an increasing extent, the man who commands the largest expense account—and who brings the largest grants to his university becomes an untouchable, with whom even the University president and trustees do not argue.

These are large problems and there is no easy solution. But the problem must be faced. It must be clearly recognized by the public, for only then will it lose some of its power to stand in the way of public good.

As you listen to the present controversy about pesticides, I recommend that you ask yourself—Who speaks?—And Why?

To Walter P. Nickell

January 14, 1963

Dear Walter:

Your letter is tremendously interesting, especially because I am now on the sidelines of a community fight in Norfolk, Virginia, where the citizens are trying to escape treatment with dieldrin (3 lbs./acre) for the white-fringed beetle. The officials are of course denying that any significant loss of wildlife or pets will result. I gave your name Saturday, to a reporter for the Norfolk *Virginian-Pilot.* This particular reporter has done a very good job of covering the subject with apparent sympathy for the point of view of the citizens and his paper has had several excellent editorials. I thought it would be well worthwhile to cover the situation you mention in southern Michigan as well as that around East St. Louis where a similar program is apparently resulting in the death of enormous numbers of wildlife, including quail.

I am glad to hear of your program in getting at least some curtailment of the spraying for Dutch elm disease. I continue to get literature from a company which claims to have an effective treatment for the fungus disease. I know nothing about the validity of their claims, but I see that they offer a three year guarantee against the death of any tree they treat. I shall enclose some of the literature on the chance that you might want to look into it.

This is interesting about the Michigan United Conservation Clubs. Have you any indication at present as to any action they may plan to take as a result of their interest? It sounds as though they could be very effective.

In Norfolk an interesting aspect that has come to light is the fact that the private property owner has no means of preventing the state from treating his property and that the state disclaims any responsibility for damage that may be done or for the loss of pets or wildlife.

Do keep me informed as to whether the farmers' petition has any effect on curtailing this program in Michigan. As soon as I can, I will send you copies of one or two talks I have given recently, for any possible interest they may have.

Sincerely yours

Dr. Walter Nickell
Cranbrook Institute
Bloomfield Hills, Michigan

To Dorothy Freeman

Wednesday A.M., March 27

Dear One,

When I got up this morning and made my usual tour of the house, opening and raising blinds, etc., I uttered a loud "Oh!" when I opened the blind of the side study window. One large clump of daffodils is suddenly in full bloom—there must be 6 or 8 blossoms. Well, spring must be just about on schedule, in spite of our record cold winter. I remember that three years ago, when I entered the hospital March 31, we had our first daffodils.

Darling, there is so much I want to say to you. It is hard for me (and perhaps for you) to see why I have been unable to find time and mood to respond to some of the things you have written. But the days do seem absurdly full, especially during that week and more of daily hospital trips and trying also to keep up correspondence. Evenings in bed I'm too weary to write—just read, doze, and talk on the phone!

There are three things I want to talk about as I can, my plans for Roger, your writing, and your dear letter in which you talked about your new thoughts on immortality. Of course, darling it touched me deeply that this new conviction has come to you through me. I have never formulated my own belief and feeling in words and am not sure I can now. Of two things I am certain. One is the kind of "material immortality" of which I wrote in the concluding paragraphs of "Undersea," and which is expressed in one of Charles Alldredge's poems which I am sending you. That is purely a biologist's philosophy. For me it has great meaning and beauty—but it is not wholly satis-fying. Then, the immortality through memory is real and, in a personal way, far more satisfying. It is good to know that I shall live on even in the minds of many who do not know me, and largely through association with things that are beautiful and lovely. When E. B. White wrote me last summer that he would always think of me when he heard his hermit thrushes, I told him I could think of no more lovely memorial. And I know, darling, that as between you and me, the one who goes first will always speak to the other through many things—the songs of the veeries and hermits, and a sleepy white throat at midnight—moonlight on the bay—ribbons of waterfowl in the sky. But, as you ask, is that enough? No, it isn't. For one thing, the concept of nothingness is hard to accept. How could that which is truly one's self cease to exist? And if not, then what kind of spiritual existence can there be? If we try to form a definite concept we are, of course, only guessing, but it seems to me that if we say we do not know and can't even imagine, this doesn't mean we disbelieve in personal immortality. Because I cannot understand something doesn't mean it doesn't exist. The marvels of atomic or nuclear physics, and the mathematical concepts of astronomy are wholly beyond my ability to grasp. Yet I know these concepts deal with proven realities, so it is no more difficult to believe there is some sort of life beyond that "horizon," and to accept the fact we cannot now know what it is. Perhaps you remember what I wrote in "Help Your Child—" about the old Swedish oceanographer Otto Pettersson—how, as he neared the end of his long life, he said to his son that in his last moments he would be sustained "by an infinite curiosity as to what was to follow." To me, that sort of feeling is an acceptable substitute for the old-fashioned "certainties" as to heaven and what it must be like. I know that we do not really "know" and I'm content that it should be so. At least, that is my present feeling. I'm glad, darling, that you opened up the subject and I hope we can talk about it. It is not a gloomy subject and certainly does not relate especially to my own situation, for this is something we all share—a normal part of life. Barney's comparison of the life-death relationship to rivers flowing into the sea is to me not only beautiful but somehow a source of great comfort and strength.

Now, darling, this must be all if I'm to get this mailed this morning as I want to do. When I can, I'll get to the other subjects. Meanwhile, dear, I do so deeply wish I could lift from you the burden of sadness that I know lies over your thoughts of me now. I do hope we may have a truly happy summer, full of fun and happy sharing of many of the things dearest to us both. Let us try to fill these weeks with lovely memories, darling.

Goodbye for now, and remember how much I love you, always—

                                                        Rachel

I'm so sorry about the uneasiness over Stan. I know so well how you both must have felt. I suppose, with all the surgery that has been done, some upsets are inevitable and I do hope it means no more than that.

Love, always
R.

1963

To Dorothy Freeman

Monday, April 1

Dearest,

I'm writing in bed this morning, planning to give my bones a good rest today. I had Elliott both Saturday and Sunday and this involved a good deal of going in and out, but the weather was so lovely I'd probably have done so anyway. Jeanne will not be here today, and perhaps not for several. Her father returned early from California because he has to have surgery for a hernia, so she is taking him to a surgeon today. The break comes at a good time for me—it is wonderful not to have the hospital trip today.

I'm sure you know how I have grieved with you and for you since the news came on Saturday. I know so well how you feel. I can remember all the anguish at Muffie's sudden passing as clearly as if it had been yesterday. Now, when Jeffie speaks to me in the night my first thought is "Oh, poor Dorothy and Stan!" But I know, too, the great comfort it is to you to have the incessant questions of those four days answered, and to know that his death was a peaceful one, in his own home. I hope it might be the same for Jeffie, and under the circumstances I must hope it may happen before the time comes when I am unable to care for him. But how much people miss when they have not known the companionship of such precious creatures. And what a store of memories you have, all aided by those wonderful slides and photographs. In time the aching emptiness will be eased, and you can enjoy all those memories almost without pain. But oh darling, I'm so sorry for you both now.

A long interruption to read the mail—then lunch—now I'm sitting in the study while a mockingbird serenades me through the chimney. Outside we have a soft April rain, bringing new green into the lawn. The willows are now a green mist, against the red haze of the maples. The colors of spring are almost as varied as autumn.

I'm sending you (under separate cover, for I discover a critical shortage of stamps, and it's heavy) a copy of the House of Lords debate inspired by *Silent Spring*. You won't want to wade through all its heavy verbiage, sprinkled with "noble Lords" but it is interesting to sample and see the amount of time spent on it. In fact, it gives me an odd feeling to know I stirred it up. The Punch cartoon, sent me today by Elizabeth Beston, is a gem, I think. Please return it in your next letter.

I think I'll call you immediately after the program Wednesday. I imagine there will be a deluge of calls and I want to speak to you first. Well, I just hope I don't look and sound like an utter idiot. When I remember my state of extreme exhaustion those two days, plus the huskiness of voice, I can't be too optimistic. Last night Roger had a CBS channel on and called to me there was a preview of *S.S.*—I missed most of it, but the shots of spraying were rather appalling.

Now I guess I mustn't write much longer. I still haven't written Barney about the Krebiozen, and I must. I know he won't like it and I guess that's why I keep putting it off. Dr. Healy will be out all week, the Clinic tells me, which poses a problem. I do want to get started. We have a new "test area" in the pains in the pelvic girdle, I'm sure, even though nothing shows as yet. I think I'll ask Dr. Biskind if he thinks it would be risky to let a nurse administer it. If these pains would yield to it, I would begin to recover some of the optimism I've lost.

Darling, I must tell you that the roses are still beautiful! They are of course fully out, and have been for some days, but not a petal has fallen. I think that is wonderful—a week tomorrow!

We'll be talking soon. Till then, my dearest love, always—

Rachel

1963

She Started It All—Here's Her Reaction

One of the most important documents to come from governmental printing presses in many years is the report, "Use of Pesticides," issued last week by the President's Science Advisory Committee. It marks the end of an era of complacency in which the government had too long refused to take seriously the hazards accompanying the use of toxic chemicals in the control of insects and other unwanted species—a hazard the President's chief science adviser now characterizes as potentially more serious than nuclear fallout.

It is important to remember that the report is not in itself a solution of the pesticide problem: it is rather a blueprint for a solution. Until it is implemented we will be little better off than we are today. If, however, its recommendations are put into effect we shall have taken a long step forward in our search for a sane policy for controlling harmful or unwanted species without also destroying ourselves.

I think no one can read this report and retain a shred of complacency about our situation. Here are some of the things this board of scientists has established:

Pesticide residues are no longer confined to the areas of original application. By various means—aerial drift, transport in running water, movement in the bodies of living animals or indirectly through biological food chains—they have been dispersed so widely that few if any places remain uncontaminated. The most startling example cited in the report is the finding of pesticide residues in ocean fish taken far at sea and in the liver oils prepared from these fish. By what processes—through food chains or the dynamics of ocean circulation—can this contamination of far ocean regions have taken place? No one knows, and the lack of knowledge is disquieting. Facts like this show how little significance is to be attached to reports, widely circulated in recent months, to the effect that only a very small percentage of the land surface is ever treated. It is the ultimate distribution of the residues that is important.

Man's more immediate and intimate environment is also contaminated. Not only may he ingest poisonous residues on his food, but his clothing, blankets and rugs may carry poisons used in moth-proofing, while from the sprays he commonly uses in household and garden he breathes in or absorbs poisonous chemicals. The panel seems to regard these last-named contacts as representing a more serious hazard than food residues and points out, moreover, that the regulation of these uses (to the extent they are regulated at all) is by the United States Department of Agriculture, whose staff is not trained in evaluating hazards to health.

Wildlife losses, even in carefully planned programs, have sometimes approached 80 per cent of songbird populations or have included most of an entire year's production of young salmon. Such valuable marine species as shrimp, crabs, and oysters are killed by the most minute traces of pesticides. Wildlife, which comes into immediate and constant contact with sprays, is not now mentioned in the low governing registration of chemicals, which affirms the intention to protect man and his domestic animals. This omission must be remedied.

The panel does not rest with stating the hazards: it presents recommendations of several kinds. Some call for immediate remedial action for our protection until comprehensive programs can be worked out. Others are long-range.

I have long felt uneasy about the system of tolerances by which permissible limits of food contamination are established. I am happy to see that the panel shares this concern and has called for a thorough review of tolerances and a re-examination of the data on which they are based. Seven commonly used chemicals are cited as being in particular need of re-evaluation. (It is interesting that the Food and Drug Administration, anticipating this recommendation, announced a week before the release of the report that such a review had been undertaken.) The panel also calls for publication of all scientific data used in establishing tolerances; this would be a welcome and much-needed ventilation of procedures that hitherto have been conducted in a vacuum of secrecy.

I am particularly pleased to see recognition given to what I consider the chief problem for man—the effect of long-continued exposure to small amounts of highly toxic chemicals. The widely publicized fact—if fact it be—that more people die of aspirin overdosage than of acute pesticide poisoning is quite beside the point, as the committee recognizes. Acute poisonings can usually be diagnosed and treated; not so the illness that comes on gradually and with obscure symptoms. The panel found available clinical studies quite inadequate to predict the long-term effects of exposure even to DDT, the most studied of all the pesticides.

It is sobering to read the statement that "physicians are generally unaware of the wide distribution of pesticides, their toxicity, and their possible effects on human health." This is amply borne out by my own observations and by many letters in my files. Indeed, the tendency in the medical schools today is to minimize the teaching of all toxicology, with the result that neither the young doctors beginning practice nor their older colleagues have sufficient awareness of the toxic hazards in our environment and of means of diagnosing and treating the poisonings they cause. The panel could find no research in progress that would aid in the diagnosing of pesticide poisoning.

The chemicals that leave long-lasting residues—in soil, air, and water, and in the bodies of men as well as animals—give most cause for concern. Within recent months the newspapers have carried many disturbing reports of aldrin residues in potatoes, of heptachlor or dieldrin in the bodies of migratory game birds, of DDT in the fish in city reservoirs. There has been a growing feeling on the part of many scientists that the chlorinated hydrocarbons would have to be abandoned because of this residue problem. The President's advisers now recommend immediate action to reduce the use of "persistent pesticides," suggesting that they now be reserved for use only against insect carriers of disease, and that the ultimate goal should be their complete elimination.

Many critics of pesticide policies, including myself, have sensed on the part of government agencies an unfortunate tendency to go along even with programs they did not approve of, rather than make effective protest. The insect control agencies seemed to be in the saddle. One may hope this will now change, and that control programs will be put into perspective against other interests involved.

The Federal pest control review board is made up of representatives of the very agencies in pest control, and so, in effect, has been asked to pass judgment on its own actions. While it has occasionally modified a program, it has never recommended that one be discontinued. This is pointed out by the panel, which now brands the government's mass eradication programs directed against the gypsy moth, fire ant, Japanese beetle and white-fringed beetle as "failures" based on unrealistic concepts.

In releasing the report to the public, the President announced he had directed the government departments to "implement the recommendations" made by the panel. It will be important to observe how long it takes for this direction of the President to bear fruit. According to press reports during the past few days, plans are going forward as usual for Federal-state gypsy moth spraying in various New York counties and perhaps elsewhere.

Long-range recommendations of the panel include setting up programs for gathering much needed data on pesticide storage in the human body, and for continuous monitoring of residues in soil, air, water, man, wildlife and fish. Legislative changes are proposed to give greater attention to safety, rather than merely to efficacy, before a chemical can be registered for use. Strong support is given to expansion of research, not only on less toxic and less persistent chemicals, but also on biological methods of control. Other research is urged to throw light on now obscure phases of the effect of pesticides on man, such as effects on reproduction and the interaction of pesticides with commonly used drugs.

One of the most refreshing recommendations is the call for full information to the public in the form of prompt reports on the actual effects of spray programs and the candid disclosure of the hazards of pesticides, where formerly only their benefits had been stressed.

This excellent report alone does not solve our problem. It must now be translated into action. This is the task of the government agencies, which, given the will, could act quickly. It is also the task of the Congress and state legislatures, where action will inevitably be slower. It is important to remember that pressures which opponents of reform know how to apply will continue unabated. The now awakened public must see that its views are also made known. These are social problems of the greatest public importance. The decisions affect every individual. As the panel points out: "In the end society must decide, and to do so it must obtain adequate information on which to base its judgments." The panel itself has made a notable contribution to this end.

New York Herald Tribune, May 19, 1963

Environmental Hazards: Control of Pesticides and Other Chemical Poisons

(Statement before the Subcommittee on Reorganization and International Organizations of the Committee on Government Operations, Washington, D.C., June 4, 1963)

THE CONTAMINATION of the environment with harmful substances is one of the major problems of modern life. The world of air and water and soil supports not only the hundreds of thousands of species of animals and plants, it supports man himself. In the past we have often chosen to ignore this fact. Now we are receiving sharp reminders that our heedless and destructive acts enter into the vast cycles of the earth and in time return to bring hazard to ourselves.

The problem you have chosen to explore is one that must be solved in our time. I feel strongly that a beginning must be made on it now,—in this session of Congress. For this reason I was delighted when I heard, Mr. Chairman, that you were planning to hold hearings on the whole vast problem of environmental pollution.

Contamination of various kinds has now invaded all of the physical environment that supports us—water, soil, air, and vegetation. It has even penetrated that internal environment within the bodies of animals and of man. It comes from many sources: radioactive wastes from reactors, laboratories and hospitals, fallout from nuclear explosions, domestic wastes from cities and towns, chemical wastes from factories, detergents from homes and industries.

When we review the history of mankind in relation to the earth we cannot help feeling somewhat discouraged, for that history is for the most part that of the blind or short-sighted despoiling of the soil, forests, waters and all the rest of the earth's resources. We have acquired technical skills on a scale undreamed of even a generation ago. We can do dramatic things and we can do them quickly; by the time damaging side effects are apparent it is often too late, or impossible, to reverse our actions. These are unpleasant facts, but they have given rise to the disturbing situations that this Committee has now undertaken to examine.

I have pointed out before, and I shall repeat now, that the problem of pesticides can be properly understood only in context, as part of the general introduction of harmful substances into the environment. In water and soil, and in our own bodies, these chemicals are mingled with others, or with radioactive substances. There are little understood interactions and summations of effect. No one fully understands, for example, what happens when pesticide residues stored in our bodies interact with drugs repeatedly taken. And there are some indications that detergents, which are often present in our drinking water, may affect the lining of the digestive tract so that it more readily absorbs cancer-causing chemicals.

In attempting to assess the role of pesticides, people too often assume that these chemicals are being introduced into a simple, easily controlled environment, as in a laboratory experiment. This, of course, is far from true.

My own studies in this field of environmental pollution have been confined largely to pesticides and I am glad, Mr. Chairman, that you have chosen to begin with this highly important problem.

It seems to me that the most significant knowledge that has developed within the past year has been the piling up of evidence about the wide dispersal of pesticide chemicals, far beyond the point of application. I should like to cite some examples to illustrate this spreading contamination.

To begin on a small scale, we accept as fact the often repeated statements that it is not the deliberate intention to spray reservoirs. Yet studies by the Massachusetts Division of Fisheries and Game during the past year, covering to date 11 reservoirs that serve as public water supplies, show that fish in these reservoirs are heavily contaminated with DDT. The average amount found in the fish from all waters examined in the Sudbury, Assabet, and Concord regions of Eastern Massachusetts was 35.4 p.p.m.; the maximum concentration of 96.7 p.p.m. was found in two places, including the Framingham Reservoir, a source of drinking water for a large area. It might be pointed out that this is nearly 14 times the legal tolerance for DDT in foods.

Although it is not difficult to imagine the paths by which domestic water supplies become contaminated, there are now examples of a different sort that defy easy or comfortable explanation. Such, for example, is the situation on Prince of Wales Island in southeastern Alaska. I am told by the Fish and Wildlife Service that its biologists have sampled resident fish in four drainage systems on this island and have found DDT, sometimes with its metabolites, in two of them. There is no record of applications of DDT on this island. The nearest town other than small native villages, is more than 50 miles away.

An even more remote region, not far below the Arctic circle, has been yielding extraordinary data to the Fish and Wildlife Service for several years. This is the Yellowknife region on the Upper Yukon River, in the Northwest Territory of Canada. It is an important waterfowl breeding area, wild, remote from any human settlements. No spraying of insecticides is known to have occurred within several hundred miles. Yet DDT and its metabolites have been found for several years both in the eggs of waterfowl and in their young. This alone might have been explained by the fact that the waterfowl are migratory and could easily have picked up the poison during their sojourn in the United States. Transfer to the eggs and young could then have followed. But there is no such explanation for the fact that native vegetation in this same area has now been found to contain residues.

The most disturbing of all such reports, however, concerns the finding of DDT in the oil of fish that live far at sea. Such residues have been found in fish caught off both coasts of North America, as well as off South America, Europe, and Asia. The species concerned include halibut living on the floor of the Pacific Ocean, and tuna, a fish of the open ocean that rarely comes close to land. Oil from some of these fish have contained DDT in concentrations exceeding 300 p.p.m.

All this gives us reason to think deeply and seriously about the means by which these residues reach the places where we are now discovering them. I must emphasize that no one can answer this question with complete assurance today, but I should like to call your attention to certain known facts that do have a bearing on the problem.

The ways by which pesticide residues may be transported over long distances are basically three: by air, by water, and in the bodies of living organisms, either indirectly through food chains or directly.

A report last year by the U.S. Department of Agriculture established the fact that aerial spraying comprises about 22% of the total acreage sprayed in the U.S. Studies by Professor George Woodwell of the University of Maine (and which confirm earlier studies by Canadian biologists) shows that of the DDT used in forest spraying, less than half falls directly to the soil. Of each 0.5 lb. released by the spray plane approximately 0.2 lb. reaches its target. The remainder is presumably dispersed as small crystals in the atmosphere. These minute particles are the components of what we know as "drift"—the phenomenon that plagues every householder who receives contaminating spray from his neighbor across the street, or from his Government's spray planes several miles away. We are now beginning to wonder how vast the reach of "drift" may be. It was known a decade ago that the herbicide 2,4-D could drift as far as 15 or 20 miles in quantities sufficient to damage vegetation. The drift of insecticides is less readily observed, but when the matter is properly studied I predict we shall discover some startling facts. It appears that little application has been made of our knowledge of atmospheric movements. Various factors influence the direction and speed of air currents. Among these is convection, or the upward flow of air which takes place when the ground temperature exceeds that of the air. Conceivably, this force could lift the very fine particles of spray materials to an altitude at which strong horizontal winds could come into play, effecting transport for long distances. We know this happens with other materials. Scientists of the Woods Hole Oceanographic Institute have studied the behavior of salt nuclei, drawn from the surface of the ocean and lifted high into the atmosphere. These tiny particles are carried great distances—at least as far as 400 miles. And we know that the upper atmosphere transports a whole assemblage of living objects—seeds, pollen spores, tiny spiders and insects—and through such transport oceanic islands are colonized. It is therefore a speculation that should be tested that the upper atmosphere may be carrying chemical particles as well as radioactive debris, and that the pesticide contamination of such remote places as those I have mentioned may be the result of a new kind of fallout.

Another factor that may contribute to atmospheric contamination is the tendency of DDT to be evaporated from the surface of water. Therefore aerial spraying may not be the sole source of chemical pollution in the atmosphere. Various studies by the Public Health Service over a period of years have clearly established the fact that rains washing over sprayed lands carry pesticides as runoff into ponds, streams, and rivers. From here, we may assume, there is further transport into the sea and into the atmosphere.

Little thought seems to have been given to the possibility of transport in dust. Yet, on a small scale, we had a vivid example of this last April, when health officials on Long Island charged that the airborne dust from potato fields, carrying arsenic and other insecticide residues, was a menace to public health. This dust had compelled the closing of a public school on several occasions, because it clogged the ventilation system. On a broader scale, it is only reasonable to assume that dust from heavily sprayed lands, especially in some areas where conditions are right, may carry insecticides for exceedingly long distances. The Dust Bowl of the 1930's gave us our most dramatic demonstration of the long range transport of soil particles, but this is a phenomenon that goes on regularly in varying degree. When we remember that insecticides remain in soil for long periods, varying from months to a decade or more, the probability of this type of dispersal is increased.

A final and especially interesting means of pesticide transportation is that which occurs in living animals, whether directly or indirectly. Direct transportation may occur over many hundreds of miles, as when woodcock carry heptachlor from southern wintering grounds in the area of fire ant treatment all the way to breeding areas in the Canadian maritime provinces. A less obvious but exceedingly important method of transportation by living organisms is that which occurs when a chemical passes from one link to another in a natural food chain, usually becoming concentrated as it goes. We now have a number of impressive demonstrations of this phenomenon. Several have been studied by biologists in California.

At Big Bear Lake, for example, toxaphene, a chlorinated hydrocarbon, was applied at a dosage of only 0.2 p.p.m. Later it was found that the minute plankton organisms in the lake had picked up this chemical and had concentrated it to a level of 73 p.p.m. The buildup continued through the food chain, with fish containing 200 p.p.m. and a fish-eating bird (a pelican) containing 1,700 p.p.m. The story does not end there. Plankton organisms collected at the lake poisoned hatchery trout when fed to them. Ten months after the insecticide was applied to the lake, fish were again able to live in these waters. The lake was accordingly re-stocked with trout. However, when fillets from the trout were analyzed, they were found to contain 3 p.p.m. of toxaphene. I might add that this experience convinced the California Division of Fish and Game that toxaphene is unsuitable for rough fish control, but the experiment did provide some very instructive data on transfer of chemicals through food chains. The same sort of phenomenon has been worked out in detail at Clear Lake, California.

I should like to add a word about the concentration or buildup of the chemicals. There is nothing surprising about this—especially about the initial concentration by the plankton. Aquatic organisms are well known to have marked ability to extract minerals and other substances from the water and concentrate them. Marine organisms in particular can do this. For example, the percentage of silica in rivers is 500 times that in the sea, because marine diatoms withdraw so much to construct their shells. Huge quantities of cobalt are extracted from sea-water by lobsters and mussels, and of nickel by various mollusks, yet human chemists recover these elements only with difficulty. Oysters concentrate zinc at a level of about 170,000 times that in the surrounding water. It should come as no surprise, therefore, to find some of these marine invertebrates collecting and concentrating such chemicals as DDT. As Secretary Udall reported to you recently, oysters exposed to levels of only one part per billion for one week then contained 132,000 parts per billion in their tissues. The implications for the human being who likes to eat oysters—or other forms of marine life—are obvious. A current publication by two Fish and Wildlife Service biologists contains this statement: "In the sea, there is the possibility of a continuous re-cycling and concentration of the more stable pesticidal compounds until they pose a real threat to man's own welfare."

All the foregoing evidence, it seems to me, leads inevitably to certain conclusions. The first is that aerial spraying of pesticides should be brought under strict control and should be reduced to the minimum needed to accomplish the most essential objectives. Reduction would, of course, be opposed on the grounds of economy and efficiency. If we are ever to solve the basic problem of environmental contamination, however, we shall have to begin to count the many hidden costs of what we are doing, and weigh them against the gains or advantages.

The second conclusion that seems apparent is that a strong and unremitting effort ought to be made to reduce the use of pesticides that leave long-lasting residues, and ultimately to eliminate them. This, you will remember, was one of the recommendations of the President's Science Advisory Committee. I strongly concur in this recommendation, for I can see no other way to control the rapidly spreading contamination I have described.

There are several other recommendations I would like to suggest, bearing on various specific aspects of the immensely complex pesticide problem. These are as follows:

1. I hope this committee will give serious consideration to a much neglected problem—that of the right of the citizen to be secure in his own home against the intrusion of poisons applied by other persons. I speak not as a lawyer but as a biologist and as a human being, but I strongly feel that this is or should be one of the basic human rights. I am afraid, however, that it has little or no existence in practice.

I have countless letters in my files describing situations in which a person has been subject to personal injury or to the loss of pets or valuable horses or other domestic animals because poisons from a neighbor's spraying invaded his property. Residents of Norfolk, Virginia, have informed me that they were told last winter that the State had the authority to apply poisons to their land but assumed no responsibility for injury that might result. It is a matter of record that dairy farmers in New York State suffered contamination of their land by Federal-State spraying for gypsy moths, with the inevitable result that their milk later contained illegal residues and was condemned by the State as unfit for market.

Under such circumstances, what is the citizen to do? You may recall the opinion of the United States Court of Appeals in the case in which a group of Long Island citizens sought an injunction to prevent a repetition of the spraying to which they had been subjected. Since no date for repeated spraying had been set the court could not grant an injunction, but it did make a significant ruling which I should like to insert in the record:

> ". . . it would seem well to point out the advisability for a district court, faced with a claim concerning aerial spraying or any other program which may cause inconvenience and damage as widespread as this 1957 spraying appears to have caused, to inquire closely into the methods and safeguards of any proposed procedures so that incidents of the seemingly unnecessary and unfortunate nature here disclosed, may be reduced to a minimum, assuming, of course, that the government will have shown such a program to be required in the public interest."

I have been informed by affected citizens in New York State that the current gypsy moth spraying has been done with no advance notice whatever. Some of these people learned of the spraying quite by chance two or three days before the planes began their work. They were told by their attorneys that in this limited time no appeal to the courts was possible. It is clear, therefore, that the intent of the Court as indicated above is thwarted in such cases.

As a minimum protection, I suggest a legal requirement of adequate advance notice of all community, state, or Federal spraying programs, so that all interests involved may receive hearing and consideration before any spraying is done. I suggest further that machinery be established so that the private citizen inconvenienced or damaged by the intrusion of his neighbor's sprays may seek appropriate redress.

2. In another area, I hope this Committee will give its support to new programs of medical research and education in the field of pesticides. I have long felt that the medical profession, with of course notable individual exceptions, was inadequately informed on this very important environmental health hazard. It was sobering to have the President's science advisors confirm this view by saying, "Physicians are generally unaware of the wide distribution of pesticides, their toxicity, and their possible effects on human health." The Panel also found a complete lack of any federally sponsored research to develop methods of diagnosing pesticide poisoning, especially when this takes the form of chronic, rather than acute illness. I am told that

in the medical schools today, because of the many subjects to be taught, the attention given the whole field of toxicology is greatly reduced. Yet this is happening at a time when toxic substances are being introduced into the environment at a rate never before approached.

The plight of the person affected by these poisons is pitiful. Many case histories have come to me in letters. As a rule these people can find no physician who understands their problem. Indeed, I remember several cases in current medical literature in which the physician, even though told of the patient's exposure to such relatively common insecticides as malathion or lindane, had never heard of the chemical and did not know the appropriate treatment. About ten years ago the American Medical Association had a special committee on pesticides which from time to time published authoritative information on the toxicology of these chemicals. I have seen none of these reports for several years. I do not know whether the committee is still functioning; if it is, it is hard to see why the American Medical Association last fall recommended that physicians seek information to allay their patients' fears, not from unbiased scientific literature, but from one of the pesticide trade organizations.

I should like to emphasize, however, that many individual physicians are aware of the hazard and of the need for research in this field. Some of the most interesting letters I receive are from doctors. In what I believe to be the first recognition of this problem by a medical organization, the Illinois Medical Society on March 17th of this year approved a resolution directing attention to delayed and indirect effects of pesticides and calling for a thorough study of the problem. I should like to introduce a copy of this resolution into the record at this point.

RESOLUTION

STUDY AND EVALUATION OF TOXICANTS

WHEREAS the total consequences to man and his renewable resources from the present widespread and often unrestrained dissemination of toxic substances into the environment are only vaguely known and some effects cannot yet even be surmised; and

WHEREAS   the indirect and untoward effects of pesticides, insecticides, rodenticides and kindred chemicals
are frequently long delayed, difficult to trace and apparent safe minimal accumulations in air, soil, water, fiber,
food and all tissues can in time accrue to harmful or even
lethal levels; and

WHEREAS   these toxicants often have a profound latent
effect on flora and fauna not originally intended for suppression or eradication; and

WHEREAS   these toxicants are among the most potent
ever known and such new incompletely evaluated substances are being developed annually; and

WHEREAS   these lethal agents can be purchased by
anyone, anywhere without adequate controls to guard
against their misuse;

NOW THEREFORE, BE IT RESOLVED   that the Board of
Trustees of the Illinois State Medical Society go on record
that efforts to manipulate ecologic balances by governmental agencies, private industry and individuals through
the use of toxicants and radiation needs urgent and con-
scientious study for the development of wise and effective
controls; and

BE IT FURTHER RESOLVED   that in the opinion of the
Board of Trustees of the Illinois State Medical Society
the present state of knowledge dictates a policy of caution, inquiry, maturity of judgement and statesmanship;
and

BE IT FURTHER RESOLVED   that the Director of the Illinois Department of Public Health through the Bureau of
Hazardous Substances and Poison Control be requested
to undertake a study of all toxicants, current and future
sold or used in Illinois, and prepare a report for appropriate distribution.

(Approved by Board of Trustees of the Illinois State Medical Society on March 17, 1963 in Chicago, Illinois.)

3. I should also like to see legislation, possibly at the state level, restricting the sale and use of pesticides at least to those capable of understanding the hazards and of following directions. To me it is shocking that these chemicals can be bought and applied by illiterate and even by mentally deficient persons. We place much more stringent restrictions on the sale of drugs—which at least are not sprayed from powerful machines! Someone wrote me recently about a man who was thought to have contracted hepatitis from a spray he had been using, making the pertinent observation that the man could buy the chemicals that made him ill with no restrictions, but had to have prescriptions to buy the drugs to cure him.

4. I should like to see the registration of chemicals made a function of all agencies concerned rather than of the Department of Agriculture alone. The deficiency in the present law has been pointed out in the report of the President's Science Advisory Committee. Many of the miscellaneous uses of chemicals, as in mothproofing, floor waxes, household sprays, and garden pesticides, have a direct relation to human health. It seems not only logical but necessary that the Department of Health, Education and Welfare should participate in decisions regarding the registration of chemicals so used. Similarly, many, probably the majority of pesticides are used at some time in such a manner that they affect wildlife and commercial and recreational fishery resources. The Department of the Interior needs to have a voice in the registration and labeling of such chemicals.

I have already trespassed upon your time and patience, and I shall mention only two more recommendations.

5. It seems to me that our troubles are unnecessarily compounded by the fantastic number of chemical compounds in use as pesticides. As matters stand, it is quite impossible for research into the effect of these chemicals on the physical environment, on wildlife, and on man to keep pace with their introduction and use. It is hard to escape the concluson that the great proliferation of new chemicals is dictated by the facts of competition within the industry rather than by actual need. I

should like to see the day when new pesticides will be approved for use only when no exisiting chemical or other method will do the job.

6. In conclusion, I hope you will give full support to research on new methods of pest control in which chemicals will be minimized or entirely eliminated. You have heard from Secretary Freeman what some of this work is. One of the outstanding values of biological controls is that they are specifically adapted to a particular species or groups of species. Therefore, since our problems of pest control are numerous and varied, we must search, not for one super-weapon that will solve all our problems, but for a great diversity of armaments, each precisely adjusted to its task. To accomplish this end requires ingenuity, persistence, and dedication, but the rewards to be gained are great.

To Dorothy Freeman

September 10

Dear One,

This is a postscript to our morning at Newagen, something I think I can write better than say. For me it was one of the loveliest of the summer's hours, and all the details will remain in my memory: that blue September sky, the sounds of wind in the spruces and surf on the rocks, the gulls busy with their foraging, alighting with deliberate grace, the distant views of Griffiths Head and Todd Point, today so clearly etched, though once half seen in swirling fog. But most of all I shall remember the Monarchs, that unhurried westward drift of one small winged form after another, each drawn by some invisible force. We talked a little about their migration, their life history. Did they return? We thought not; for most, at least, this was the closing journey of their lives.

But it occurred to me this afternoon, remembering, that it had been a happy spectacle, that we had felt no sadness when we spoke of the fact that there would be no return. And rightly—for when any living thing has come to the end of its life cycle we accept that end as natural.

For the Monarch, that cycle is measured in a known span of months. For ourselves, the measure is something else, the span of which we cannot know. But the thought is the same: when that intangible cycle has run its course it is a natural and not unhappy thing that a life comes to its end.

That is what those brightly fluttering bits of life taught me this morning. I found a deep happiness in it—so, I hope, may you. Thank you for this morning.

Rachel

1963

The Pollution of Our Environment

(Kaiser Foundation Hospitals and Permanente Medical
Group Seventh Annual Symposium, "Man Against
Himself," San Francisco, October 18, 1963)

THANK YOU, Dr. Keene. I am very happy, indeed, to be here tonight, and I consider it a special privilege to present the opening lecture in this symposium.

I suppose it is rather a new, and almost a humbling thought, and certainly one borne of this atomic age, that man could be working against himself. In spite of our rather boastful talk about progress, and our pride in the gadgets of civilization, there is I think a growing suspicion—indeed, perhaps an uneasy certainty—that we have been sometimes a little too ingenious for our own good. In spite of the truly marvelous inventiveness of the human brain, we are beginning to wonder whether our power to change the face of nature should not have been tempered with wisdom for our own good, and with a greater sense of responsibility for the welfare of generations to come.

The subject of man's relationship to his environment is one that has been uppermost in my own thoughts for many years. Contrary to the beliefs that seem often to guide our actions, man does not live apart from the world; he lives in the midst of a complex, dynamic interplay of physical, chemical, and biological forces and between himself and this environment there are continuing, never-ending interactions. I thought a good deal about what I could say most usefully tonight on the subject assigned to me—the pollution of our environment. Unfortunately, there is so much that could be said. I am afraid it is true that, since the beginning of time, man has been a most untidy animal. But in the earlier days this perhaps mattered less. When men were relatively few, their settlements were scattered; their industries undeveloped; but now pollution has become one of the most vital problems of our society. I don't want to spend time tonight giving a catalog of all the various kinds of pollution that today defile our land, our air, and our waters. I know that this is an informed and intelligent audience, and I am sure all these facts are known to you. Instead, I would like to present a point of view about pollution—a point of view which seems to me a useful and necessary starting point for the control of an alarming situation. Since the concept of the environment and its relation to life will underlie everything I have to say (and indeed, I think it is central to this whole symposium), I should like in the beginning to remind you of some of the early history of this planet.

I should like to speak of that strange and seemingly hostile environment that, nevertheless, gave rise to an event possibly unique in our solar system: the origin of life. Of course, our thoughts on this must be speculative; but nevertheless there is fairly wide agreement among geologists, astronomers, geochemists, and biologists, about the conditions that must have prevailed just before life appeared on earth. They were of course very different from those of the present day. Remember, for example, that the atmosphere probably contained no oxygen; and because of that there could be no protective layer of ozone in the upper atmosphere. As a result, the full energy of the sun's ultraviolet rays must have fallen upon the sea; and there in the sea, as we know, there was an abundance of simple chemical compounds. These included carbon dioxide, methane, and ammonia, ready at hand for the complex series of combinations and syntheses that must have occurred. I shall not take time to describe the stages that presumably took place over long eons of time to produce, first, molecules capable of reproducing themselves; then some simple organisms, possibly resembling the viruses, and then doubtless much later organisms able to make their own food because of their possession of chlorophyll. Rather than stressing these details, I want to suggest two general thoughts: (1) So far as our present knowledge goes, nowhere else in the solar system have conditions equally hospitable to life occurred. This earth, then, presented an environment of extraordinary fitness; and life is a creation of that environment. (2) No sooner was life created than it began to act upon the environment. The early virus-like organisms must have rapidly reduced the supplies of nutrients adrift in that primitive ocean. But more important was the change that took place as soon as plants began the process of photosynthesis. A byproduct of this process was the release of oxygen into the atmosphere. And so, gradually, over the millions and billions of years, the nature of the atmosphere has changed;

and the air that we breathe today, with its rich proportion of oxygen, is a creation of life.

As soon as oxygen was introduced into the atmosphere, an ozone layer began to form, high up; this shielded the earth from the fierce energy of the ultraviolet rays and the energy needed for the creation of new life was withdrawn.

From all this we may generalize that, since the beginning of biological times, there has been the closest possible interdependence between the physical environment and the life it sustains. The conditions on the young earth produced life; life then at once modified the conditions of the earth, so that this single extraordinary act of spontaneous generation could not be repeated. In one form or another, action and interaction between life and its surroundings have been going on ever since.

This historic fact has, I think, more than academic significance. Once we accept it we see why we cannot with impunity make repeated assaults upon the environment as we now do. The serious student of earth history knows that neither life nor the physical world that supports it exists in little isolated compartments. On the contrary, he recognizes that extraordinary unity between organisms and the environment. For this reason he knows that harmful substances released into the environment return in time to create problems for mankind.

The branch of science that deals with these interrelations is Ecology; and it is from the viewpoint of an ecologist that I wish to consider our modern problems of pollution. To solve these problems, or even just to keep from being overwhelmed by them, we need, it is true, the services of many specialists, each concerned with some particular facet of pollution. But we also need to see the problem as a whole; to look beyond the immediate and single event of the introduction of a pollutant into the environment, and to trace the chain of events thus set into motion. We must never forget the wholeness of that relationship. We cannot think of the living organism alone; nor can we think of the physical environment as a separate entity. The two exist together, each acting on the other to form an ecological complex or an ecosystem.

There is nothing static about an ecosystem; something is always happening. Energy and materials are being received, transformed, given off. The living community maintains itself in a dynamic rather than a static balance. And yet these concepts, which sound so fundamental, are forgotten when we face the problem of disposing of the myriad wastes of our modern way of life. We behave, not like people guided by scientific knowledge, but more like the proverbial bad housekeeper who sweeps the dirt under the rug in the hope of getting it out of sight. We dump wastes of all kinds into our streams, with the object of having them carried away from our shores. We discharge the smoke and fumes of a million smokestacks and burning rubbish heaps into the atmosphere in the hope that the ocean of air is somehow vast enough to contain them. Now, even the sea has become a dumping ground, not only for assorted rubbish, but for the poisonous garbage of the atomic age. And this is done, I repeat, without recognition of the fact that introducing harmful substances into the environment is not a one-step process. It is changing the nature of the complex ecological system, and is changing it in ways that we usually do not foresee until it is too late.

This lack of foresight is one of the most serious complications, I think. I remember that Barry Commoner pointed out in a masterful address to the Air Pollution Conference in Washington last winter, that we seldom if ever evaluate the risks associated with a new technological program before it is put into effect. We wait until the process has become embedded in a vast economic and political commitment, and then it is virtually impossible to alter it.

For example, surely it would have been possible to determine in the laboratory how detergents would behave once released into public water supplies, to foresee their nearly indestructible nature. Now, after years of use in every woman's dishwasher and washing machine, the process of converting to "soft" detergents will be a long and a costly one.

So our approach to the whole problem is shot through with fallacies. We have persisted too long in the kind of thinking that may have been appropriate in the days of the pioneers, but is so no longer. The assumption that the rivers, the atmosphere, and the sea are vast enough to contain whatever we pour into them—I remember not long ago, I heard a supposedly able scientist, the director of one of our agricultural institutions, talk glibly about the "dilution of the pollution," repeating this magical phrase as though it provided the answer to all our problems. It does not, for several reasons.

One reason, as I expect Dr. Brown will tell us tonight, is that there are entirely too many of us; and so our output of pollutants of all kinds has become prodigious. Another reason is the very dangerous nature of much of the present-day pollution. Substances that are highly capable of entering into biological reactions with living organisms. The third very important reason is that the pollutant seldom stays where we put it, and seldom remains in the form in which it was introduced.

Let us look at a few examples. The most serious problem related to modern synthetic pesticides, in my opinion, is the fact that they are becoming long-term, widespread contaminants of the environment. Some of them persist in soil for ten years or more, entering into what surely is one of the most complex and delicately balanced of all ecological systems. They have entered both surface and ground waters; they have been recovered not only from most of the major river systems but in the drinking water of many communities. Their importance as air contaminants is only beginning to be recognized. I remember this past summer there was a freak mishap in the State of Washington, which provided a rather dramatic illustration: a temperature inversion kept a very dangerous chemical which was sprayed from the air from settling on the crops that were being sprayed. Instead, the chemical remained in a drifting cloud for some hours and before the incident was over several cows had died of poisoning and some thirty people had been hospitalized. Then there was the incident in Long Island last winter, when several schools had to be closed because of dust from the potato fields—dust that was carrying insecticides and blowing through the screens of the school windows.

Less dramatic than those examples, but probably more important in the long run, is the fact, seldom remembered, that, for example, of all the DDT sprayed from the air less than half falls directly to the soil or to the intended target. The remainder is presumably dispersed in small crystals in the atmosphere. These minute particles are the components of what we know as "drift," or the dispersal of pesticides far beyond the point of application. This is a subject of great importance and one on which few studies have been made. We don't even know the mechanics or the mechanisms of drift. We certainly need to find out.

A few months ago, wide publicity was given to a release purporting to show that only a very small percentage of the land surface of the United States is sprayed with pesticides in any year. I don't necessarily quarrel with the statement; it may or may not be correct. But I do quarrel very seriously with the interpretation, which implies that the pesticide chemicals are confined to very limited areas; to the areas where they are applied. There are a number of reports, from many different sources, which show how inaccurate that is. The Department of the Interior, for example, has records of the occurrence of pesticide residues in water fowl, in the eggs of the water fowl, and in associated vegetation in far arctic regions hundreds of miles from any known spraying. The Food and Drug Administration has revealed the discovery of pesticide residues in quite substantial amounts in the liver oils of marine fishes taken far at sea, fishes of species that do not come into inshore waters. How do those things happen? We do not know. But we must remember that we are dealing with biological systems and cyclic movements of materials through the environment.

Take, for example, some of the recent demonstrations of what happens when pesticides enter a natural food chain. They progress through it in a fashion that is really explosive. You have several examples here in the State of California, at the Tule Lake and Klamath National Wildlife Refuges. Water entering the refuges from surrounding farms is carrying in residues of insecticides. These have now become concentrated in food chain organisms and in recent years have resulted in a heavy mortality among fish-eating birds.

Then, at Big Bear Lake in San Bernardino County, toxaphene was applied to the lake at a concentration of only 0.2 of 1 part per million. But notice how it was built up. Four months later it was concentrated in plankton organisms at a level of 73 parts per million. Later, residues in fish reached 200 parts per million. In a fish-eating bird, a pelican, 1700 parts per million.

And at Clear Lake, not far from here, efforts to control the gnat population have had a long and a troubled history. Beginning in 1949, the chemical DDD was applied to the lake in very low concentrations. It was later picked up by the plankton, by plankton-eating fish, and by fish-eating birds. The maximum application to the water itself was only 1/50 part per million; yet in some of the fishes the concentration reached 2500 parts per million. The western grebes which nested on the shore of the lake and are fish-eaters almost died out. When their tissues were analyzed they were found to contain heavy concentrations of the chemical. A very interesting phenomenon was that five years after the last application of the chemical, although the water of the lake itself was free of the poison the chemical apparently had gone into the living fabric of the lake; all of the resident plants and animals still carried the residues and were passing them on from generation to generation.

One of the most troublesome of modern pollution problems is the disposal of radioactive wastes at sea. By its very vastness and seeming remoteness the sea has attracted the attention of those faced with the problem of disposing of the byproducts of atomic fission. And so the ocean has become a natural burying-place for contaminated rubbish and for other low-level wastes of the atomic age. Studies to determine the limits of safety in this procedure for the most part have come after rather than before the fact, and disposal activities have far outrun our precise knowledge as to the fate of these waste products.

If disposal of radioactive wastes at sea is to be safe, the material must remain approximately where it is put, or else it must follow predictable paths of distribution, at least until the decay of the radioactive substances has reduced them to relatively harmless levels. The more we know about the depths of the sea, the less do they appear to be a place of calm where deposits may remain undisturbed for centuries. There is far greater activity at deep levels than we formerly suspected. Below the known and charted surface currents there are others which run at their own speeds, in their own directions, and with their own volume. There are powerful turbidity currents that rush down over the continental margins. Even on the ocean floor, at great depths, moving waters are constantly sorting over the sediments, leaving the evidence of their work in ripple marks.

All of these activities, plus the long recognized upwelling of water from the depths and the opposite, downward sinking of great masses of surface water result in a gigantic mixing process. When we dump radioactive wastes in the sea we are introducing them into a dynamic system. But this transport by the sea is only part of the problem, because marine organisms also play an important part in concentrating and distributing radioisotopes. We still need to learn a great deal about the processes involved when radioactive materials are introduced, through fallout, into the marine environment. The studies that have been made reveal movements of great complexity between sea water and the hordes of plankton creatures, between the plankton and the organisms higher in the food chain, between the sea and the land and from the land to the sea.

The most important fact about this is that the marine organisms bring about a marked distribution, both vertical and horizontal, of the radioactive contaminants. As the plankton make regular migrations, sinking into deep water in the daytime and rising to the surface at night, with the organisms go the radioisotopes they have absorbed, or that may adhere to them. As a result, the contaminants are made available to other organisms in new areas; and as they are taken up by larger, more active animals, they are subject to transport over long horizontal distances; migrating fishes, seals, whales may distribute radioactive materials far beyond the point of origin.

All these facts have important meaning for us. They show that the contaminant does not remain in the place deposited, or in its original concentration, but rather becomes involved in biological activities of an intensive nature.

It is surprising, then, that so little thought seems to have been given to the biological cycling of materials in one of the most crucial problems of our time! The understanding of the true hazards of radiation and fallout. There have been situations in the news in recent months that are perfect illustrations of our lack of application of the ecological understanding that we have. I think one of the best examples of what I mean is taking place now in the arctic regions in both eastern and western hemispheres. Only two or three years ago it was reported that both the Alaskan Eskimos and the Scandinavian Lapps are carrying heavy burdens of both Sr^{90} and Ce^{137}. This is not because fallout is especially heavy in these far northern regions; indeed, it is lighter there than in areas of heavier rainfall somewhat farther south. The reason is that these native peoples occupy a terminal position in a unique food chain. This begins with the lichens of the arctic tundras; it continues through the bones and the flesh of the caribou and the reindeer, and at last ends in the bodies of the natives, who depend heavily on these animals for meat. Because the so-called "reindeer moss" and other lichens receive nutrients directly from the air, they pick up large amounts of the radioactive debris of fallout. Lichens, for example, have been found to contain 4 to 18 times as much Sr^{90} as sedges, and 15 to 66 times the Sr^{90} content of willow leaves. They are long-lived, slow-growing plants; so they retain and they concentrate what they take out.

Cesium^{137} also travels through this arctic food chain, to build up high values in human bodies. As you remember, cesium has about the same physical half-life as Sr^{90}, although its stay in the human body is relatively short, only about 17 days. However, its radiation does take the form of the highly penetrating gamma rays, thus making it potentially a hazard to the genes. About 1960 it was reported that Norwegians and also the Finnish and Swedish Lapps were carrying heavy body burdens of Ce^{137}. Then, during the summer of 1962, a team from the Hanford Laboratories in Washington went up into the arctic and measured the levels of radioactivity in about 700 natives in 4 different villages above the arctic circle. They found that the averages for Ce^{137} were about 3 to 80 times the burden in individuals who had been tested at Hanford. In one little village, where caribou is a major item of diet, the average burden of Ce^{137} was 421 nanocuries; the maximum burden was 790. The counts for 1963, which extended over a wider geographic area in Alaska, are said to have been still higher.

This situation almost certainly existed from the beginning of the bomb tests; yet somehow it does not seem to have been anticipated, or at least it was not widely discussed and acted upon, though the Scandinavian countries have been rather active in their investigations.

Another example which has become familiar to many of us in recent months is provided by radioactive iodine. This must always have been an important constituent of fallout, so we wonder why its significance has been largely ignored until very recently. Probably the answer lies in its very short half-life, which is only about 8 days, and in the assumption that decay would have rendered it harmless before it could affect human beings.

But the facts, of course, are otherwise. Radioactive iodine is a component of the lower atmospheric fallout and so, depending upon weather conditions it may reach the earth so early that much of its radioactivity is retained. Its distribution may be spotty, also, because of wind, rain, or other weather conditions. So we have the occurrence of the so-called "hot spots."

But we are not primarily concerned with the amount of radioactive iodine on the ground. It is not believed that we absorb significant amounts through the skin or even by inhalation. What is important is the entrance of this material into the food chains. From that point the route to the human body is short and direct. From contaminated pasture grass to the cow, from fresh cow's milk to the human consumer; and once in the body, the iodine finds its natural target, the thyroid gland. It follows that small children, with their small thyroids and their relatively large intake of milk, are endangered more than are adults.

Only a few years ago, it was declared by a scientist testifying before the Committee on Atomic Energy that radioiodine from worldwide fallout is not a problem of concern to humans, and it is not expected that it will become a problem in the future. At the time this prediction was made, there was no national system for sampling. Most of the sampling done since then seems to have suffered from various defects. For example, data on milk supplies for large cities have little meaning, because such milk is a mixture of collections from various areas and the occasional high levels of contamination are easily obscured. Until the summer of 1962 no attempt seems to have been made to collect fallout data and milk contamination data at the same place and time. It appears also that much of the monitoring data reported by the AEC refers to measurements of gamma-ray intensity from the ground, or of beta radioactivity near the ground or in the air. However, as we have seen, what is important is not the radioactive source outside the body, but the entrance of the radioactivity into the food chain and so into the human body.

In the summer of 1962, the Utah State Department of Health began to make its own evaluation of this problem and quickly decided that a hazardous situation existed. All of the five bomb tests carried out in Nevada in July 1962 had carried radioactive iodine into Utah. As exposures began to exceed the yearly radiation protection guide, the state recommended protective measures. Of course, for radioactive iodine, these are very simple: cows may be transferred to stored feed; contaminated milk may be diverted to processing plants for use in ways that will assure an appropriate lapse of time before it is consumed. Knapp, of the AEC's Division of Biology and Medicine, made other observations in Utah, examining single samples of milk rather than dealing in averages, or in composite samples. And these studies bore out the contention that high levels of radioactive iodine were occurring in certain areas. The Utah situation is probably not unique. A few months ago the Committee for Nuclear Information testified before the Joint Committee on Atomic Energy, and declared that a number of local populations, especially in Nevada, Utah, Idaho, and probably other communities scattered throughout the continental United States have been exposed to fallout of medically unacceptable proportions, especially in the cases of children who drink fresh, locally produced milk. The evidence provided by the Committee, as well as that collected in the recently released Knapp report, would seem to support this conclusion. Yet as recently as May of this year the Public Health Service stated that I^{131} doses from weapons testing have not caused undue risk to health.

My reason for reviewing these facts, which I am sure are familiar to most of you, is simply to emphasize that we have not yet become sophisticated enough to view these matters as the ecological problems which they are. Of course there are various ways of studying the problem; there are various angles from which it must be approached, and what I am suggesting does not necessarily preclude other approaches. But I think that the ecological aspect of it must be considered. We must remember that we have introduced these things into dynamic systems that comprise our environment, and it is not enough to monitor the entrance of the contaminant into the environment at that single point. We must be prepared, with the best understanding of all concerned—the physician, the biologist, the ecologist—to follow the contaminants through whatever path they take, through physical and biological systems. This demands more extensive studies than any that have been undertaken, more comprehensive monitoring programs, and more realistic evaluations.

In my opinion, we have been too unwilling to concede the possibility of hazard or of the actual existence of hazard. We have been too unwilling to give attention to the preparation of countermeasures to cope with the hazardous situations when they do arise. Perhaps not now, but perhaps in the future that will arise. Indeed, a report to the Surgeon General by the National Advisory Committee on Radiation in 1962 revealed that except in the case of I^{131}, no effective countermeasures exist. In the climate of euphoria that is generated by repeated assurances that all is well, there is little public support and there is little money for the kind of research that needs to be done. I, for one, would like to see the public considered to be capable of hearing the facts about the hazards that exist in the modern environment. I should like to have them considered capable of making intelligent decisions as to prudent and necessary measures that ought to be taken.

Currently, in this specific area of radiation hazards, I think there is a certain danger that we will feel that the recent test ban treaty makes the whole fallout problem obsolete. This, in my opinion, is not true. The longer lived isotopes will remain in the upper atmosphere for years to come, and we are still destined to receive heavy fallout from past tests. Another very important point is that underground tests have been known to produce atmospheric contamination through venting in the past, and they will almost certainly continue to do so.

The third point is that environmental contamination by radioactive materials is apparently an inevitable part of the atomic age. It is an accompaniment of the so-called "peaceful" uses of the atom as well as of the testing of weapons. This contamination will come about occasionally by accidents, and perpetually by the disposal of wastes.

Underlying all of these problems of introducing contamination into our world is the question of moral responsibility. Responsibility, not only to our own generation, but to those of the future. We are properly concerned about somatic damage to generations now alive; but the threat is infinitely greater to the generations unborn; to those who have no voice in the decisions of today, and that fact alone makes our responsibility a heavy one.

I recently read some calculations made by Professor H. J. Muller. His general conclusion was that the amount of somatic damage from radiation as it is distributed today is far less than the damage which this same radiation, received and transmitted by the present generation, will inflict upon posterity. His further conclusion was that hereditary damage should be the chief touchstone in the setting of permissible or acceptable dose limits. But apparently we have a long way to go and much enlightenment to gain before any agreement can be reached on standards of this kind.

The question of genetic damage from harmful elements in the environment is one that particularly interests me. Elsewhere I have made the suggestion that pesticide chemicals should be viewed with great suspicion as possible agents of genetic damage to man. This suggestion has been challenged by some on the grounds that there is no proof that these chemicals are having such an effect. I don't believe we should wait for some dramatic demonstration before making a thorough study of the potential genetic effect of all chemicals that are widely introduced into the human environment. By the time such a discovery is made otherwise, it will be too late to eradicate them. Some of the chemicals that are now in use as herbicides and insecticides do have mutagenic effects on lower organisms. Others have the ability to cause chromosome damage or a change in chromosome number, and as you know this type of chromosomal abnormality may be associated with a wide variety of congenital defects in man, including mental retardation. I think we should test the pesticide chemicals on several of the organisms that reproduce rapidly and so lend themselves to genetic experiments. If the chemicals then prove mutagenic, or otherwise disruptive of genetic systems in a variety of test organisms, then I think we should withdraw them from use. I am not impressed with the argument that they might not have similar effects in man. After all, the science of genetics was founded when an obscure Austrian monk performed some experiments on garden peas; and the basic hereditary laws he discovered have proved generally applicable in both plants and animals.

Again, another fact of far-reaching significance, that influences in the external environment can cause mutations, was discovered by Professor Muller in experiments on an insect; yet few doubt its applicability to man. Indeed, one of the most striking phenomena in biology is the basic similarity of genetic systems throughout the living world. Yet again and again, in this whole field of environmental influences in relation to life, and this includes our theme of pollution and its impact on life, we meet a strange reluctance to concede that man is, himself, susceptible to harm. It may be admitted freely, for example, that an agricultural chemical entering a river could kill thousands of fish; but it will be denied that this chemical could do any harm to the person who might drink the water. Reports of the decimation of whole populations of birds are shrugged off with the thought that it can't happen to us. If we carry this view to its logical conclusion, it would make a mockery of all the elaborate testing, involving millions of laboratory animals; yet I have been astonished to discover how frequently it crops up, if not stated directly, then at least as the implied basis for an official point of view or decision, or perhaps more often for the lack of any decisive action. I wonder sometimes whether this attitude may not have a deep significance which is relative to our theme tonight. It seems to me to imply a sort of rejection of our past—a reluctance or an unreadiness to accept the fact that man, like all other living creatures, is part of the vast ecosystems of the earth, subject to the forces of the environment.

As I look back through history I find a parallel. I ask you to recall the uproar that followed Charles Darwin's announcement of his theories of evolution. The concept of man's origin from pre-existing forms was hotly and emotionally denied, and the denials came not only from the lay public but from Darwin's peers in science. Only after many years did the concepts set forth in *The Origin of Species* become firmly established. Today, it would be hard to find any person of education who would deny the facts of evolution. Yet so many of us deny the obvious corollary: that man is affected by the same environmental influences that control the lives of all the many thousands of other species to which he is related by evolutionary ties.

I find it quite fascinating to speculate what hidden fears in man, what long-forgotten experiences, have made him so loath to acknowledge first, his origins and then his relationship to that environment in which all living things evolved and coexist. The Victorians at last freed themselves from the fears and superstitions that made them recoil in shock and dismay from Darwinian concepts. And I look forward to a day when we, also, can accept the facts of our true relationship to our environment. I believe that only in that atmosphere of intellectual freedom can we solve the problems before us now.

Thank you.

To Dorothy Freeman

Monday, October 21
Return voyage of the *Victory Chimes*

Dear One,

Now we are over Ohio and considering the speed of jet flight I haven't long to write. But until we are east of Denver I can't bear to do anything but *look*. Then there was lunch and now we are nearly home, with the prospect of being only about 10 minutes late due to unexpected headwinds.

We flew over Yosemite, or rather just to the north, but from our altitude it wasn't possible to form any conception of it. I can tell you its peaks were snow clad as were the mountains west of Denver. Lake Tahoe was in sight from the opposite side of the plane as we passed Yosemite.

Strange how absolutely alien and uninhabitable the west appears from the air—it might easily be the interior of Asia, so little relation does it seem to have to the world I know. I guess one should drive across it by several routes to know it. But I have been so impressed, thinking about water in relation to the landscape—or especially the lack of it!

No doubt I'll talk to you tonight and repeat anything I may say here, but for the record I'll say I fell quite in love with San Francisco and if I had another life to live would like to spend at least a few years of it there. Between the Browers (Sierra Club) and Marie's friends we saw so much that ordinary sightseeing wouldn't cover. Of course one of the highlights was Muir Woods. There was a wheelchair there and a most obliging ranger took the "controls" and gave me a good long ride along the paths. Such a wonderful freshness in the air there!

As you know, I had wanted so much to get to Pacific Grove to see the butterflies and to Carmel. I could have had two chances to go—Sunday with the Browers or today, if I could have stayed over, with Marie's friends, but I decided the drive was too long for me and that it just wouldn't be wise to undertake it. But I did see something of the Monarchs. Yesterday Llewellyn and Carolyn picked us up at 10 o'clock, drove us around the city a while, through Golden Gate Park to the ocean and to the Cliff House for lunch. (Four sea lions asleep on rocks.) Then we went to their house in Belvedere—just a heavenly outlook over the water—and I went to bed. But all the time I lay there, Monarchs were flying past the window, headed south toward Pacific Grove. I counted for a while and I'm sure a couple of hundred must have passed while I watched, for about two hours.

It proved to be quite a chore to get to the Ballroom where the symposium was held. Besides, I stole Sat. A.M. for Muir Woods. So actually I heard none of the Saturday sessions except the last talk, by Medawar. The chairs were terribly uncomfortable and an all-day session must have been pure torture. My hosts were most attentive. Dr. Sheain's secretary, an attractive little Oriental, was waiting at the hotel with a very handsome wheelchair. Then flowers began to arrive—tiny red roses from Dr. Sheain, bronze & yellow chrysanthemum from Dr. Keene, beautiful long-stemmed red roses from Paul, and later a corsage of my favorite tiny orchids from a local admirer, an old gentleman I've had some correspondence with. Very lovely, but it seemed such a waste this morning to walk out and leave them.

It was wonderful that Marie could time her own trip this way, for I couldn't possibly have gone without her or someone in her place! I was far more dependent than I expected to be—indeed, I guess I was about at my worst physically. I'm not sorry I went, but am slightly appalled at my temerity in crossing the country in this condition. But now I'm almost back, and apparently am going to beat the hurricane to Washington! The captain tells us the overcast and rain are scheduled for later this evening.

My internal time clock never did get adjusted to S.F.—I always woke between 5 and 6. Now I've just pushed my watch on 3 hours.

Your *wonderful* farewell to Maine letter was waiting at the Fairmont. Getting it revived memories; reading and re-reading it have been pure delight. Now we are coasting down toward Dulles!

All my love
Rachel

1963

To Dorothy Freeman

Saturday afternoon, October 26

Darling,

Your Thursday letter was such a joy to have—I had hoped for some word but didn't expect such a good, full letter. Now of course I'm eager for the sequel to "As though the swans weren't enough—." It was as though you had planned to end this installment in a tantalizing mid-sentence!

That was a wonderful experience—more swans (Yes, they must have been whistlers; the trumpeters are only in the West.) than I ever saw, even at Mattamuskeet. But it made me long to return there, to hear the music of the thousands of Canada geese. How about a trip there in January?

I'm so happy about the outcome of Thursday's visit to your doctor, and I do hope nothing further will be necessary.

You ask about the redwoods. Of course this brief, tantalizing visit only served to answer some of the major questions in my mind and to make me eager to return, under other conditions. I longed to wander off, *alone*, into the heart of the woods, where I could really get the feeling of the place, instead of being surrounded by people! And confined to a wheelchair! I was so grateful to the Browers for taking me, and to the ranger for his hospitality and his fund of information, but the thing that would have made my enjoyment complete I couldn't have.

The setting is interesting and quite different from the way I imagined it. Once over the Golden Gate Bridge one climbs up and up into those smooth, brown hills, so much of the road lined with eucalyptus trees. (Ansel Adams refuses to photograph them, we learned, because they aren't native!) Then a long winding descent, one hairpin curve after another, into the canyon where the redwoods are. It is deep enough that despite their height they don't present good targets for lightning—this is one reason why fires, at least from natural causes, are no great hazards. Nevertheless, one sees great, burned out stumps here and there, looking fresh enough to have resulted from a fire last year; yet the ranger said there had been no fire in Muir Woods for at least 150 years. It was fascinating to see the circles of young trees surrounding the old ones. There is very little reproduction from seed in these coastal redwoods (but otherwise, I understood, in the "big trees" of the Sierra country) but instead young trees sprout up around the base of a mature tree. There was a marvelous freshness in the air, though I couldn't detect a distinctive odor. The under story of these woods is chiefly the California laurel—a huge tree. When an old tree falls, a long line of young ones grows up from the

trunk, so ⸺⸺⸺⸺⸺ . Then if one of these topples over the same process ensues. The ranger pointed out three generations in one place—the "grandmother" being a very old tree. Perfectly huge oxalis is a predominant ground cover and there are marvelous ferns, sword especially.

It is now Monday, and if I'm to send this on its way today I must hurry along. (Slept all morning, as usual!) The Scotts from Pittsburgh were here yesterday afternoon—they are in town for some conservation meetings.

Do you realize how many perfectly wonderful bird experiences you have had in recent years? I know, because you have been so generous in sharing them with me. How I long to get out somewhere to get the feeling of these fall migrations! I think you will enjoy Irston Barnes's column—don't return it. Everyone speaks of the great abundance of grackles, blackbirds, and the like but no one seems sure of the reason.

This morning's mail brought a copy of Sec. Udall's book on conservation, *The Quiet Crisis*. I thought it was just a publisher's copy but when I opened it there was an inscription in a bold, decisive hand: "For Rachel—An educator-crusader for conservation whose stones have wide ripples! Sincerely, Stewart." Now I want to read this book, which seems to be a history of the conservation movement. Not surprisingly, my reading this past week has a western flavor—John Muir, the Teale anthology, the American Heritage book I mentioned. George Stewart did the section on the Great Basin, which has led me to a re-reading of *Storm*. Now there is much I want to learn about winds, clouds, air movements and, of course, rain.

Did I send you the reprint Betsy Reimers supplied? I meant to just before I left for Calif. but am not sure I did. I really wish you would ask Dr. B. to read it—there is nothing quackish about it—it seems to be important new work. It seems to have provided the answer to various troubles Paul had, including the several times he passed out completely. I believe the remedy is as simple as substituting mono-saccharide for poly.

Our leaves are going fast now, but one of the big maples is still a blaze of color—gold and scarlet—and the dogwoods are a rich wine red.

This piece of paper is smudged but they are Jeffie's tracks, so you won't mind, I'm sure. That reminds me to say that Dr. Monahan finally sent me his bill and the pathologists' report, which confirmed the diagnosis of pancreatitis. As to the bill, I had been prepared for at least $100, knowing how much time he had given little Moppet, for 10 days, to say nothing of the expensive medication. He wrote a note saying he was charging only for the food and medicines—"to include time would make the bill entirely too high." The result: $42! Of course I won't let him do it, but where except in Maine could that happen?

Now I must stop, dear. Your letters keep me feeling so close to you. Yes, Saturday's did come today, as you would know, I guess, from my comment about the blackbirds.

All my dearest love goes with this—now and always—

Rachel

P.S. The Dutch *Silent Spring* arrived today, the title no doubt a literal translation but sounding considerably less attractive— *Dode Lenten*! There has been a terrible uproar there, with Dr. Briejèr probably my only defender.

1963

To Walter C. Bauer

November 12, 1963

Dr. Walter C. Bauer
Committee for Nuclear Information
6504 Delmar Boulevard
St. Louis 30, Missouri

Dear Dr. Bauer:

I was much interested to read in your September–October Bulletin that CNI is planning to extend the scope of its interests and activities to include environmental contaminants other than radioactive substances. I have long admired your organization and have repeatedly referred to it as a model when I am asked about setting up a similar organization for the study of pesticide problems. This suggestion has come to me repeatedly during the past year and in fact, the need has become increasingly apparent. My personal correspondence on this subject has been tremendous, coming from people with genuine interest and concern who wish advice about their own problems or suggestions as to how they may play an effective role in bringing about improvement of the situation.

It is far beyond my own capacity to handle this correspondence in the way it needs to be handled and I had about come to the conclusion that we must take steps to set up some central organization to cope with it. Before proceeding with our plans however, I should like very much to know how much you feel your own organization will be able to do in the field of pesticides. It may well be that there is no need for another organization. If on the other hand, this subject will occupy a minor role then I should like very much to have an opportunity to consult with someone on your staff so that we may have the benefit of some of your experience in setting up a committee on pesticides. I shall look forward with much interest to hearing from you at your earliest convenience.

Under separate cover I am sending in my application for membership and may I say that I think your review "War and the Living Environment" was one of the most important contributions your organization has made. I recently spoke in San Francisco at a symposium in which my topic was "Pollution of the Environment." I wish I had seen this paper first. I did however, refer to the valuable work done by CNI in publicizing the problems posed by radioactive iodine.

Sincerely yours,

To C. J. Briejèr

December 2, 1963

Dear Dr. Briejèr:

I have your most interesting and moving letter before me but I must reply only briefly because I am getting ready to go to New York in the morning. As you can imagine, we are stunned here by the terrible events and can scarcely think clearly about their impact on the future.

I am very glad to hear that the report of the President's Science Advisory Committee will be published in translation in Holland, and especially that you are to write a preface for it. I can perhaps give you a few details in answer to your questions. You may remember that the excerpts from *Silent Spring* appeared in *The New Yorker* magazine during June, and that the book itself was published at the end of September. Mr. Kennedy was known to be a regular reader of *The New Yorker* and had apparently read or at least was familiar with my material before book publication because at a press conference held late in August he was asked whether any steps were being taken in the government in regard to growing public concern about pesticides. He replied that special consideration was being given to the problem "especially in view of Miss Carson's book" and announcement was made either that day or very shortly thereafter of a special committee of scientists to review the problem. As you know, the work of the committee (actually, two committees were involved eventually) occupied about 8 months and some of the members told me that during this period Mr. Kennedy often inquired about the progress of the committee and urged speed in getting out the report. When it was published in May it was accompanied by a letter from the President (which I think appears in the printed report) stating that he had instructed the various government departments involved to take immediate steps to implement the report. I think there is no question about his unusual grasp of problems such as this and his deep concern that steps be taken to improve conditions.

I am much interested in all you tell me and shall write again soon when I have a little more time.

Sincerely,

Speech Accepting the Audubon Medal
of the National Audubon Society
(New York, December 3, 1963)

I COME BEFORE YOU tonight to receive the Audubon Medal with mingled feelings of pride and humility. The roll call of names of those who have received this medal before me includes many whose place in the history of American conservation is both illustrious and secure. I have known at first hand the character and achievements of most of these men; my years of government service were carried out under two of them. To be adjudged worthy to join this distinguished company is an honor I receive, not lightly, but with deep appreciation and gratitude. I am mindful, too, that a prophet is not always honored in his own country—but the National Audubon Society *is* my country—I belong to it—and from its own strength and devotion to the cause of conservation I have drawn strength and inspiration, as have many thousands of others.

For we all are united in a common cause. It is a proud cause, which we may serve secure in the knowledge that the earth will be better for our efforts. It is a cause that has no end: there is no point at which we shall say, "Our work is finished." We build on the achievements of those who have gone before us; let us, in turn, build strong foundations for those who will take up the work when we must lay it down.

Recently I read Secretary Udall's book, *The Quiet Crisis*, which is a fine summary of the history of conservation in America. It is a history that on the one hand inspires dismay, on the other a measure of hope. It inspires dismay at the ruthless stripping of timber lands, the relentless slaughter of wildlife, the reckless raid on the resources of a rich and beautiful land—but a land not inexhaustible. It inspires hope and courage because even the darkest hours have brought forth men who, though their numbers were but a handful, nevertheless contrived to save for us enough that we may at least glimpse the grandeur of this continent as it was a few centuries ago.

Over the decades and the centuries, the scenes and the actors change. Yet the central theme remains—the greed and the shortsightedness of the few who would deprive the many of their rightful heritage. It is a theme supported by the false assurances that whatever is financially profitable is good for the nation and for mankind. These assurances were offered in the days of the timber barons and the land grabbers; they are heard today.

If the crisis that now confronts us is even more urgent than those of the early years of the century—and I believe it is—this is because of wholly new factors peculiar to our own time. These are, first of all, the phenomenal growth of the human population, threatening to over-run its own environment in a way that can bring only deep concern to thoughtful students of population problems. The second factor is a corollary of the first: that as people and their demands increase, there is a smaller share of the earth's resources for each of us to use and enjoy. There is less clean water, less uncontaminated air; there are fewer forests, fewer unspoiled wilderness areas. The third reason is the introduction of new and dangerous contaminants into soil, water, air, and the bodies of plants and animals as our new technology spreads its poisons and its discarded wastes over the land.

And so we live in a time when change comes rapidly—a time when much of that change is, at least for long periods, irrevocable. This is what makes our own task so urgent. It is not often that a generation is challenged, as we today are challenged. For what we fail to do—what we let go by default, can perhaps never be done.

I take courage, however, in the fact that the conservation effort has a broader base than ever before. There is more organized effort; there are many more individuals who are conscious of conservation problems and who are striving, in their own communities or on the national scene, to solve these problems. The educational programs carried on by the National Audubon Society have played a distinguished role in bringing this about.

And so the effort must and shall go on. Though the task will never be ended we must engage in it with a patience that refuses to be turned aside, with determination to overcome obstacles, and with pride that it is our privilege to contribute so greatly.

Chronology

1907–12   Born Rachel Louise Carson on May 27, 1907, in Springdale, Pennsylvania, the third child of Robert Warden
Carson, forty-three, and Maria Frazier (McLean) Carson,
thirty-eight. Father, eldest of six children of Irish immigrants, works as travelling insurance salesman; mother,
a graduate of Washington Female Seminary and former
schoolteacher, keeps house and gives piano lessons. Mar-
ried in 1894, they had daughter Marian Frazier in 1897
and son Robert McLean in 1898, moving to Springdale in
1900. In 1910, father subdivides their wooded, sixty-four-
acre property, advertising lots for sale to supplement his
irregular earnings.

1913–17   Attends School Street School in Springdale; mother encourages outdoor nature study. Reads Beatrix Potter, Kenneth Grahame, and Gene Stratton-Porter; later describes
herself as a "solitary child" who was "happiest with wild
birds and creatures as companions." Sister Marian marries
in 1915; the next year, her husband is arrested for desertion.
In 1917 is confirmed in Sunday school classes at Cheswick
Presbyterian Church. Brother Robert enlists in Army Air
Service.

1918–19   Writes story "A Battle in the Clouds," based on one of
brother's letters; it is published in children's magazine *St.
Nicholas* in September 1918. Contributes additional stories
in January, February, and August of next year. Brother,
returning from France, works at electrical repair company.
Sister Marian's divorce finalized.

1920–22   Takes high school classes at School Street School. Continues to write and publish in children's magazines, beginning a ledger of literary submissions, acceptances, and
rejections. Sister Marian remarries.

1923–24   Transfers to high school in Parnassus, Pennsylvania, in
class of 1925. Plays basketball and field hockey. Sister Marian has daughter, Virginia; brother Robert marries.

1925   Graduates from high school in May, first in her small class, and is accepted at Pennsylvania College for Women in Pittsburgh. Wins state tuition scholarship; family borrows money, sells land and china to help pay for room and board. Arriving at school in September, hopes to major in English and fulfill literary aspirations. Writes in composition class: "I love all the beautiful things of nature, and the wild creatures are my friends." Mother frequently visits campus on weekends. Sister Marian has second daughter, Marjorie Louise; brother Robert's wife has daughter Frances.

1926   Becomes goalkeeper on field hockey team and plays basketball as substitute. Publishes "The Master of the Ship's Light," a sea story, in *The Englicode*, literary supplement of student newspaper *The Arrow*. In June is selected for freshman honors list. Returns to crowded family home over summer, brother and sister-in-law, now-separated sister, and young nieces having moved in. Joins staff of *The Arrow* in September. Finds mentor in biology teacher Mary Scott Skinker.

1927   Story "Broken Lamps" wins college literary prize. Decides to add a science minor to her English major. Writes more for *The Englicode*.

1928   Changes major to biology, hoping to work toward master's degree. In May, is accepted to Johns Hopkins University graduate program in zoology but is unable to afford graduate tuition; completes senior year at Pennsylvania College for Women instead. Tutors high school students in Springdale over summer. Corresponds frequently with Mary Scott Skinker, now at Marine Biological Laboratory in Woods Hole, Massachusetts. Cofounds a science club; takes histology, genetics, organic chemistry, and physics. Reapplies to Johns Hopkins in December.

1929   Graduates magna cum laude in June. After a month in Springdale and a hiking trip in the Shenandoah Valley with Skinker, spends six weeks as beginning investigator at Marine Biological Laboratory in Woods Hole. In October begins graduate study at Johns Hopkins with a one-year full-tuition scholarship.

1930   Rents a house in Stemmers Run, Maryland, within commuting distance of Baltimore; parents, sister, and nieces move in. Works over summer as teaching assistant for an undergraduate biology class at Hopkins, taught by Grace

Lippy; continues as Lippy's assistant for the next four summers. Unable to afford increased second-year tuition, becomes a part-time student, taking part-time job as laboratory technician at the Institute for Biological Research in Hopkins' School of Hygiene and Public Health.

1931   Abandons preliminary work on African *Anomalurus* squirrels, a potential thesis subject, after specimens prove too difficult to obtain. Brother Robert, working as a radio technician, moves in with family. Begins dissertation research on catfish. In September, obtains a teaching assistantship at University of Maryland Dental and Pharmacy School in College Park, Maryland.

1932   Completes thesis, "The Development of the Pronephros During the Embryonic and Early Larval Life of the Catfish (*Ictalurus punctatus*)" and in June receives master's degree from Johns Hopkins. After summer teaching assistantship, returns to Marine Biological Laboratory at Woods Hole for six weeks. Back in Baltimore, begins experiments at Hopkins on salt tolerance in eels, intending to pursue doctorate. In lieu of long-overdue student loan payments, gives Pennsylvania College for Women title to mortgaged family property in Springdale.

1933   Continues teaching biology at University of Maryland. Sister Marian, now diabetic, increasingly unable to support herself and her children; father's health declines.

1934   Formally withdraws from doctoral program at Hopkins and seeks full-time employment as biology instructor. Unsuccessfully submits poems and short fiction to literary magazines.

1935   Passes federal civil service exams for various positions including junior parasitologist, junior wildlife biologist, and junior aquatic biologist, but finds none available. Father dies on July 6. In October, takes part-time job writing radio scripts for Bureau of Fisheries educational series "Romance Under the Waters." Elmer Higgins, head of Division of Scientific Inquiry, attempts to find permanent job for her.

1936   Publishes articles based on Bureau of Fisheries research ("It'll Be Shad-Time Soon" and "Numbering the Fish of the Sea") in the Baltimore *Sun*. In April, completes introductory pamphlet on marine life for Bureau; Higgins recommends she submit her work, "The World of Waters," to *The Atlantic*. Hired in July as junior aquatic biologist, works with assistant bureau chief Robert Nesbit in Baltimore field office on study of Chesapeake Bay fishes.

1937   Sister Marian Williams dies of pneumonia in January. With mother, takes responsibility for care of nieces Virginia, twelve, and Marjorie, eleven; they move to a larger house in Silver Spring, Maryland, in July. Writes articles on Chesapeake fisheries for the Baltimore *Sun*, the Charleston *News and Courier*, and the *Richmond Times Dispatch*. "Undersea," a revision of "The World of Waters," is published in *The Atlantic* in September; after reading the essay, Simon & Schuster editor Quincy Howe and historian Hendrik Willem van Loon encourage her to write a book about marine life.

1938   Meets with Howe and Van Loon in January at the latter's home in Greenwich, Connecticut, and with Howe at Simon & Schuster the next day to discuss further book plans. Hoping to make an undersea dive, asks Van Loon for an introduction to naturalist and ocean explorer William Beebe. Attends North American Wildlife Conference in Baltimore. Reviews books for *The Atlantic*, and continues to write on wildlife and marine subjects for Baltimore *Sun*. In July, takes vacation with family in Beaufort, North Carolina, visiting U.S. Fisheries Station and exploring Outer Banks.

1939   In June, is transferred to Bureau's field office in College Park, Maryland, continuing work as Nesbit's research assistant. Writes brochures for series "Our Aquatic Food Animals." Publishes articles in Baltimore *Sun* on trout hatcheries, starlings, and shad. In August, spends ten days at Fisheries Biological Station in Woods Hole; visits nearby wetlands, birdwatching with friend and fellow researcher Dorothy Hamilton. Moves into a larger house in Silver Spring. Writes mornings, evenings, and weekends; reads passages aloud with mother, who helps with typing.

1940   Sends preliminary book chapters to Quincy Howe, who offers contract in June. Commissions Baltimore *Sun* artist Howard Frech to work on illustrations. Returns to Woods Hole in July, visiting naturalist Henry Beston's "outermost house" at Eastham, Massachusetts, with Hamilton.

Completes *Under the Sea-Wind*, sending manuscript to Simon & Schuster on December 31, her deadline.

1941   *Under the Sea-Wind* is published on November 1. William Beebe, Howard Zahniser, and others review the book favorably, and it is offered as a main selection of the Scientific Book Club, but in the wake of the December 7 attack on Pearl Harbor, it attracts little further attention and sales fail to meet expectations.

1942   Promoted to assistant aquatic biologist in May, begins working in the Bureau of Fisheries' offices at the Department of the Interior in D.C. Edits *Progressive Fish-Culturist*; writes press releases and pamphlets for the Fish and Wildlife Service. In August is transferred to Chicago, the Bureau of Fisheries forced to give up office space to other wartime agencies. Moves with mother to Evanston, Illinois. Begins work on "Food from the Sea" series.

1943   Returns to the D.C. area in May, newly promoted to associate aquatic biologist in Office of the Coordinator of Fisheries; rents house in Takoma Park, Maryland. *Food from the Sea: Fish and Shellfish of New England* and *Food from Home Waters: Fishes of the Middle West* published.

1944   Publishes "The Bat Knew It First," on bat sonar, in *Collier's* magazine, and a feature article on milkweed. *Fish and Shellfish of the South Atlantic and Gulf Coasts* appears; chapters from *Under the Sea-Wind* are selected for William Beebe's anthology *The Book of Naturalists*. Unsuccessfully seeks position as science editor at *Reader's Digest*.

1945   Has appendix removed. Moves back to Silver Spring. Article about oceanarium in Marineland, Florida, appears in April *Transatlantic Review*. Proposes article on DDT research at Patuxent, Maryland, for *Reader's Digest*. Visits Maryland coast with artist Shirley Briggs, a new colleague; they later join others on a trip to Hawk Mountain Sanctuary in Pennsylvania. Asks William Beebe about possible employment in public education at the New York Zoological Society, and inquires about editorial positions at *Audubon* magazine. "Sky Dwellers," condensed from a longer article about chimney swift migration patterns, appears in *Coronet*. Begins friendly correspondence with ornithologist Ada Govan, author of *Wings at My Window*.

1946   Plans series of Fish and Wildlife Service booklets on national wildlife refuge system, "Conservation in Action." In April, with Shirley Briggs, visits recently established refuge near Chincoteague, Virginia, researching first booklet. Takes monthlong summer vacation with mother in Boothbay Harbor, Maine. Writes Briggs: "My greatest ambition is to be able to buy a place here and manage to spend a great deal of time in it—summers at least!" *Under the Sea-Wind* goes out of print, having sold only about 2,000 copies. In September, for second "Conservation in Action" booklet, travels to Parker River in northern Massachusetts with colleague Kay Howe, a designer and illustrator. Researches refuge history at Massachusetts Audubon Society library in Lincoln, Massachusetts. Contemplating a new book of her own, meets with Harvard marine biologist and oceanographer Henry Bigelow. Receives $1,000 prize from *Outdoor Life* magazine for "Conservation Pledge": "I pledge myself to preserve and protect America's fertile soils, her mighty forests and rivers, her wildlife and minerals, for on these her greatness was established and her strength depends."

1947   Makes additional trips with Howe for "Conservation in Action": in February to Mattamuskeet National Wildlife Refuge in eastern North Carolina, and in September to refuges in Montana and Utah and fish hatcheries in Oregon. Also joins in several local Audubon Society outings. In May, while birding near Seneca, Maryland, meets Louis Halle, author of *Spring in Washington*; later has lunch with Halle and nature writer Edwin Way Teale.

1948   Publishes "The Great Red Tide Mystery" in February *Field and Stream*. Meets with literary agents recommended by friend Charles Alldredge; decides to work with Marie Rodell, a mystery novelist and editor then setting up her own New York firm. *Guarding Our Wildlife Resources*, part of the "Conservation in Action" series, appears in October. Flies to Chicago to visit Mary Scott Skinker, who dies of cancer in December. Drafts chapters for new sea book. Is elected to board of Washington, D.C., Audubon Society.

1949   Travels to New York in April, birding on Long Island with Edwin and Nellie Teale and visiting William Beebe at the New York Zoological Society. Receives writing fellowship to work on book tentatively titled "Return to the Sea." Oxford University Press editor Philip Vaudrin asks to see manuscript; after meeting him in Washington, she signs a contract. Makes two summer research trips: with Shirley Briggs, tours the Florida Everglades and briefly attempts an undersea dive; with Marie Rodell, sails from Woods Hole to Georges Bank on an Atlantic fisheries survey aboard *Albatross III*. Plans another book, for which she would write the introduction, collecting Louis Agassiz Fuertes' illustrations of Mexican birds; Oxford declines the project. Discussing it with Houghton Mifflin editor Paul Brooks, agrees instead to write an Atlantic coast guidebook. Takes leave of absence from Fish and Wildlife Service beginning in October.

1950 Decides on *The Sea Around Us* as title for Oxford book (rejecting others, including "Out of My Depth" and "Carson at Sea"). Meets with climatologists and oceanographers at Harvard and Yale; sends a draft chapter to wave expert Walter Munk at Scripps Institution of Oceanography in San Diego, and another to *Kon-Tiki* author Thor Heyerdahl. Enlists Kay Howe as illustrator. Submits manuscript to Oxford at end of June; in August, William Shawn proposes that a three-part condensation appear in *The New Yorker*. Has surgery in September to remove a breast tumor, biopsied as nonmalignant. Recuperates for a week at Nags Head, North Carolina. Applies for Guggenheim Fellowship to support research on Atlantic coast guidebook. Plans for Fuertes volume frustrated by artist's daughter. Travels to Cleveland in December to accept American Academy of Arts and Sciences–Westinghouse science writing prize for essay "The Birth of an Island," published earlier in *The Yale Review*. Proofs of *The Sea Around Us* arrive.

1951 Awarded Guggenheim Fellowship in March, obtains further leave of absence from Fish and Wildlife Service. Meets with Shawn to discuss *New Yorker* serialization of *The Sea Around Us*; it appears in the magazine on June 2, 9, and 16. Oxford hosts book party at National Press Club in Washington, D.C. Published July 12, the book is a surprise runaway best seller; it is later offered as alternate Book-of-the-Month Club selection. Spends time in Maine with Marie Rodell and Fish and Wildlife colleague Bob Hines, who agrees to illustrate guidebook, now titled "Guide to Seashore Life on the Atlantic Coast." Writes liner notes for new recording of Debussy's *La Mer*. Speaks at *New York Herald Tribune* luncheon in October, presenting hydrophone recordings of shrimp, fish, and whale sounds, and later at benefit for National Symphony Orchestra. In December sells film rights to *The Sea Around Us*. Oxford publishes new edition of first book, *Under the Sea-Wind*.

1952   Receives Henry Grier Bryant medal from Geographical Society of Philadelphia and, in New York, National Book Award for nonfiction, giving acceptance speeches. Niece Marjorie has son, Roger Allen Christie, on February 18. Accepts John Burroughs Medal in New York in April. Researching new book, now tentatively titled "Rock, Sand, and Coral: A Beachcomber's Guide to the Atlantic Coast," visits Myrtle Beach, South Carolina; Simons Island, Georgia; and Florida Keys. Receives honorary doctorates from Oberlin, Drexel Institute of Technology, and her alma mater. Formally resigns from Fish and Wildlife Service. In June, new edition of *Under the Sea-Wind* is made available as an alternate selection of Book-of-the-Month Club and is serialized in *Life*; it joins *The Sea Around Us* on best-seller lists. Spends July at Marine Biological Laboratory in Woods Hole; is elected to MBL Corporation. Visits Nantucket with Edwin and Nellie Teale. Purchases property on Southport Island, Maine, and hires contractor to build cottage. Declines invitation to join four-month South Pacific research expedition, citing family and publishing obligations. Reads script for RKO documentary *The Sea Around Us*, finding it "really dreadful." Visits Sanibel and Marco Islands, Florida, with Rodell.

1953   In January, accepting New York Zoological Society gold medal, praises society's Coney Island Aquarium project. Takes long March vacation in Myrtle Beach, South Carolina. Receives honorary doctorate from Smith College. Moves into new cottage on Southport Island, naming it "Silverledges"; niece Marjorie and grandnephew Roger follow later. In July meets Southport neighbors Dorothy and Stanley Freeman. Presents paper "The Edge of the Sea" at American Academy of Arts and Sciences symposium in Boston. RKO film *The Sea Around Us* wins Academy Award for Best Documentary Feature. Signs contract with Harper & Brothers for book on evolution, to be titled "Origin of Life." Writes and calls Dorothy Freeman frequently, sometimes enclosing intimate letters ("apples") within letters to Freeman and family. ("Our brand of 'craziness' would be a little hard for anyone but us to understand," she later explains.)

1954   Writes letter to *The Washington Post*, published on April 22, arguing that dismissal of Fish and Wildlife Service director Albert M. Day constitutes "an ominous threat to the cause of conservation." Presents material from book-in-progress, now titled *The Edge of the Sea*, as a lecture at Cranbrook Institute of Science in Bloomfield Hills, Michigan; later addresses Theta Sigma Phi Matrix Table Dinner in Columbus, Ohio. With Dorothy Freeman, visits Henry Beston and wife Elizabeth Coatsworth at home in Nobleboro, Maine; they exchange visits in subsequent years. *The New Yorker* asks to publish chapters from *The Edge of the Sea*. Over summer in Maine, takes photographs of shore creatures with Stanley Freeman; uses their slides to illustrate December lecture at Audubon Society dinner in D.C.

1955   Sends nearly completed manuscript of *The Edge of the Sea* to Paul Brooks at Houghton Mifflin, who suggests minor revisions. Hires friend Dorothy Algire to help complete taxonomic appendix for the book. Dorothy Freeman arrives in Maryland in March for a weeklong visit; *The Edge of the Sea* is dedicated to Freeman and her husband. Speaks at publisher's sales conference in Boston. Spends summer in Southport. After first serial publication in *The New Yorker* on August 20 and 27, *The Edge of the Sea* appears on October 26. Attends book party at New York's 21 Club. Elected honorary fellow of Boston Museum of Science. Asked to write television script about clouds, meets in New York with meteorologist and producer.

1956   *The Edge of the Sea* nominated for National Book Award. Program "Something About the Sky" is broadcast in March. "Help Your Child to Wonder," article for *Woman's Home Companion*, is published in July; Marie Rodell and Paul Brooks suggest she expand it as a book. In April visits correspondent Curtis Bok and wife Nellie Lee Bok in Radnor, Pennsylvania; later, conceiving plan to purchase forest land near her Southport cottage, turns to Bok for counsel. Receives Achievement Award of the American Association of University Women in Washington. Allows juvenile adaptation of *The Sea Around Us* to proceed, assigning royalties to niece Marjorie. Is invited by Maria Leiper at Simon

& Schuster to edit nature anthology. Organizes a Maine chapter of The Nature Conservancy, becoming honorary chairman.

1957   Marjorie dies of pneumonia on January 30. Takes responsibility for care of grandnephew Roger, now five. Builds larger home in Silver Spring, occupying it in July, and adds extensions to Southport cottage. Abandons plan to purchase "Lost Woods" in Southport, finding the property too costly. Postpones editing of nature anthology. Writes "Our Ever-Changing Shore" for *Holiday* magazine.

1958   In February, studying dangers of insecticide use, decides to write a magazine article; drafts memorandum on "the horrifying facts about what is happening." Writes to E. B. White, encouraging him to report for *The New Yorker* on a Long Island lawsuit against Department of Agriculture gypsy moth spraying program. Contacts Marjorie Spock and Polly Richards, plaintiffs in the suit, who share trial transcript and extensive research. Working with Paul Brooks, initially envisions a "small, quick" book on insecticide problem, "How to Balance Nature"; Rodell enlists *Newsweek* science editor Edwin Diamond as project collaborator. After discussions with William Shawn at *The New Yorker*, decides longer treatment of subject is called for. Withdraws from collaboration with Diamond and hires research assistant, Bette Haney. Over summer, befriends Beverly Knecht, a young blind woman hospitalized with diabetic complications; they call and write each other often. Meets Spock and Richards in Southport, and has productive visit with Robert Rudd, University of California professor at work on *Pesticides and the Living Landscape*. Corresponds with physicians, geneticists, agronomists, ornithologists, entomologists, and others on aspects of new subject. Mother dies on December 2, after a stroke.

1959   Studies health effects of insecticides in libraries of National Cancer Institute and National Institutes of Health. Writes introduction for new edition of *The Sea Around Us*. Attends February meetings of National Wildlife Federation in New York; declines invitation to present preliminary findings. With other members of D.C. Audubon Society, sends letter to Agriculture Secretary Ezra Taft Benson, protesting film *The Fire Ant on Trial* and ant-eradication efforts; finds her access to USDA entomologists blocked. Throws party for Lois Crisler, author of 1956 memoir *Arctic Wild* and new friend. Tentatively titles book "Man Against the Earth," hoping for publication in spring 1960; hires another research assistant, Jeanne Davis. Addresses outdoor meeting of Quaint Acres Community Association in Silver Spring, convincing residents to reject a proposed spraying program. In October, presents research-in-progress at meeting of Audubon and allied societies at The Brookings Institution in Washington, D.C. Attends November Food and Drug Administration hearings on the "Great Cranberry Scare," the carcinogenic herbicide aminotriazole having been detected in the year's cranberry crop.

1960   Is diagnosed with duodenal ulcer in January. In April, discovering lumps in breast, has radical mastectomy; is told falsely that no malignancy has been found. Serves on Natural Resources Committee of Democratic Advisory Council, making recommendations about environmental priorities in a future Democratic administration. After summer in Maine, also serves on Women's Committee for New Frontiers, meeting at Georgetown home of Senator John F. Kennedy and Jacqueline Kennedy. Makes several visits to USDA laboratory of Edward F. Knipling, investigating biological alternatives to chemical insecticides. Rodell suggests *Silent Spring* as book title. In November finds swelling on rib near operation site. After initial radiation treatments seeks additional medical opinion, writing to George Crile at Cleveland Clinic: "I want to do what must be done, but no more. I still have several books to write, and can't spend the rest of my life in hospitals." Travels to Cleveland, where metastasized cancer is confirmed.

1961   Attends Reception for Distinguished Ladies at National Gallery of Art during Kennedy Inauguration. Undergoes radiation treatments later in January and continuing into March; suffers staphylococcus infection and septic arthritis as complications. In April, recovering, writes Marjorie Spock: "I have been working quite steadily on The Book again." Meets with Brooks in June, in Washington, to discuss chapter revisions, and again in New York in September; he hires Louis and Lois Darling as illustrators. Spends summer in Maine, photographers Charles Pratt and Alfred Eisenstaedt among few visitors. Learns that Beverly Knecht has died. Steady writing progress interrupted in November by painful iritis and loss of vision.

1962   Sends fifteen of seventeen chapters of *Silent Spring* to
            Rodell in January. William Shawn calls to offer praise
            ("you have made it literature"). Confides to Freeman:
            "suddenly the tensions of four years were broken and I
            got down and put my arms around [cat] Jeffie and let the
            tears come." Travels to Cleveland in March for further
            radiation treatments. Finishes another chapter and works
            on bibliography; is fitted for wig. In May, meets Supreme
            Court Justice William O. Douglas at dinner of National
            Parks Association Trustees and Conservation Forum;
            he later praises *Silent Spring* as "the most revolutionary
            book since *Uncle Tom's Cabin*." Flies to Los Angeles in
            June to deliver commencement address at Scripps Col-
            lege, stopping in Denver to see Lois Crisler. A condensed
            Silent Spring appears in *The New Yorker* on June 16, 23,
            and 30, prompting unprecedented reader response. *The
            New York Times* publishes article "'Silent Spring' Is Now
            Noisy Summer"; President Kennedy refers to "Miss Car-
            son's book" at a press conference. Published by Hough-
            ton Mifflin on September 27, *Silent Spring* is chosen as
            October main selection of the Book-of-the-Month Club
            and reaches top of best-seller lists; Carson is subject of *Life*
            magazine profile, "The Gentle Storm Center." Speaks at
            annual meeting of National Parks Association in Washing-
            ton, D.C., at meetings of National Council of Women and
            before Women's National Press Club. Attends informal
            "Kennedy Seminar" at home of Interior Secretary Stewart
            Udall; is interviewed by Eric Sevareid for *CBS Reports*.
            Returns to hospital for additional radiation at end of
            December; collapses in a Chevy Chase department store
            while shopping for Roger's Christmas gifts.

1963   Begins letter to be read after her death, to Dorothy Free-
            man. Receives Albert Schweitzer medal from Animal Wel-
            fare Institute. Meets with President's Science Advisory
            Committee; its subsequent report, *Use of Pesticides*, con-
            firms findings and recommendations of *Silent Spring*. Has
            "difficult days" in March, suffering through pain and nausea
            after radiation; bone metastasis evident. Investigates now-
            discredited Krebiozen as alternative therapy. Hour-long
            special "The Silent Spring of Rachel Carson" airs on *CBS
            Reports* on April 3, with an estimated audience of 10–15
            million. Appears on *Today* show late in May, and before
            Congress, twice, in June. Receives "Woman of Conscience"

award from National Council of Women. After summer in Maine, travels to San Francisco with Marie Rodell to speak at symposium "Man Against Himself." Visits Muir Woods, in a wheelchair, with Sierra Club director David Brower. Works with Marie Rodell to organize her papers. Attends dinner of American Geographical Society and is inducted into American Academy of Arts and Letters.

1964   Attends Connecticut funeral of Stanley Freeman, who dies of a heart attack on January 14. Signs new will in February, establishing trust for Roger; leaves papers to Yale University and bequests to many, including The Nature Conservancy and the Sierra Club. Undergoes final radiation treatments at Cleveland Clinic in March and April. Dorothy Freeman visits. Dies of a heart attack on April 14. After a large public funeral at the Washington National Cathedral and smaller memorial service at All Souls Unitarian Church in Washington, D.C., her ashes are divided, half buried by brother Robert near mother's grave, half scattered by Dorothy off Southport Island.

Note on the Texts

This volume contains the complete text of Rachel Carson's *Silent Spring* (1962), including Lois and Louis Darling's original illustrations, and presents a selection of Carson's letters, speeches, and other writings on environmental themes. A Library of America volume in preparation will gather her "sea trilogy": *Under the Sea-Wind* (1941), *The Sea Around Us* (1951), and *The Edge of the Sea* (1955). The text of *Silent Spring* has been taken from the first edition, published by Houghton Mifflin. The texts of Carson's other works—sixty-three items written from 1945 to 1963—have been taken from a variety of manuscript and printed sources, as described below. Approximately half of these other works are believed to be published here in their entirety for the first time.

Silent Spring. Carson had been concerned about the ecological consequences of the use of DDT and other pesticides for many years before *Silent Spring.* In July 1945, for instance, she attempted unsuccessfully to interest the editor of *Reader's Digest* in an article on the subject. But according to her own account, her decision to write the book was precipitated by a January 1958 letter from Olga Owens Huckins, who maintained a bird sanctuary in her Massachusetts backyard. Huckins's letter, relating the disastrous effects of aerial spraying on local songbirds, sharply refocused Carson's attention and prompted her to act. Within only a few weeks, she made "valuable contacts" and "discovered many leads"; prepared a detailed memorandum on "the horrifying facts about what is happening" for her agent Marie Rodell; encouraged E. B. White to report for *The New Yorker* on upcoming court proceedings in an antipesticide lawsuit, *Murphy v. Benson*; and decided to write a magazine article herself, perhaps for *Ladies' Home Journal.* "I feel I should do something," she wrote her editor Paul Brooks, at Houghton Mifflin, on February 21.

Though she was reluctant to postpone other works in progress, Carson soon reconceived her potential article as a "small, quick" book, "How to Balance Nature," to which she would contribute an opening chapter or chapters. A collaborator, the *Newsweek* science editor Edwin Diamond, was to assist in excerpting transcripts from *Murphy v. Benson* and in writing additional material. But the collaboration, advanced by Brooks and Rodell, was short-lived. On April 1, William Shawn at *The New Yorker* indicated that if Carson were to extend her initial findings as a book-length series of articles, his magazine would be inclined to publish them; thereafter, she understood the project as her own, and not a small or quick one. It would appear first in *The New Yorker* and then as a Houghton Mifflin hardcover, like *The Edge of the Sea* before it.

By February 1959 Carson had a new tentative title, "Man Against the Earth," and a sense that "all the pieces of an extremely complex jig-saw puzzle are at last falling into place." Brooks optimistically scheduled publication for the spring of 1960. Carson's health—a mastectomy in April 1960 and subsequent cancer diagnosis, followed by radiation treatments—forced her to set all such schedules aside, but she persisted in her work, writing whenever she could find the strength to do so. She felt ready to send a typescript to Shawn, with all but two of seventeen chapters complete, in January 1962. Rodell and Brooks had read and met with her about successive drafts along the way; she also solicited feedback, on matters of fact particularly, from a wider group of specialists in different aspects of her subject, each commenting on individual chapters. Rodell proposed an epigraph from Keats's "La Belle Dame sans Merci" that made the title *Silent Spring*, her earlier suggestion, newly appealing to Carson.

The New Yorker published *Silent Spring* in three parts, on June 6, 23, and 30, 1962. Houghton Mifflin followed on September 27. Carson was involved in the editing of both texts, which differ significantly, though they were prepared from her identical initial typescripts. Shawn and others at *The New Yorker* condensed her work with her consent and oversight, reflected in page proofs now among her papers at the Beinecke Library. Editors at Houghton Mifflin suggested fewer revisions to her typescript text, and she similarly reviewed and approved page proofs for the book, which includes an extensive list of sources necessarily omitted in *The New Yorker*. Carson is not known to have made or desired any alteration to her work after publication, though it went through additional printings at Houghton Mifflin and was also published in London, by Hamish Hamilton, in 1963. The text of *Silent Spring* in the present volume is that of the 1962 Houghton Mifflin first printing.

Other Writings on the Environment. Following *Silent Spring*, the present volume gathers sixty-three of Carson's letters, speeches, and contributions to periodicals that document the writing and reception of her book and reflect her interest in conservation and the natural world more generally. These items, written from 1945 to 1963, are presented in approximate chronological order of composition. A list of the sources from which the text of each has been taken appears below. With the few exceptions noted, the texts of items published during Carson's life-time have been taken from the source in which each first appeared. The texts of eighteen otherwise unpublished letters to Dorothy or Dorothy and Stanley Freeman have been taken from *Always, Rachel: The Letters of Rachel Carson and Dorothy Freeman, 1952–1964*, edited by the Free-mans' granddaughter Martha Freeman and published by Beacon Press in 1995. The texts of the remaining items have been taken from Car-son's original manuscripts. Some of Carson's unpublished works were collected posthumously in *Lost Woods: The Discovered Writings of Rachel Carson*, edited by Linda Lear and published by Beacon Press in 1998. Others, marked with an asterisk (*) below, are believed to appear in their entirety in the present volume for the first time.

*To Harold Lynch, July 15, 1945: Typescript, Rachel Carson Papers (YCAL MSS 46), Beinecke Rare Book & Manuscript Library, Yale University, New Haven, Connecticut. (Hereafter abbreviated *Beinecke*.)

*To Shirley Briggs, September 28, 1946: Typescript, *Beinecke*.

*To Raymond J. Brown, October 15, 1946: Typescript, *Beinecke*.

*To Edwin Way Teale, February 2, 1947: Typescript, Edwin Way Teale Papers (1981.0009), Archives and Special Collections, Thomas J. Dodd Research Center, University of Connecticut Libraries, Storrs, Connecticut. (Hereafter abbreviated *U Conn*.)

*To Maria Carson, September 21, 1947: Typescript, *Beinecke*.

*To Shirley Briggs, September 29, 1947: Manuscript, Shirley A. Briggs Papers (IWA 0197), Iowa Women's Archives, University of Iowa Libraries, Iowa City, Iowa.

Mr. Day's Dismissal: *Washington Post*, April 22, 1953.

*To Fon Boardman, c. September 1953: Typescript, *Beinecke*.

To Dorothy Freeman, September 28, 1953: *Always, Rachel: The Let-ters of Rachel Carson and Dorothy Freeman, 1952–1964*, ed. Martha Freeman (Boston: Beacon Press, 1995), 5–7. (Hereafter abbreviated *AR*. *AR*'s footnotes and square-bracketed explanatory interpola-tions have been omitted in the present volume; Carson's signature, and the dates of her letters, have been printed in roman rather than italic type.)

To Dorothy Freeman, January 21, 1954: *AR*, 16–17.

The Real World Around Us: Typescript, *Beinecke*.

To Dorothy Freeman, June 1, 1954: *AR*, 42–43.

*To Edwin Way Teale, August 16, 1955: Manuscript, *U Conn*.

Help Your Child to Wonder: *Woman's Home Companion*, July 1956. (Among the future projects she envisioned before her death, Car-son hoped to write a book-length version of this article. *The Sense of*

Wonder, published posthumously by Harper & Row in 1965, adds photographs by Charles Platt and others to the magazine text.)

To Stanley and Dorothy Freeman, August 8, 1956: *AR*, 186–87.

To Curtis and Nellie Lee Bok, December 12, 1956: Typescript, William Curtis Bok and Nellie Lee Holt Box Papers (Collection 3096), Historical Society of Pennsylvania, Philadelphia, Pennsylvania.

*To DeWitt Wallace, January 27, 1958: Typescript, *Beinecke*.

To Dorothy Freeman, February 1, 1958: *AR*, 248–49.

*To Marie Rodell, February 2, 1958: Typescript, *Beinecke*.

*To E. B. White, February 3, 1958: Typescript, *Beinecke*.

To Dorothy Freeman, June 12, 1958: *AR*, 257–58.

*To Edwin Way Teale, October 12, 1958: Typescript, *U Conn*.

*To Marjorie Spock, November 17, 1958: Manuscript, *Beinecke*.

To Marjorie Spock, December 4, 1958: Manuscript, *Beinecke*.

*To William Shawn, February 14, 1959: Typescript, *Beinecke*.

"Vanishing Americans": *Washington Post*, April 10, 1959.

*To Beverly Knecht, April 12, 1959: Typescript, *Beinecke*.

*To R. D. Radeleff, May 20, 1959: Typescript, *Beinecke*.

*To J. I. McClurkin, September 28, 1959: Typescript, *Beinecke*.

*To Grace Barstow Murphy, November 16, 1959: Typescript, *Beinecke*.

To Dorothy Freeman, November 19, 1959: *AR*, 290–91.

*To Morton S. Biskind, December 3, 1959: Typescript, *Beinecke*.

To Dorothy Freeman, December 7, 1959: *AR*, 291–92. (Dated "Monday night" in Carson's original manuscript and tentatively dated December 7, 1959, by the editor of *Always, Rachel*.)

*To Marjorie Spock and Polly Richards, April 12, 1960: Manuscript, *Beinecke*.

*To C. Girard Davidson, June 8, 1960: Typescript, *Beinecke*.

*To Marjorie Spock, September 27, 1960: Typescript, *Beinecke*.

To Dorothy Freeman, October 12, 1960: *AR*, 310–11.

*To Clarence Cottam, January 4, 1961: Typescript, *Beinecke*.

*To Paul Brooks, June 26, 1961: Typescript, *Beinecke*.

To Dorothy Freeman, January 6, 1962: *AR*, 390–91.

To Dorothy Freeman, January 23, 1962: *AR*, 393–94.

*To Frank E. Egler, January 29, 1962: Typescript, *Beinecke*.

*To Lois Crisler, February 8, 1962: Manuscript, *Beinecke*.

To Dorothy Freeman, April 5, 1962: *AR*, 402.

To Dorothy Freeman, May 20, 1962: *AR*, 404–6.

Of Man and the Stream of Time: *Scripps College Bulletin*, July 1962.

*Form Letter to Correspondents, August 2, 1962: Typescript, *Beinecke*.

On the Reception of *Silent Spring*: Typescript, *Beinecke*. (Written for the Book Review section of *The New York Times*, and published there in part in "What's the Reason Why: A Symposium of

Best-Selling Authors," December 2, 1962. Title supplied for the present volume.)

Speech to the Women's National Press Club: Typescript, *Beinecke*.

*Speech Accepting the Schweitzer Medal: Typescript, *Beinecke*.

Speech to the Garden Club of America: Typescript, *Beinecke*. (A version published in the May 1963 *Bulletin of the Garden Club of America* as "A New Chapter to *Silent Spring*" differs in small ways from Carson's typescript.)

*To Walter P. Nickell, January 14, 1963: Typescript, *Beinecke*.

To Dorothy Freeman, March 27, 1963: *AR*, 446–47.

To Dorothy Freeman, April 1, 1963: *AR*, 450–51.

She Started It All—Here's Her Reaction: *New York Herald Tribune*, May 19, 1963.

Environmental Hazards: Control of Pesticides and Other Chemical Poisons: Typescript, *Beinecke*. (Congressional testimony published in government documents and reprinted elsewhere during Carson's lifetime, with changes to the typescript text.)

To Dorothy Freeman, September 10, 1963: *AR*, 467–68.

The Pollution of Our Environment: Typescript, *Beinecke*.

To Dorothy Freeman, October 21, 1963: *AR*, 481–82.

To Dorothy Freeman, October 26, 1963: *AR*, 484–86. (A small sketch included in Carson's original manuscript letter, not reproduced in *AR*, is printed here in facsimile by courtesy of the Edmund S. Muskie Archives and Special Collections at Bates College, which holds Dorothy Freeman's papers.)

*To Walter C. Bauer, November 12, 1963: Typescript, *Beinecke*.

*To C. J. Briejer, December 2, 1963: Typescript, *Beinecke*.

*Speech Accepting the Audubon Medal of the National Audubon Society: Typescript, *Beinecke*.

The texts of the forty items taken from Carson's manuscripts and typescripts in the list above are newly prepared clear texts. Carson's holograph corrections and cancellations have silently been accepted. Slips of the pen, missed keystrokes, and inadvertently misplaced or omitted punctuation have silently been corrected. Titles of books and periodicals, biological taxa, and the names of vessels have been printed in italic type. In one instance—her December 1962 speech to the Women's National Press Club—Carson's single extant typescript was specially prepared for use at the lectern, and contains paragraphing, capitalization, underlining, and spacing intended to aid oral presentation. The present volume regularizes the lectern features of the typescript of this speech. Other speeches have been preserved in her papers in both conventional and lectern forms, and the conventional forms are followed here.

This volume presents the texts of the original printings and manuscripts chosen for inclusion here, but it does not attempt to reproduce features of their typographic design, such as the display capitalization of chapter openings, or holographic features, such as variations in the length of dashes. The texts are reprinted without change, except for the correction of typographical errors, slips of the pen, and other manuscript and typescript accidentals noted above. Spelling, punctuation, and capitalization are often expressive features, and they are not altered, even when inconsistent or irregular. The following is a list of such errors corrected, cited by page and line number: 55.33–34, putrifaction; 158.32, salmonberries and; 166.21, are; 208.27, enviroment; 316.9, black-and blue; 321.27, is; 321.30, flaskes; 322.28, or the; 323.25, Mushbach's; 324.11, picnicing; 325.5, patter; 327.25, snoeshoe; 328.37, springs,; 334.19, interpretor; 334.30 and 32, direr; 336.4, ansser; 336.8, alsmost; 338.19, *Wind* was; 341.1, mountains's; 341.19, spalsh; 343.9, expense; 343.22, travled; 344.11, limstone; 344.25, unforgetable; 345.12, ords; 345.37, Petterson; 346.5, Petterson; 346.7, curiousity; 346.22, juman; 348.9, Rochard; 348.21, myslef; 349.1, articicial; 351.9, a siege; 369.11, pomdering; 369.20, stemographic; 376.26, engagen in; 377.11, resistent; 377.26, first air; 377.32, damage. Some; 378.3, cripped; 378.30–31, Sterility; 378.37, carconigenic; 381.16, acrea; 382.10 (and *passim*), Doves; 384.28, Straus; 385.2, studies) I; 388.3, it as; 388.20, inditement; 389.10, indicting; 389.15, inadvertant; 389.35, is; 390.6, alternative; 397.15 (and *passim*), Fleming; 398.15, Allan; 398.23, away; 401.2, Americal; 405.22, indisturbed; 431.12–13, reaction—; 434.39, lead; 436.39, setup; 439.20, hippopotomi; 443.26, state; 444.29, insecticide,; 447.9, animals; 447.35, see; 449.21–22, any substantial"; 453.16, Petterson; 464.18, studies; 465.40, foodchain,; 467.15, recommendations.; 469.20, patients; 478.24, air,; 479.31–32, toxophene; 482.25, were; 492.10, oxalic; 496.31, inexhaustable.

ILLUSTRATIONS

1. Photograph by Shirley A. Briggs. Courtesy of Rachel Carson Council, Inc.
2. Photograph by Shirley A. Briggs. Courtesy of Rachel Carson Council, Inc.
3. Courtesy of Rachel Carson Council, Inc.
4. Photograph by Rex Gary Schmidt. Courtesy of the U.S. Fish and Wildlife Service/National Conservation Training Center Museum/Archives.
5. Photograph by Edwin Gray. Courtesy of Rachel Carson Council, Inc.
6. Photograph by Shirley A. Briggs. Courtesy of Rachel Carson Council, Inc.
7. Courtesy of Associated Press.
8. Courtesy of Bettmann/Getty Images.
9. Courtesy of Bettmann/Getty Images.
10. Courtesy of Associated Press.
11. Photograph by Joseph Steinmetz. Courtesy of the Joseph Janney Steinmetz/Florida Photographic Collection, State Archives of Florida.
12. Courtesy of the Beinecke Rare Book & Manuscript Library, Yale University.
13. Courtesy of the Beinecke Rare Book & Manuscript Library, Yale University.
14. Courtesy of the Beinecke Rare Book & Manuscript Library, Yale University.
15. Photograph by Erich Hartmann. Copyright © Erich Hartmann/ Magnum Photos.
16. Photograph by Alfred Eisenstaedt. Courtesy of Alfred Eisenstaedt/The LIFE Picture Collection/Getty Images.
17. *Peanuts* © 1962 Peanuts Worldwide LLC. Distributed by Andrews McMeel Syndication. Reprinted with permission. All rights reserved. *New Yorker* drawing on May 18, 1963. Courtesy of James Stevenson/The New Yorker Collection/The Cartoon Bank.
18. Courtesy of CBS Photo Archive/Getty Images.
19. Courtesy of Associated Press.

Notes

In the notes below, the reference numbers denote page and line of this volume (the line count includes chapter headings but not blank lines). No note is made for material included in standard desk-reference works. Biblical references are keyed to the King James Version. For further information about Carson's life and works, and references to other studies, see Paul Brooks, *The House of Life: Rachel Carson at Work* (Boston: Houghton Mifflin, 1972); Martha Freeman, ed., *Always, Rachel: The Letters of Rachel Carson and Dorothy Freeman, 1952–1964* (Boston: Beacon Press, 1995); Linda Lear, *Rachel Carson: Witness for Nature* (New York: Henry Holt, 1997); Linda Lear, ed., *Lost Woods: The Discovered Writing of Rachel Carson* (Boston: Beacon Press, 1998); Peter Matthiessen, ed., *Courage for the Earth: Writers, Scientists, and Activists Celebrate the Life and Writing of Rachel Carson* (New York: Houghton Mifflin, 2007); Priscilla Coit Murphy, *What a Book Can Do: The Publication and Reception of* Silent Spring (Amherst: University of Massachusetts Press, 2005); William Souder, *On a Farther Shore: The Life and Legacy of Rachel Carson* (New York: Crown, 2012); Philip Sterling, *Sea and Earth: The Life of Rachel Carson* (New York: Thomas Y. Crowell, 1970); and the website www.rachelcarson.org. For Carson's sources in *Silent Spring*, see her original "List of Principal Sources" on pages 259–311 of the present volume.

SILENT SPRING

2.1–5 Albert Schweitzer . . . the earth."] From a letter to a French apiarist whose bees had been destroyed by insecticide spraying, printed in part in the December 1956 *Bulletin* of the International Union for the Conservation of Nature and Natural Resources, a copy of which is among Carson's papers at the Beinecke Library. As published, the letter reads: "I am aware of some of the tragic repercussions of the chemical fight against insects taking place in France and elsewhere, and I deplore them. Modern man no longer knows how to foresee or to forestall. He will end by destroying the earth from which he and other living creatures draw their food. Poor bees, poor birds, poor men. . . ."

3.1–3 The sedge . . . KEATS] See John Keats's ballad "La Belle Dame sans Merci," first published in 1820.

3.4–10 I am pessimistic . . . E. B. WHITE] From a letter to *Newsweek* reporter Bruce Lee, quoted in "Typewriter Man," *Newsweek*, February 15, 1960.

14.13–14 Albert Schweitzer . . . creation."] See Norman Cousins, *Dr. Schweitzer of Lamberéné* (1960), chapter X.

20.11–13 the words of Jean Rostand . . . to know."] See Rostand's "Popularization of Science," *Science*, May 20, 1960.

96.3–5 Tomlinson's . . . no evil"] See "A Lost Wood," first collected in H. M. Tomlinson's *Out of Soundings* (1931): "Improvement had come. In the heart of the wood oaks were being felled, and by the torn roots of one was a dead hedgehog, which had been evicted from its hibernaculum into the frigid blast of reform. Unseen but near a saw was at work, and its voice was like the incessant growling of a carnivore which had got its teeth into a body and would never let go. This Easter, by all the signs, was the last the wood would see. The bluebells had been coming, expecting no evil, and, had they been allowed the grace of a few more weeks, they would have put the depth of the sky between the trees; but carts and engines had crushed them, and had exposed even their white bulbs, as though the marrow of the earth were bared."

137.5–6 a British ecologist . . . the earth.] In *The Ecology of Invasions by Animals and Plants* (1958), Charles S. Elton (1900–1991) describes aerial pesticide spraying as an "astonishing rain of death upon so much of the world's surface."

162.19–22 Lewis Carroll's . . . be seen."] See the song "Haddocks' Eyes" in chapter 8 of Carroll's *Through the Looking-Glass* (1871).

200.15–16 the Japanese fisherman . . . *Lucky Dragon*.] Aikichi Kuboyama (1914–1954) was one of twenty-three crewmembers exposed to radioactive fallout from a U.S. atmospheric hydrogen bomb test in the Marshall Islands, on May 1, 1954; he died on September 23.

241.4 Robert Frost's familiar poem] See "The Road Not Taken," first collected in *Mountain Interval* (1916).

OTHER WRITINGS ON THE ENVIRONMENT

316.1–5 *Shirley Briggs . . . Kay*] Shirley Briggs (1918–2004) and Katherine Howe (1919–2009) both worked as illustrators for the U.S. Fish and Wildlife Service and collaborated with Carson on the "Conservation in Action" booklet series (1947–50), about national wildlife refuges.

317.22 Kitty] Catherine Birch (1910–2002), a Washington, D.C., physician and clinical instructor in otolaryngology; she and Carson enjoyed walks in Rock Creek Park and sailing on the Potomac.

319.6 my contribution] Carson won second place, and a $1,000 prize, for a short essay on conservation ("Why America's Natural Resources Must Be Conserved") and a "Conservation Pledge" for children: "I pledge myself to preserve and protect America's fertile soils, her mighty forests and rivers, her wildlife and minerals, for on these her greatness was established and her strength depends."

319.10   *Edwin Way Teale*] Carson and Teale (1899–1980) exchanged letters and occasional visits from the late 1940s until her death; this is the first of their letters known to survive. His many books include *Near Horizons: The Story of an Insect Garden* (1942), *The Lost Woods: Adventures of a Naturalist* (1945), *Days without Time: Adventures of a Naturalist* (1948), and *North with the Spring: A Naturalist's Record of a 17,000 Mile Journey with the North American Spring* (1951).

319.26   Dr. Hildebrand] Samuel F. Hildebrand (1883–1949), senior ichthyologist at the U.S. Bureau of Fisheries.

320.30   Dr. Sharp] Ward M. Sharp (b. 1904), refuge manager for the Red Rock Lakes National Wildlife Refuge, in southwestern Montana.

321.5–7   Driftwood Valley . . . in the book.] See *Driftwood Valley* (1946), an account of life in the Driftwood Valley Country of central British Columbia by Theodora C. Stanwell-Fletcher (1906–2000).

321.27   the Peterson card] A flashcard (or possibly a notecard) associated with *A Field Guide to the Birds* (1934) by Roger Tory Peterson (1908–1996).

322.4   Robert] Robert McLean Carson (1899–1988), Rachel's brother.

323.25   the Mushbachs] George E. Mushbach (1881–1963), superintendent of the National Bison Range in western Montana, and his wife Gertrude F. Mushbach (1881–1956).

323.32   "Bob, Son of Battle"] Film of 1947, also titled *Thunder in the Valley.*

325.10–12   the head of the Park Service . . . replaced.] Conrad L. Wirth (1899–1993) led the National Park Service from December 1951 to January 1964.

327.1   *To Fon Boardman*] The original typescript from which the text of Carson's letter has been taken bears a marginal pencil annotation in her hand, its last words incompletely legible: "Without having read anything he had written (for Sand County Almanac came while I was too busy with the Sea to do [with reading it?])." Boardman (1911–2000) was director of marketing at Oxford University Press, which published Carson's *The Sea Around Us* in 1951.

327.3–4   Aldo Leopold's *Round River*] A collection of the early journals of the ecologist Aldo Leopold (1887–1948), edited by his son Luna B. Leopold and published by Oxford University Press in 1953.

328.19   *Dorothy Freeman*] Freeman (1898–1978), a former home economics teacher, wrote to welcome Carson when she learned in the fall of 1952 that the two would soon be summer neighbors on Southport Island, Maine; they developed a close lifelong relationship. *Always, Rachel: The Letters of Rachel Carson and Dorothy Freeman, 1952–1964* (1995), edited by Martha Freeman, collects some of their extensive correspondence.

330.28   the *Register*'s] The *Boothbay Register*, a weekly newspaper.

330.34   the movie] *The Sea Around Us* (1953), a documentary based on Carson's book of the same title.

331.25 Roger] Roger Christie (b. 1952), Carson's grandnephew and, after her niece's death in 1957, adopted son.

332.9 Martha] Martha Freeman, Dorothy Freeman's granddaughter, born in 1953.

332.15 "regular" letter] In their mutual correspondence, Carson and Freeman made a distinction between "regular" or "family" letters, intended to be shared with others, and "apples," meant to be private; in some instances a "regular" letter would contain an "apple." (See, for instance, the beginning of Carson's letter to Freeman of January 6, 1962, on page 411 of the present volume.)

332.32 Stan] Stanley L. Freeman (1879–1964), Dorothy's husband.

332.37 Mr. Cloos] Hans Cloos (1885–1951), German geologist whose 1947 memoir *Gespräch mit der Erde* had recently been published in English as *Conversation with the Earth* (1953).

336.19–23 Emily Dickinson . . . must be.] See Dickinson's "I never saw a Moor."

337.39–40 Hendrik Willem van Loon] Van Loon (1882–1944) was a prolific and popular author most widely known for his young adult history *The Story of Mankind* (1921).

339.16 Marie] Marie Rodell (1912–1975), Carson's literary agent.

342.26 Shirley] Shirley Briggs (see note 316.1–5).

345.36–346.7 His son . . . what is to follow."] See "An African Interlude," chapter 17 of *Westward Ho with the Albatross* (1953) by Hans Pettersson (1888–1966).

347.33–38 an editorial writer . . . beyond price."] See "Endangered Sanctuary," *New York Times*, October 5, 1952.

348.9–15 Richard Jefferies . . . mere endurance."] See Jefferies' essay "The Pageant of Summer," collected in *The Life of the Fields* (1884).

350.19–20 "of a spirit . . . Mr. Halle said] Louis J. Halle, Jr. (1910–1998), describes the song of the veery in his *Spring in Washington* (1947): "It is a soft, tremulous, utterly ethereal sound, swirling downward and ending again. Heard in the gloom of twilight, back and forth across the marshes, it gives the impression that this is no bird at all but some spirit not to be discovered."

351.9 Nellie] Nellie I. Teale (1900–1993), Edwin's wife.

351.22 Bob's illustrations] Robert W. Hines (1912–1994) illustrated Carson's book *The Edge of the Sea* (1955).

353.5 Mrs. Newman] Josephine Oliver Newman (1878–1968), a Maine naturalist.

353.8 Marjie] Marjorie Williams Christie (1925–1957), Carson's niece.

363.24–33   Otto Pettersson . . . to follow."] See note 345.36–346.7.

366.31   *Curtis and Nellie Lee Bok*] Curtis Bok (1897–1962), a judge and novelist, wrote Carson in praise of *The Edge of the Sea* in October 1955, and the two then corresponded extensively. Carson visited Bok and his wife Nellie Lee (1901–1984) the following April, at their home in Radnor, Pennsylvania.

369.27   the Singing Tower and Sanctuary] Principal features of a park built by Curtis Bok's parents Edward W. Bok (1863–1930) and Mary Louis Curtis Bok (1876–1970) near Lake Wales, Florida; it opened in 1929.

373.3   ————] The beginning of Carson's letter is not known to have been preserved.

373.4   the book] In 1953, Carson had agreed to write a book for the World Perspectives series, edited by Ruth Nanda Anshen and published by Harper & Row. At one point tentatively titled "Remembrance of Earth," it remained unfinished.

374.5   the "new heaven and the new earth"] See Revelation 21:1.

374.7–9   arrogance . . . Frank Lloyd Wright's use of it.] Carson wrote Freeman on January 1, 1958, in a letter begun the previous day: "One of the things I've had in the back of my mind to comment on was a TV interview with Frank Lloyd Wright, of which I happened to hear a little weeks ago. Something he said (while it didn't really surprise me) stuck in my mind with a sense of shock. It was to the effect that long ago he had to choose between 'honest arrogance' and 'hypocritical humility'—and had chosen honest arrogance. It somehow crystallized my belief that a large share of what's wrong with the world is man's towering arrogance—in a universe that surely ought to impose humility, and reverence."

375.1   *Marie Rodell*] See note 339.16.

380.5–6   Cass Canfield and Paul Brooks] Canfield (1897–1986) was chairman of the executive committee of Harper & Brothers. Brooks (1909–1998), an editor-in-chief at Houghton Mifflin, worked with Carson on *Silent Spring*.

382.10   Joan Daves—Fairfield Osborn] Daves (1919–1997) was a partner in Marie Rodell's literary agency. Osborn (1887–1969), a prominent environmentalist, was the author of *Our Plundered Planet* (1948), *The Limits of Earth* (1953), and other works.

382.11   Mr. Shawn] William Shawn (1907–1992), editor of the *The New Yorker* from 1952 to 1987.

382.30   Dr. Newton Harvey] E. Newton Harvey (1887–1959), an authority on bioluminescence.

382.32   my proposed Harper book] See note 373.4.

382.34   Vincent Schaefer] Schaefer (1906–1993) was an inventor of cloud seeding.

382.35–36   little horror story . . . Dick Pough] Richard H. Pough (1904–2003) helped to establish Hawk Mountain Sanctuary and The Nature Conservancy, serving as president of the latter organization. The "horror story" to which Carson refers is not known to have appeared in *The New Yorker*.

383.11–12   Elizabeth Lawrence] An editor at Harper & Brothers.

383.19   Mrs. Pinkham's] Izetta Pinkham, whose husband owned the general store in West Southport, Maine.

383.30   *The Return of the Native*] Novel by Thomas Hardy (1840–1928), first published in 1878.

384.19–21   One of my delights . . . "51" volunteers.] See Wayland J. Hayes, Jr., et al., "The Effect of Known Repeated Oral Doses of Chlorophenothane (DDT) in Man," *Journal of the American Medical Association* 162.9 (1956): 890–97, and *Silent Spring*, page 170 in the present volume.

384.28   as slick . . . Admiral Strauss was in his] As chairman of the Atomic Energy Commission from 1953 to 1958, Lewis Strauss (1896–1974) discounted evidence of the dangers of fallout from atmospheric nuclear bomb testing. When that evidence became incontrovertible, he urged further testing to develop a "clean" bomb.

384.38   Mr. Broley's reports on the eagle.] In "The Plight of the American Bald Eagle" (*Audubon*, July–August 1958), Charles L. Broley (1879–1959) blamed DDT for the precipitous declines he had observed, after two decades of bird banding and field study, in Florida bald eagle populations.

385.18–19   John Kieran] Kieran (1892–1981), a radio and television personality, was the author of *John Kieran's Nature Notes* (1941), *An Introduction to Birds* (1946), *Footnotes on Nature* (1947), and later *A Natural History of New York* (1959).

385.31   *Marjorie Spock*] An organic farmer living in Brookville, Long Island, Spock (1904–2008) joined others in May 1957 to seek an injunction against the U.S. government, which planned to spray parts of Long Island with DDT in an attempt to control Dutch elm disease. After this effort failed, a larger group of plaintiffs sued for damages; their case reached the Supreme Court in March 1960. Spock and her friend Mary T. ("Polly") Richards (1908–1990), who largely funded these legal efforts, shared transcripts, articles, and other information with Carson as she was writing *Silent Spring*.

387.17   Albert Schweitzer's "reverence for life."] See chapter 13 ("First Activities in Africa, 1913–1917") of Schweitzer's autobiography *Out of My Life and Thought*, originally published in 1931 as *Aus meinem Leben und Denken*.

391.23–24   a noted British ecologist . . . the earth."] See note 137.5–6.

391.37–392.5   Professor George Wallace . . . reproductive failure."] See George J. Wallace, "Insecticides and Birds," *Audubon*, January–February 1959.

393.1 *Beverly Knecht*] Knecht, a young, blind woman hospitalized with diabetes and its complications, sent Carson a letter after hearing a recorded version of *The Edge of the Sea*, and the two corresponded extensively. Knecht died in 1961.

393.4–6 Housman's . . . cherry hung with snow.] See the untitled second poem ("Loveliest of trees, the cherry now") in *A Shropshire Lad* (1896) by A. E. Housman (1859–1936).

393.19 Dr. Biskind] Morton S. Biskind (1906–1981), a Connecticut physician and authority on insecticides and public health.

393.33–36 I shall dedicate . . . the earth] See note 2.1–5.

394.8–10 The next book . . . with children.] The article to which Carson refers ("Help Your Child to Wonder," *Woman's Home Companion*, July 1956) is printed on pages 354–64 of the present volume. It was published in book form after her death, with photographs by Charles Pratt, as *The Sense of Wonder* (1965).

394.10–11 the Sea Anthology] Carson had recently agreed to edit an anthology for Harper & Brothers titled "Magic of the Sea"; it was not ultimately published.

397.13–17 *Grace Barstow Murphy* . . . Long Island group] Murphy (1888–1975) coordinated activities of the Committee Against Mass Poisoning, formed in June 1957 to publicize the dangers of pesticide use. Her husband Robert Cushman Murphy (1887–1973), an ornithologist, was lead plaintiff in *Murphy v. Benson*, a lawsuit seeking an end to aerial spraying on Long Island. (See also note 385.31.)

397.18 Sec. Flemming's . . . the sprayers] On November 9, 1959, U.S. Secretary of Health, Education, and Welfare Arthur S. Flemming (1905–1996) announced at a press conference that some of the 1958–59 cranberry crop contained unsafe levels of the herbicide aminotriazole, a known carcinogen, and warned consumers not to buy or use cranberry products. His statement precipitated a national "cranberry scare" or "cranberry crisis."

398.15–17 Allen Morgan . . . Russ Mason.] Allen H. Morgan (1925–1990) and C. Russell Mason (1895–1983), new and former executive directors of the Massachusetts Audubon Society.

398.18–19 Dr. Sears's] Ecologist Paul B. Sears (1891–1990).

398.22 the big Cranberry Meeting] At a public meeting in Washington, D.C., on November 18, 1959, government, consumer, and cranberry industry representatives discussed the "cranberry crisis" and worked out a plan that would enable uncontaminated berries to be labelled and sold.

399.27 Ida] Ida B. Sprow (1920–2011), Carson's housekeeper from November 1956 until her death.

399.34   *Morton S. Biskind*] See note 393.19.

399.38   Mrs. Hunter] Probably Beatrice Trum Hunter (1919–2017), later the author of *The Natural Foods Cookbook* (1961).

401.26–27   Mrs. D.] Jeanne Davis (1915–2004), Carson's assistant and secretary from September 1959 until her death.

402.6   Dr. Hueper] Wilhelm C. Hueper (1894–1978), director of the Environmental Cancer Section of the National Cancer Institute from 1938 to 1964.

402.20   *Polly Richards*] See note 385.31.

403.24   Mr. Butler] Paul M. Butler (1905–1961), chairman of the Democratic National Committee.

405.36–406.1   the article in the Rochester . . . Mr. King] See Floyd King's "Outdoors" column ("Do You Object to Poisoning?") in the Rochester *Democrat and Chronicle*, July 3, 1960.

406.1–2   Dr. Hueper] See note 402.6.

406.34   Dr. Pfeiffer] Ehrenfried Pfeiffer (1899–1961), an authority on biodynamic agriculture.

407.15–16   Margaret Louise Hill] Hill served as chairman of the Conservation Committee of the Texas Ornithological Society.

408.20   *Clarence Cottam*] After a career with the U.S. Bureau of Biological Survey and the U.S. Fish and Wildlife Service, Cottam (1899–1974) became director of the Welder Wildlife Foundation in 1955.

408.32   Dr. Decker] George C. Decker (1900–1993), professor of entomology at the University of Illinois.

411.27   an apple] See note 332.15.

412.13   Jeanne] See note 401.26–27.

412.32–33   the *Help Your Child . . . Nature*" book.] See notes 373.4 and 394.8–10.

413.4   Lois's] Lois Crisler (1896–1971), author of *Arctic Wild* (1958).

415.1   *Frank E. Egler*] Egler (1911–1996), an ecologist and conservationist, had campaigned against overuse of herbicides in roadside and right-of-way vegetation management.

415.25   Olaus Murie] Murie (1889–1963) was the author of *The Elk of North America* (1951) and other books.

416.7–8   a paper by Goodrum . . . Wildlife Conference.] See Phil Goodrum and Vincent H. Reid, "Wildlife Implications of Hardwood and Brush Controls," *Transactions of the North American Wildlife Conference* (1956).

417.5   *Lois Crisler*] See note 413.4.

417.34   Dr. Hard] Frederick Hard (1898–1981), president of Scripps College in Claremont, California.

419.12   Clarence Cottam] See note 408.20.

419.15   Justice Douglas] William O. Douglas (1898–1980), associate justice of the Supreme Court and advocate of wilderness protection.

420.4   poor Senator Neuberger] Maurine Neuberger (1907–2000), an Oregon Democrat.

422.12–18   E. B. White . . . dictatorially."] See note 3.4–10.

423.6–15   John Muir . . . hanging the wicked."] See "Cedar Keys," chapter 6 of Muir's *A Thousand-Mile Walk to the Gulf* (1916).

424.4–7   Francis Thompson . . . troubling of a star.] See Thompson's poem "The Mistress of Vision," XXI, first collected in *New Poems* (1897).

425.2–3   Albert Schweitzer . . . life."] See note 387.17.

425.36–39   In 1955 . . . urban sprawl.] See *Man's Role in Changing the Face of the Earth* (1956), edited by William L. Thomas.

426.28–32   the English essayist Tomlinson . . . yet discovered."] See the concluding paragraphs of "The Little Things" (1944) by H. M. Tomlinson (1873–1958), collected in his book *The Face of the Earth* (1950).

434.12   the FIFRA] The Federal Insecticide, Fungicide, and Rodenticide Act (1947).

439.11–12   his own account . . . Reverence for Life.] See note 387.17.

451.1   *Walter P. Nickell*] In *Silent Spring*, Carson describes Nickell (1903–1973) as "one of the best-known and best-informed naturalists" in Michigan; see page 81 in the present volume.

452.25   "Undersea,"] Published in the *Atlantic Monthly* in September 1937.

452.26   Charles Alldredge's poems] Alldredge (1911–1963), with whom Carson had been friendly since the early 1940s, died on March 10. His poems were collected posthumously in *Some Quick and Some Dead* (1964).

453.16–18   Otto Pettersson . . . follow."] See note 345.36–346.7.

453.26   Barney's] George "Barney" Crile, Jr. (1907–1992), head of general surgery at the Cleveland Clinic, was one of the physicians who treated Carson's breast cancer.

454.10   Elliott] A handyman.

454.12   Jeanne] See note 401.26–27.

455.21   Krebiozen] A now-discredited alternative cancer treatment.

474.5   Dr. Keene] Clifford Keene (1910–2000), vice president and general manager of Kaiser Foundation Hospitals.

489.3   the *Victory Chimes*] A three-masted schooner that sailed out of Booth-bay Harbor, Maine, on vacation cruises.

489.24   the Browers] David Brower (1912–2000), executive director of the Sierra Club beginning in 1952, and his wife Anne Hus Brower (1913–2001).

490.9   the last talk, by Medawar.] Peter Medawar (1915–1987), British immunologist.

492.13–14   The Scotts from Pittsburgh] Ruth Jury Scott (1909–2003) and her husband J. Lewis Scott (1901–1968), both active in environmental and conservation groups.

492.20   Irston Barnes's column] Barnes (1904–1988) wrote a weekly column, "The Naturalist," for *The Washington Post*. In "Autumn Brings a Change in Mood," on September 29, 1963, he describes the fall migrations to which Carson refers.

492.30–31   John Muir . . . American Heritage book] *The Wilderness World of John Muir* (1954), edited by Edwin Way Teale, and *The American Heritage Book of Natural Wonders* (1963), edited by Alvin M. Josephy.

492.31–33   George Stewart . . . *Storm.*] Stewart (1895–1980), a U.S. Forest Service ecologist, published the novel *Storm* in 1941.

492.35–39   the reprint Betsy Reimers . . . Paul had] Reimers, one of Carson's neighbors in Silver Spring, Maryland, supplied a reprint that had helped her husband Paul; Carson hoped that it would be useful for Stanley Freeman and his physician, Dr. Boles.

493.23–24   Dr. Briejér] Cornelis Jan Briejér (1901–1986), director of the Plantenziektenkundige, the Dutch Plant Protection Service.

494.24–25   "War and the Living Environment"] See *War and the Living Environment* (1963), edited by Robert H. Wurtz.

495.1   the terrible events] President John F. Kennedy was assassinated on November 22, 1963.

Index

Acetylcholine, 32–33
Adams, Anselm, 491
Adenosine triphosphate, 177–78, 180–82, 201
Adipose tissue, 168
Adirondack Mountains, 138
Adrenal gland, damage to, 49–50
Aerial spraying, 80, 82, 85–86, 95–96, 99–100, 103, 137–51, 371–72, 376–78, 434–35, 465–68, 478
Aerosol sprays, 198–99, 442
Afghanistan, 234
Africa, 206, 221, 224, 231, 234–35, 245, 439
Agricultural Experiment Station, Alabama, 144
Agricultural Experiment Station, Louisiana, 151
Agricultural Experiment Station, Mississippi, 151
Agricultural Experiment Station, New York, 141
Agricultural Extension Service, 416
Agricultural Research Service, 395
Agriculture and Markets Department, New York, 140
Agriculture Department, Illinois, 84, 444
Agriculture Department, Michigan, 80
Agriculture Department, U.S., 16, 59, 81, 84–85, 90, 103, 125, 138–51, 155, 222, 239, 243–47, 249–50, 378–81, 389–90, 395, 397, 404, 434, 442, 445, 457, 464, 471
Agriculture Ministry, U.K., 38, 110–11
Air pollution, 137–51, 156, 404, 429, 443, 445, 459, 461, 463–65, 470, 474, 477–78
Air Pollution Conference, 477
Alabama, 42, 93–94, 125–26, 142, 144–47, 149, 151, 244–45
Alabama Cooperative Wildlife Research Unit, 147
Alabama Polytechnic Institute, 144
Alaska, 109, 158–59, 463, 481

Albatross III (fishing boat), 338–41
Aldrin, 30–31, 56, 80, 82–83, 87, 109, 111–12, 181, 236, 433, 458
Alexander, Peter, 185, 248
Alfalfa, 27, 69–70, 253–54
Algae, 54, 329–30
Algire, Dorothy Hamilton, 6
Alkyl phosphates, 31–36
Alldredge, Charles, 452
Alligators, 343–44
American Cancer Society, 193
American Chemical Society, 401
American Medical Association, 155, 168, 388, 436, 449, 469
American Museum of Natural History, 70, 93, 235
American Society of Ichthyologists and Herpetologists, 125
Aminotriazole, 161, 196–97
Amitrol, 39
Ammonia, 475
Anchorage, Alas., 159
Animal Welfare Institute, 439
Annapolis Valley, 225–27
Antibiotics, 410
Ants, 124–25, 142–51, 222, 255–56, 376, 381, 389, 395, 459, 465
Apennine Mountains, 255
Aphids, 36, 217, 219, 254
Appalachian Mountains, 94, 106, 139
Apple orchards, 220–21, 225–27, 254
Arant, F. S., 144
Argentina, 78, 194
Aristotle, 251
Arizona, 216
Arkansas, 377
Army Chemical Corps, U.S., 44–45
Arsenic, 22–23, 38–39, 44, 51, 57–58, 67, 72, 76, 191, 194–95, 206, 220, 226, 230, 465
Assabet, Mass., 462
Associated Press, 434
Astoria, Ore., 324
Atlantic Monthly, 337
Atlantic Naturalist, 437

531